D0990036

FISHES
OF THE SEA

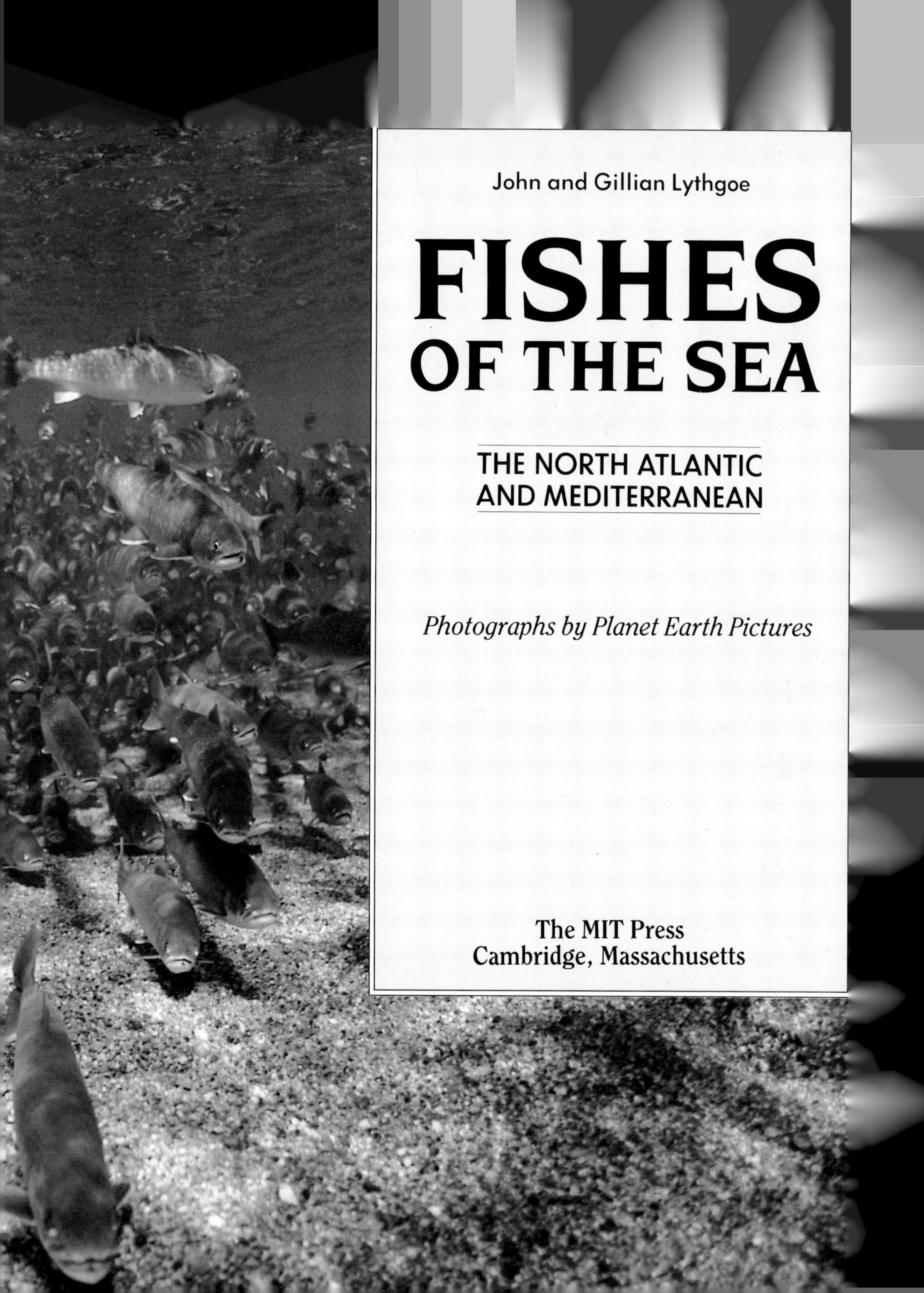

John and Gillian Lythgoe

FISHES OF THE SEA

THE NORTH ATLANTIC AND MEDITERRANEAN

Photographs by Planet Earth Pictures

The MIT Press
Cambridge, Massachusetts

ACKNOWLEDGEMENTS

Underwater photographs of living fish in their natural environment form the backbone of this book. Such photography is not easy, for it combines all the challenges of land-based natural history photography with the added problems that come from working under water. We thank all the photographers who have devoted their skill and perseverance to taking the pictures that are reproduced here.

One of the pleasures in assembling the photographs for the book is the unstinting and generous way that the photographers suggested others who might have valuable collections of pictures, or perhaps a single picture of a species we needed. Many marine biologists from both sides of the Atlantic gave us a great deal of help and we gratefully acknowledge Drs Ed Brothers, John Heise, Phil Lobell, Ellis Loew, Nigel Merrett, Peter Miller and Jon Witman. A special word of thanks goes to Antonio Soccol who introduced us to several European photographers.

Many thanks go to Jennifer Jeffrey and Carla Stubbs for their exceptional work on the index. We are particularly grateful to those photographers who made pictures available that were not finally needed. Their contribution to the final result is unseen, but none the less real for that.

John and Gillian Lythgoe

All pictures have been supplied by Planet Earth Pictures.

First MIT Press edition, 1992

This book was printed and bound in Singapore

Library of Congress Cataloging-in-Publication Data

Lythgoe, J.N.
Fishes of the sea : the North Atlantic and Mediterranean / John and Gillian Lythgoe : photographs by Planet Earth Pictures.
p. cm.
"First published in the United Kingdom in 1991 by Blandford . . . London" – T.p. verso
ISBN 0–262–12162–X
1. Marine fishes – North Atlantic Ocean – Identification. 2. Marine fishes – Mediterranean Sea – Identification. 3. Marine fishes – North Atlantic Ocean – Pictorial works. 4. Marine fishes – Mediterranean Sea – Pictorial works. I. Lythgoe, G.I. (Gillian I.) II. Title.
QL621.2.L97 1991
597.092′11—dc20 91–23354
CIP

Typeset by Bookworm Typesetting, Manchester, England

Contents

INTRODUCTION

A free-swimming fish can look quite different from a drawing made from a dead specimen, and photographs taken of the living fish in its own environment are very helpful for giving it a name. Drawings of fish used in identification usually show the fins fully extended in order to display their features. Living fish often keep one or more sets of fins folded down for most of the time giving a different profile to the fish. Patterns and colour in the living fish are not fixed; the needs for camouflage change from time to time and from place to place and the coloration of a fish at night is sometimes quite different from the same fish in the daytime. Colours and pattern are also used for display and can change from season to season, or even from second to second. Photographs of the living fish show it as it actually appears, although the use of flash photography means that many fish look more colourful than they do to a diver.

The underwater photographs are the work of many different photographers. We have tried to include photographs of as many fish in their natural environment as possible, and these should at least give an idea of the family a fish belongs to. The summary drawings and the brief descriptions that accompany them should be enough to make an accurate identification in most cases. The longer descriptions give further information that may be needed to distinguish between closely similar species.

The previous editions of the book contained some North American species but the coverage was very far from complete. In this book most species from the cold and temperate waters of North America are included. To keep the book to a reasonable size, we have not included species that primarily live in fresh water but do sometimes enter brackish water, nor have we included species that live too deep to be encountered by scuba divers and anglers. These changes have increased the number of species described by about 50 per cent.

It is not possible to define the geographical range of a fish with absolute precision because their movements depend in large part on the exact set of the ocean currents and the temperature of the water from year to year. We have not included those fish that have migrated through the Red Sea into the Mediterranean, and we have not included some fish that may occasionally migrate up the east coast of the United States from the tropical Atlantic. However, we have tried to present a reasonably complete coverage of the important game fish species including the billfish and tunas.

The geographical area covered by the book includes the cold and temperate waters of the North Atlantic and the Mediterranean (Fig.1). To the west it includes the waters of the United States and Canada northwards from about Chesapeake Bay. To the east it covers the whole of the Mediterranean, and from Gibraltar north to the Arctic including all the coasts of Europe. We have not included species that rarely come more shallow than 50 m, nor have we included open ocean species that rarely come within a few kilometres of the coast. There are, however, no fixed frontiers in the ocean, and unexpected strays do turn up, especially in years when the sea temperature is unusually high or low.

Since 1971, when this book was first published, there have been numerous changes in the scientific names and classification of fish. These changes are made in line with modern thinking about the relationships of fish, and scientific opinion does change as more research is carried out. English names of fish often vary from place to place along the coast. There are also differences between the American and European usages. For example the fish that the British call Gurnards, are called Searobins by Americans. Recently, more order has come into the English names. The American Fisheries Society publishes a list of American fishes in which only one scientific name and one English name is given. In Europe the CLOFNAM volumes of fish

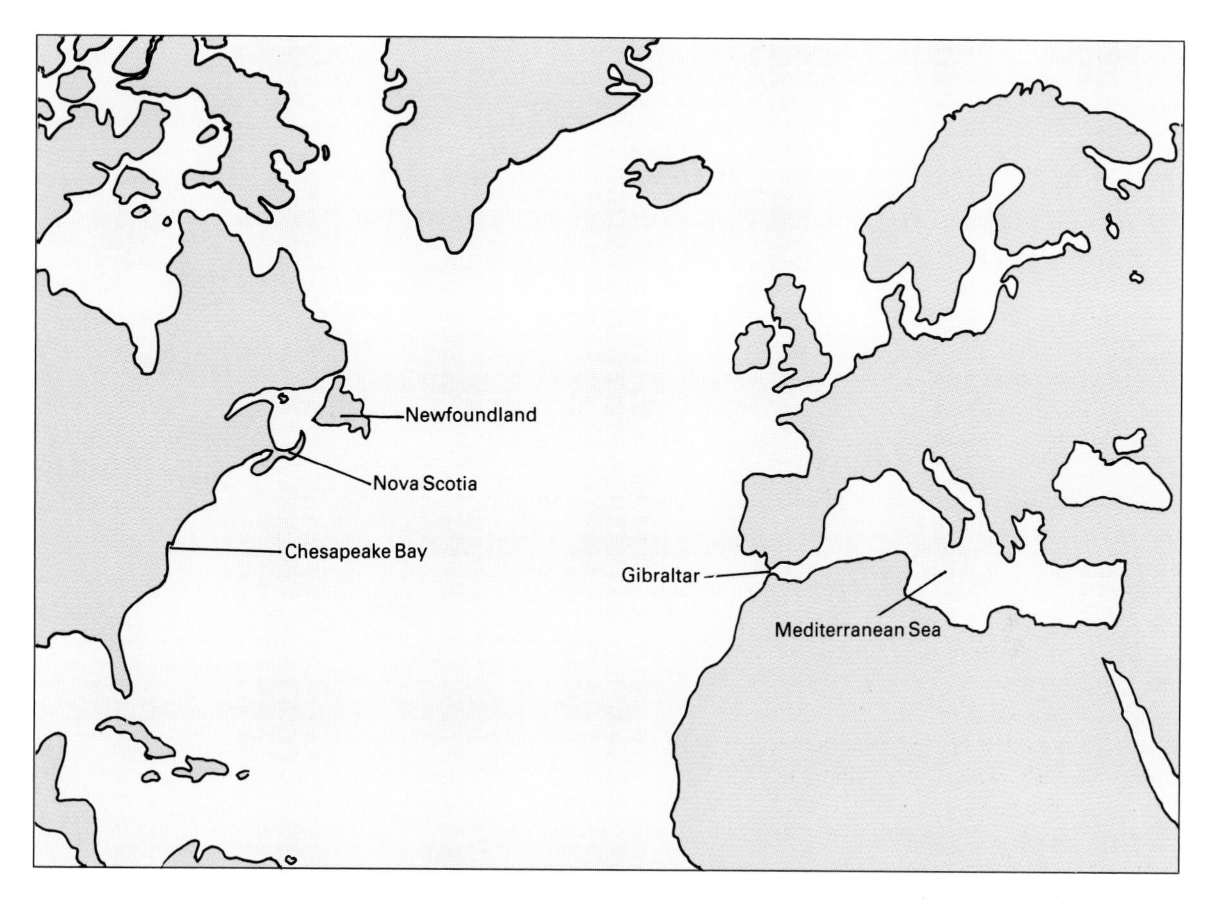

Figure 1

identification gives the names in the main European languages and these are also likely to become standard. References to these books, and other valuable books for identification, appear at the end of this introduction.

For almost every species there is an outline drawing, where the main features are indicated. Each drawing is supported by a brief description of the main diagnostic points including the geographical range, which can be a useful aid to identification. The size given for each fish is the total length from the snout to the base of the tail and it refers to a large specimen, but not a record-breaking one. The sizes of some of the Rays refer to the width across the body because the long whip-like tail makes measurements of total length rather unhelpful. In some cases the species are so similar that individual drawings are not helpful and the important points of difference are indicated in the text. Wherever possible we have avoided technical terms. This may result in some loss of precision for the specialist, but we hope that the meaning remains clear.

The parts of the fish that are mainly used for identification in the field are shown in Figs 2–5.

The fins are important features in identification. The unpaired fins include the dorsal fins, the tail fin (caudal fin), and the anal fin. The dorsal and anal fins often have two kinds of rays; the hard rays or spines and the soft rays. The number of spiny rays is shown in Roman numerals and the soft rays by conventional Arabic numerals. In some species there are so many hard rays that Roman numerals are difficult to decode and Arabic numerals are used instead. In some families, notably in the Mackerel family, there are a series of finlets behind the dorsal fins. The pectoral and

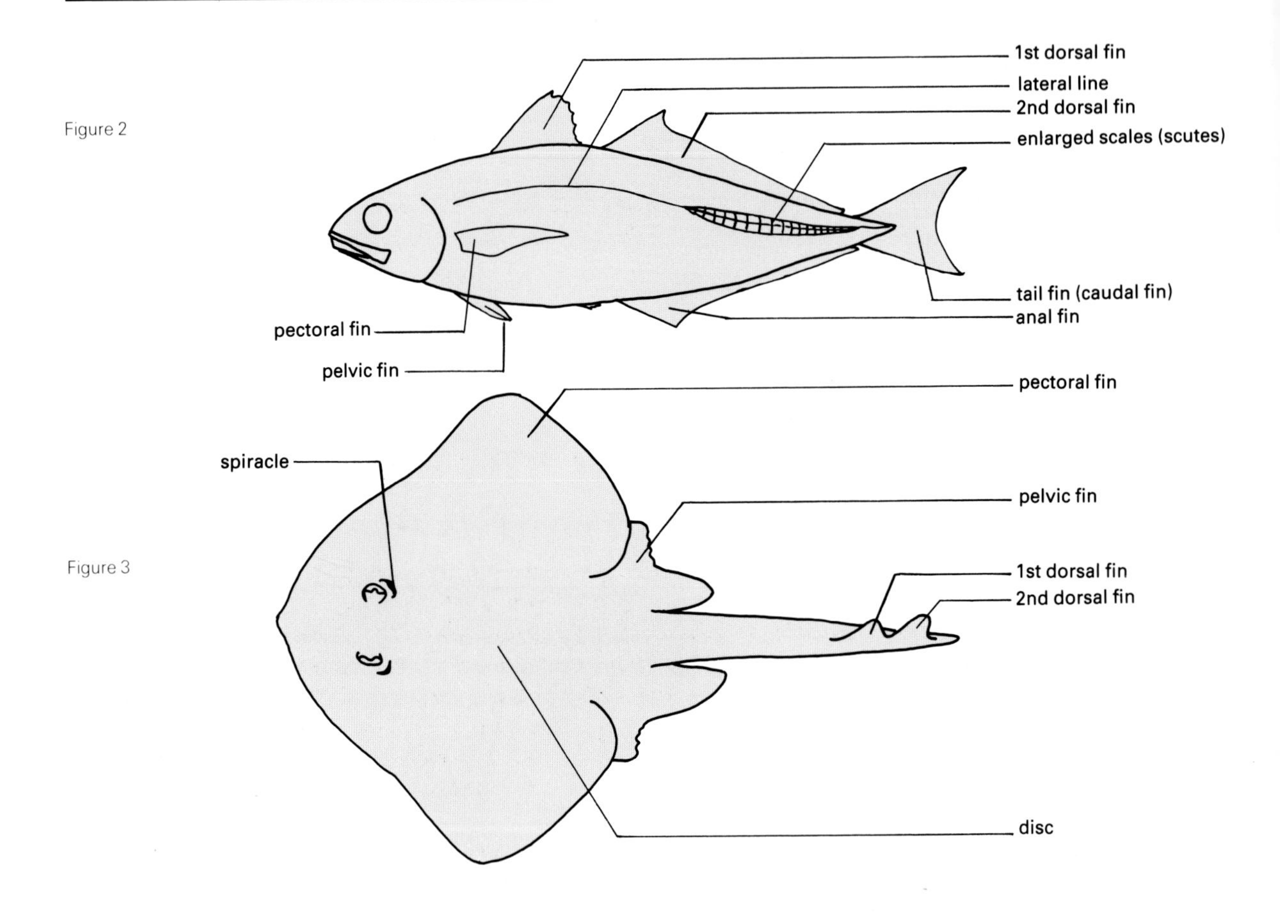

pelvic fins are paired and are comparable to the front and rear pairs of limbs in other vertebrates. The fins are often highly modified to perform other functions besides swimming. For example, the pelvic fins of the Lumpsucker are formed into a disc to anchor the fish to a rock. In the Remoras, the first dorsal fin is modified into a sucker which the Remora uses to attach itself to a shark or billfish. In fish like the Searobins and Gurnards the rays of the pelvic fins are sometimes strengthened or prolonged for support and for moving across the bottom. Modified fins may also have taste buds for the direct sampling of the bottom.

Many fish have barbels equipped with taste buds attached to the mouth or the lower part of the head which are also used for locating food. Fish that do not need to be streamlined may have fleshy outgrowths from various parts of the body, especially the head. These certainly help to break up the outline of the fish and one of their main uses are probably for camouflage.

Teeth can provide useful information about the lifestyle and feeding habits of the fish. For example the Sea Breams (Sparidae) have incisor-like teeth for biting off vegetation, canine-like teeth for holding animals, and molar-like teeth for grinding hard-shelled molluscs and other animals. Sharks can be very difficult to identify by sight alone, but if they can be examined, their teeth are important aids to identification.

Scales may cover the whole body, or part of it, or may be absent altogether. Sometimes they are ridged or grooved, and the margins may be slightly toothed. In fish such as the herrings, the scales are very easily shed, whilst in sharks they are so strongly attached that the skin can be used like sandpaper. In the Jack Mackerel

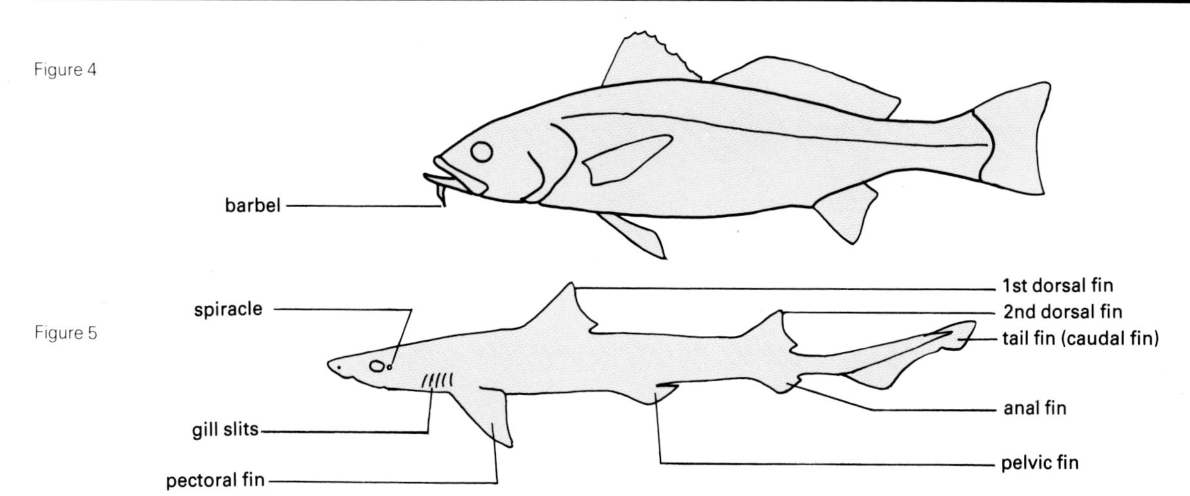

family (Carangidae) the scales along the lateral line may be enlarged to form large shield-like scutes, and their size, position and number are important in identification.

The lateral line is the external part of a series of sensory canals that run just beneath the surface and are used to detect currents and eddies in the water. It is often branched, but the main section usually runs along the mid-line of the body from the gill cover to the tail. Sharks and rays have a set of canals that open to the surface as small pores. These are part of the electrosensory system which allows a shark or ray to locate fish that are buried centimetres deep in the bottom using the electrical fields set up by its prey.

As in the previous editions the section on the gobies has been written by Dr P.J. Miller, who is a leading authority on these important but difficult fish. His contribution adds substantially to the value of the book.

Anyone who wishes to know more about the fishes of the North Atlantic and Mediterranean will find the books listed below to be useful. They contain direct references to the original scientific papers on which most of our modern knowledge is based, and in most cases the sections are written by the specialists themselves.

Fishes of the North-eastern Atlantic and the Mediterranean (1986) edited by P.J.P. Whitehead *et al*. This is a multi-author work published by the United Nations Educational Scientific and Cultural Organization.

Fishes of the Atlantic Coast of Canada (1966) by A.H. Leim and W.B. Scott. This is a one-volume book published by the Fisheries Board of Canada, Ottawa.

Atlantic Fishes of Canada (1988) by W.B. Scott and M.G. Scott. Bulletin 219 of the Canadian Bulletins of Fisheries and Aquatic Science.

Fishes of the Western North Atlantic. This is a multi-volume work for the Sears Foundation. The first volume was published in 1948, but the series has not yet been completed. Published by the Sears Foundation for Marine Research, Yale University, New Haven, USA.

A list of common and scientific names of fishes from the United States and Canada, 4th edition (1980), edited by C.R. Robins *et al*. American Fisheries Society Special Publication No 12.

Class:
CYCLOSTOMATA

The **Lampreys** and **Hagfish** are very primitive vertebrates which superficially resemble eels. They have no bones in the true sense but instead are strengthened with separated nodules of cartilage. There are no scales and the body is very slimy. Instead of jaws there is a round or slit-shaped mouth armed with horny teeth. The single nostril is on top of the head and it opens into from 1 to 15 pairs of simple gill openings. There are two orders, the *Myxinoidea* (Hagfish) and the *Hyperotreta* (Lampreys).

Order:
MYXINOIDEA

Family:
MYXINIDAE

Purely marine. The Hagfish have slit-like mouths with barbels and poorly developed fins. The eyes are degenerate.

Hagfish
Myxine glutinosa L.

North-East Atlantic; north North Sea; Mediterranean; Atlantic coast of North America; Davis Straight south to Florida.

Body eel-like; *eyes* are not visible and very degenerate: *nostril* 1 at extreme tip of the head; *barbels* (see diagram) one pair each side of the nostril and another pair, about twice as large as the others, each side of the mouth; *gill slits* are present as one pair of small round openings on abdomen; *skin* smooth, no scales; *fins* 1 fin which is narrow and starts about ¾ of the way along the body, runs around the tail and ends about ½ way along the underside. Yellowish, greyish or reddish, darker on the back, lighter underneath.

Usually 30–35 cm, sometimes up to 45 cm.

Hagfish are extremely sensitive to conditions and are only found in areas of high-medium salinity (31–34 parts per thousand), low temperature 10–13°C and low light intensity. The eyes are very poorly developed but the skin is somewhat sensitive to light. They are found on or very near muddy bottoms where they bury themselves in the mud. Their burrows are marked by small mounds with a hole at the top for the fish to enter. Rarely found above 25 m, usually between 30–600 m. They feed on any dead or dying animals by fastening themselves to the skin, often to the gills or on any injuries that the fish may have. They then devour the whole of the animal except for the skin and bones. When they find some particularly tough meat they are able to tear off the mouthful by tying their body into a knot and pulling the mouthful through the knot. Each fish has both sets of reproductive organs but only one set in any fish reaches maturity. The eggs, which are horny with many filaments at each end, are laid throughout the year. There is a row of mucus sacs along each side of the body that produce a prolific quantity of slimy mucus. A single fish is capable of filling a 4.5 litre bucket with slime and water in only a few seconds.

Hagfish *Myxine glutinosa*
Barbels around snout;
1 continuous fin around body.
Eastern and western Atlantic.
35 cm.

Order:
HYPEROTRETA
Family:
PETROMYZONIDAE

All **Lampreys** breed in fresh water but some species spend part of their adult life in the sea and are thus included here. The mouth is circular or funnel-shaped and there are no barbels. Eyes are present in the adult but are rudimentary in the larvae.

Sea Lamprey
Petromyzon marinus L.

Sea Lamprey
Petromyzon marinus
Adults with dark mottling: mouth disc (see diagram) 90 cm.

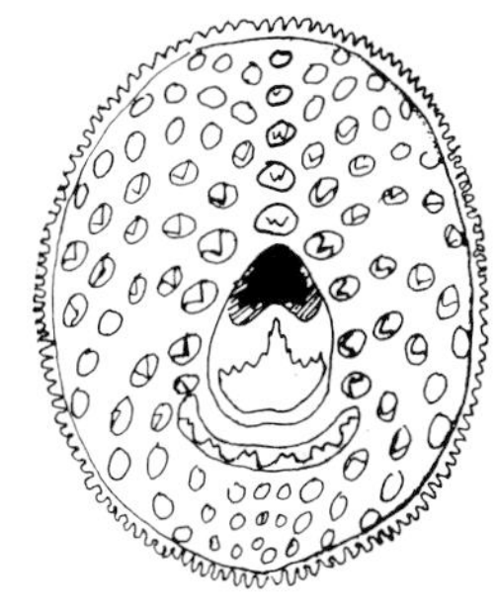

North-west European and African coasts; Iceland; Atlantic coast of North America: Labrador south to Florida; occasionally Baltic; Mediterranean.

Body eel-like, front half round and then gradually becoming flattened sideways towards the tail; *eyes* small; *mouth* a sucker with a fringed edge, oval when in use and slit-like when closed; *nostril* 1 at the level of the eye on the back; *teeth* (see diagram), the arrangement of the teeth in concentric rows with one large double-pointed central tooth is an important feature for precise identification; *gill slits* 7 pairs; *scales* none, skin smooth; *fins* 2 dorsal fins separate but sometimes nearly touching. The second fin is larger than the first and has a distinct notch between it and the tail fin.

Back olive-brown to yellow-grey with a black mottling. Underside greyish. Young fish may not have the black markings.

Up to 90 cm.

Found at all depths to 500 m and frequently near the mouths of rivers. The majority of their life is spent in fresh water. In the breeding season the male fish develop a ridge along their backs and the females a fold on their lower surface in front of the tail fin. The adults move into the rivers during the spring and make their nests on the gravel bottoms of fast-flowing streams. Activity is restricted to the night and the day is spent attached to a rock by their sucker. Spawning occurs during May and June, and up to 200,000 eggs may be laid. Both the male and female fish attach themselves to a rock by their suckers and the male twists around the female. After spawning they die. The young fish are very different from the adults and live in mud for about 6 years after which they change into the adult form and migrate to the sea.

Lampreys feed on living fish by attaching themselves to the skin of the prey, rasping through it and sucking out the blood. In some areas (the Great Lakes in Canada) they are a serious menace to fisheries. In the past they were considered a great delicacy and Henry I is said to have died from 'his having too plentiful a repast on these fishes'.

Sea Lamprey, *Petromyzon marinus* (Gilbert van Rijckevorsel).

Class:

CHONDRICHTHYES (Cartilagenous fishes)

The **Sharks**, **Skates**, **Rays** and **Chimaeras** (rabbit fishes). Included in this class are all fishes that have a basically cartilagenous skeleton, no true fin rays and a mouth underneath the head. There is a pair of nostrils and a series of exposed gill openings. The skin is covered with hard dermal denticles or placoid scales. These are tooth-like structures with a bone-like base embedded in the skin and an enamel-covered spine which is directed backwards and is exposed. It is these that give the characteristic rough surface to these fishes.

Fertilization is always internal and the mature males have special copulatory organs (claspers) modified from the pelvic fins. Each clasper has a cartilagenous skeleton and a canal running along its length. During copulation the female is grasped by the claspers and the seminal fluid is introduced into her cloaca through the canals. The young may pass their early life in one of three ways. In the oviperous type the eggs, which are usually encased in a horny capsule, are laid directly into the sea soon after fertilization. In the ovoviviperous type the eggs are not laid into the sea but are retained in the oviducts. The eggs break just before birth and the young are born alive into the sea. In the viviperous type the young are nourished through a 'placenta' in the mother, and are also born alive.

The cartilagenous fishes in this book fall within two orders, the Pleurotremata (Sharks and Monkfishes) and the Hypotremata (Rays).

Order:

PLEUROTREMATA

The **Sharks** and **Monkfish**. Front margin of each pectoral fin is free. Gill slits on the side of the head.

The Sharks are the most dangerous group of fishes. They are a serious menace to swimmers in some places, notably in Australia. However, they do not seem to attack swimmers and divers in waters further north than the Mediterranean and even there it is extremely rare for a shark to attack.

They must always be treated with great respect and the diver who gets blasé about them and tries to provoke them, or who swims too close, is asking for trouble.

Family:

HEXANCHIDAE

Medium to large sharks with 6–7 gill slits, one dorsal fin without spines, a well marked upper lobe to the tail fin and an anal fin. The eyes have no nictating membrane and the spiracle is rounded. They are viviparous.

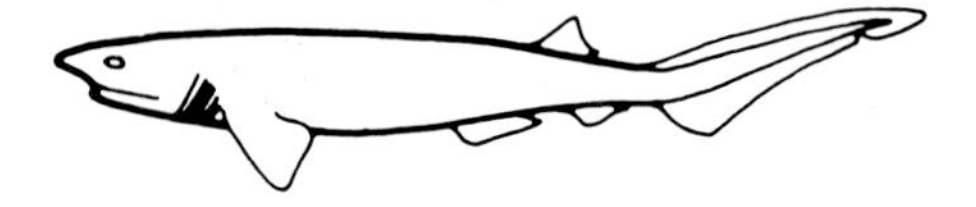

Bluntnosed Shark; Six-Gilled Shark; Grey Shark *Hexanchus griseus*

1 dorsal fin;
6 gill slits.
5 m.

Seven-Gilled Shark
Heptranchias perlo

1 dorsal fin;
7 large gill slits.
3 m.

Bluntnosed Shark; Six-gilled Shark; Grey Shark
Hexanchus griseus (Bonnaterre)

Mediterranean; temperate and subtropical east and west Atlantic.

Body slender, elliptical; *head* wider than body; *snout* rounded; *eye* large and oval with no nictating membranes; *teeth* (see diagram above left) different in each jaw, those in the upper jaw are triangular and those in the lower jaw are pointed with cusps at each side; *gill slits* 6, gradually decreasing in size from the head backwards and they are all situated in front of the pectoral fin; *fins* 1 dorsal fin situated between the pelvic and anal fins. Tail fin large and notched near end of the upper lobe.

The colour can vary from light to dark brown, grey or reddish. The underside is lighter and the lateral line is usually distinct as a lighter stripe.

Up to 5 m. The females are usually longer than the males.

Solitary deep-water sharks found down to 2000 m but usually between 100 and 1000 m. Occasionally they may move into more shallow water particularly in their northern ranges. They feed principally on fish but also on crustacea.

The young are born alive between 40–60 cm in length and in very large numbers. Up to 100 embryos have been reported and 47 confirmed. Sexual maturity is reached at about 2.1 m.

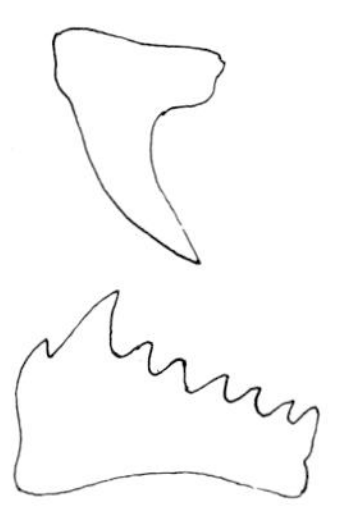

Seven-gilled Shark
Heptranchias perlo (Bonnaterre)

Mediterranean; central Atlantic from Portugal to Cuba.

Body long and rather slender; *head* narrow; *snout* pointed; *eye* large (wider than the distance between the nostrils) and oval; no nictating membranes; *teeth* (see diagram below left) different in each jaw. Those in the upper jaw have one point and those in the bottom jaw have many; *gill slits* 7, they are very large and continue under the throat. They decrease in size from the head backwards and are all in front of the pectoral fin; *fins* 1 dorsal fin between pelvic and anal fins. Tail fin large and notched near the end of the long upper lobe.

Greyish back, lighter underside. The front edges of the pectoral fins are edged with white and the tip of the dorsal fin is black. There are two white spots on the dorsal fin. The upper edge of the tail fin is black with the extreme rear edge and lower edge bordered in white.

Up to 3 m.

Found in deep water down to 300 m. They are fish-eaters and probably live near the bottom. The young are born between 25–28 cm in length and up to 20 in number.

Family:
ODONTASPIDIDAE

Medium to large **sharks** with 5 gill slits, 2 dorsal fins and an anal fin. The upper lobe of the tail fin is large. The head is flattened but the snout is conical. There is a small spiracle. Each scale has 3 keels. No nictating membrane. Viviparous.

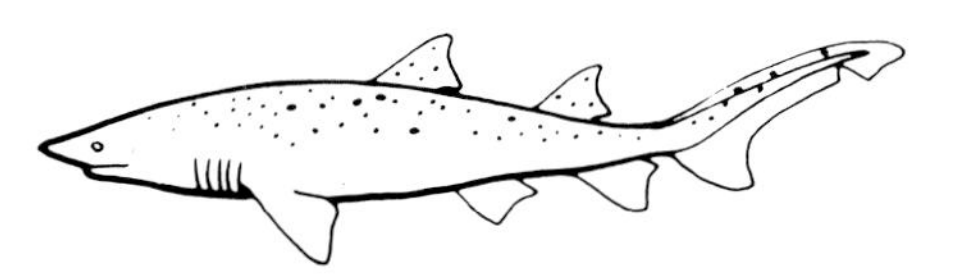

Sand Shark
Eugomphodus taurus

Grey-brown with brown to yellow irregular spots; dark edges to fins.
3 m.

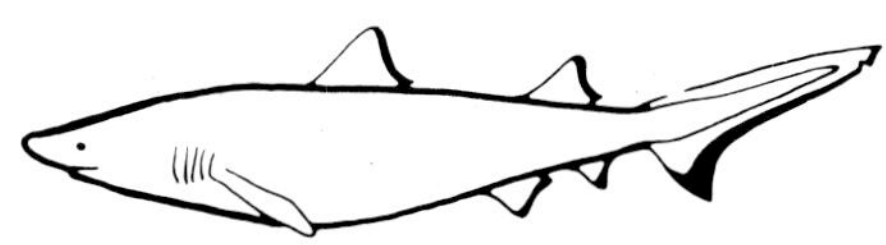

Fierce Shark *Odontaspis ferox*

Dark rear edge to fins.
2 m.

Sand Shark, *Eugomphodus taurus* (Ken Lucas).

Sand Shark

Eugomphodus taurus (Rafinesque)

Mediterranean; eastern Atlantic from Morocco to south Africa and all warm seas.

Body long and fairly stout; *head* small, flattened from above; *snout* pointed; *eyes* small and round; *teeth* (see diagram) similar in each jaw, pointed with cusps on each side; *gill slits* 5, all situated in front of the pectoral fins; *scales* small and scattered over the surface; *fins*, 2 dorsal fins nearly equal in size, the first completely in front on the pelvic fins. Tail fin large with a deep notch on the underside of the upper lobe. Pectoral fin longer than in the Fierce Shark (*Odontaspis ferox*).

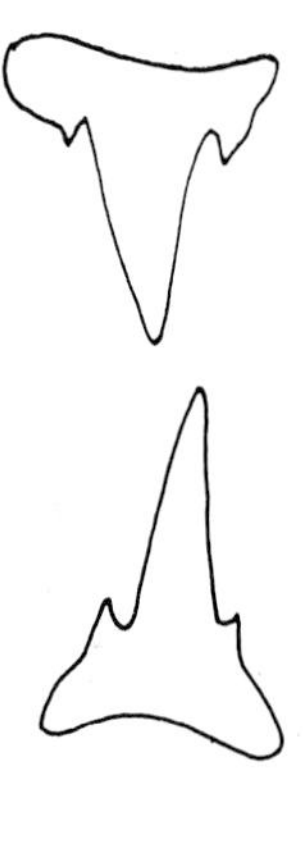

Greyish or brownish on the back becoming lighter underneath. Brownish-yellow irregular spots are scattered over the back and sides. The spots are more clearly marked in young fish than in adults.

Up to 3 m.

Coastal, near bottom 0–70 m, occasionally 200 m. They normally live over sand but spend their life continually swimming. They are able to swallow air and keep it in their stomachs which they then use as an air bladder to help them retain buoyancy in the water. Predominantly a fish- and cephalopod-eater they will attack any animal that attracts them. Viviparous, the young are born in winter.

Fierce Shark

Odontaspis ferox (Risso)

Mediterranean; east Atlantic to Madeira and the Gulf of Gascony.

Body long and fairly stout; *snout* prominent and more or less pointed; *eyes* round and small without nictating membranes; *teeth* are similar in both jaws, and are pointed with cusps on either side; *gill slits* 5, all situated in front of the pectoral fins; *fins* 2 dorsal fins, the front edge of the second fin situated in front of the anal fin. Tail fin long (total length of fish 3–3½ times tail), upper lobe elongate and notched near the end.

Dark grey to black on the back shading to light grey underneath. The first dorsal, second dorsal, anal and pelvic fins have black rear margins. The lower edge of the tail fin is also black.

Up to 2 m, occasionally 4 m.

Deep-water sharks which approach shore during the summer. Very little is known about their activities but stomach contents show that they are carnivorous.

Family:

LAMNIDAE

Medium to large oceanic **sharks**. There are 5 gill slits, 2 dorsal fins, and an anal fin which is about the same size as the second dorsal fin. The lower lobe of the tail fin nearly equals the upper giving a 'moon shape'. The snout is pointed and conical. The tail stalk is strongly flattened from above and below. There is a lateral keel on each side which extends onto the fin.

Mako

Isurus oxyrinchus Rafinesque

Mediterranean; throughout tropical and temperate Atlantic north to Scotland; western English Channel.

Body elongate and more slender than Porbeagle (*Lamna nasus*); tail stalk with keels present on either side and well developed upper and lower pits; *head* slender; *snout* pointed; *teeth* (see diagram above left) similar in each jaw, long and pointed with no cusps; *gill slits* 5, long and all situated in front of the pectoral fins; *fins*, 2 dorsal fins, the first situated behind the pectoral fins and the second, which is much smaller, originates just in front of the anal fin. Tail fin moon-shaped but with the upper lobe longer and notched.

Deep blue-grey above with an abrupt change to the whitish underside.

Up to 4 m.

Solitary and only found in shallow open water, rarely below 18 m. Often confused with the Porbeagle (*Lamna nasus*). They have often been observed jumping out of the water and are a much faster fish than the Porbeagle. They are carnivorous and feed on shoals of small fishes. Sexual maturity is reached at a length of 2 m in the male, rather longer in the female.

Longfin Mako Shark

Isurus paucus Guitart Manday

South-west coasts of Spain and Portugal, warm waters of the Atlantic; Indopacific.

Resembles the Mako (*Isurus oxyrinchus*) except for the pectoral fin, which is much longer. Little is known of the habits of this shark.

Porbeagle; Mackerel Shark

Lamna nasus (Bonnaterre)

Mediterranean; throughout central Atlantic; West Atlantic Newfoundland to Carolinas; east Atlantic north to Norway and Iceland, south to Morocco; North Sea; west Baltic; English Channel.

Body deep and robust; *eye* large; *keels* are present each side of the tail stalk and smaller ones on the tail fin; *pits* there are distinct upper and lower pits immediately in front of the tail fin; *snout* rounded; *teeth* (see diagram below left) pointed with cusps

Mako, *Isurus oxyrinchus* (Marty Snyderman).

White Shark, *Carcharodon carcharias* (Marty Snyderman).

on either side and similar in both jaws; *gill slits* 5, long but they do not reach the undersurface, they are all situated in front of the pectoral fins; *fins*, 2 dorsal fins; the 1st immediately behind the base of the pectoral fin and the 2nd, which is much smaller, situated above the anal fin. Tail fin moon-shaped but with the upper lobe notched and longer.

Dark grey to grey-blue on the back shading to white underneath; outer parts of pectoral fin dusky.

Up to 3 m, rarely 4 m.

Found from shallow water down to 200 m. Often confused with the Mako (*Isurus oxyrychus*). The Porbeagle is often encountered in colder water than other sharks. They are strong swimmers and feed principally on shoals of fish especially mackerel, which small groups of sharks sometimes follow into shore. The young are born at a length of 50–60 cm, probably in summer and only in small numbers. Considered excellent sporting fish and sometimes found on sale as 'swordfish'! However, they are an excellent food fish in their own right.

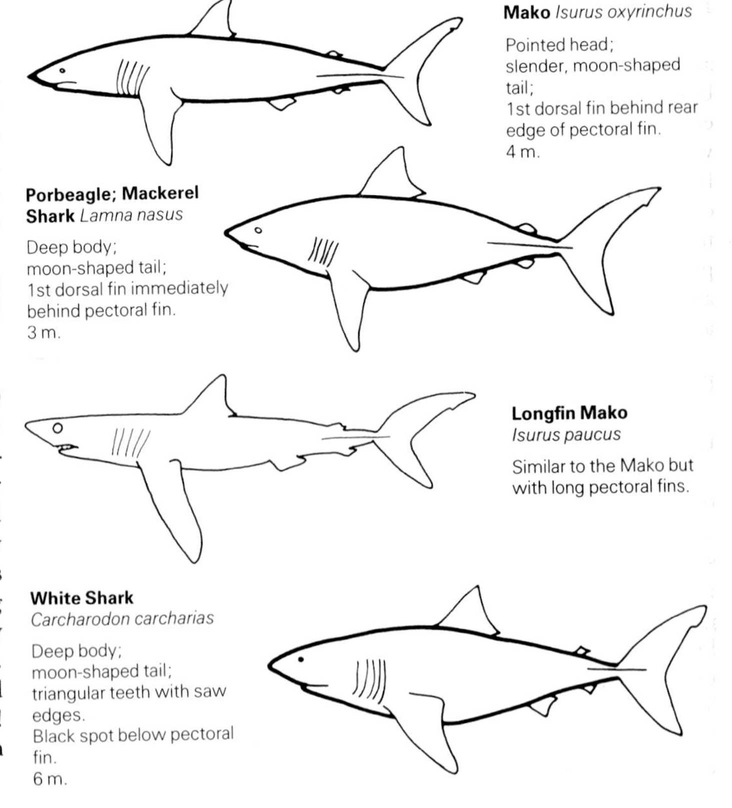

White Shark
Carcharodon carcharias (L.)
= *Carcharodon rondeletii* Müller & Henle

Mediterranean; Atlantic north to Biscay; all temperate and tropical seas.

Body deep and elongate with keels each side of the tail stalk; *snout* rounded; *teeth* (see diagram) triangular with saw edges and are characteristic of this shark; *gill slits* 5, long and all situated in front of the pectoral fins; *fins* 2 dorsal fins, 1st triangular and behind the pectoral fin; 2nd smaller and immediately in front of the anal fin. Tail fin large and moon-shaped with lobes nearly the same size; upper lobe notched. Pectoral fins large and long.

Grey-brown to slate-blue on the back, shading to light grey underneath. The pectoral fins have a dark spot at their tip on the undersurface.

Up to 10 m, more usually between 3–6 m.

Found in the open sea either singly or in small groups. White Sharks are the most feared of all the sharks and are known to attack bathers. Feed on fish, turtles, seals, refuse from ships, etc.

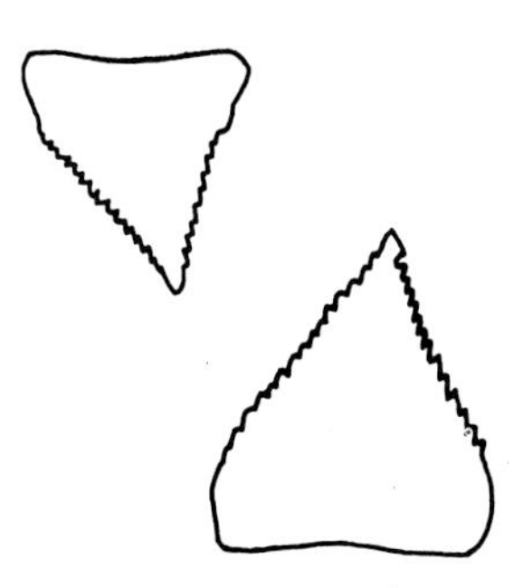

Family:
CETORHINIDAE

There are several species of **Basking Shark** in temperate seas but only one in our area. They are distinguished by having horny gill-rakers. Otherwise they resemble the Isuridae except that they have very numerous minute teeth. Viviparous.

Basking Shark
Cetorhinus maximus (Gunnerus)

Mediterranean; eastern Atlantic north to Iceland and Norway, south to Madeira.

Body deep and elongate with keels present on each side of the tail stalk; *snout* blunt. In young fish the snout is greatly enlarged and moveable; *eyes* small; *teeth* very small; *gill slits* 5, very long extending from the back to the undersurface and are characteristic of this shark; *fins* 2 dorsal fins, 1st situated between the pectoral and pelvic fins, 2nd immediately in front of the anal fin. Upper lobe of the tail fin larger and notched near rear end.

Brown or grey on the back shading to light grey underneath.

Up to 15 m, usually between 3–12 m. Can weigh over 3 tonnes.

Basking Sharks are primarily inhabitants of open water but during the summer they are thought to make a migration towards shore where they are occasionally seen by divers. They hibernate in the winter. A gregarious species usually seen in groups of between 12–15 individuals but sometimes up to 60. When seen they are usually 'basking' or swimming slowly near the surface, feeding with their mouths open and their snout and dorsal fin and tail tip just emerging above water. This habit has almost certainly given rise to 'sea-serpent' stories in the past as have their decaying carcasses washed up on beaches. Basking Sharks are plankton eaters, the gill system filters the plankton from the sea water and up to 1000 tonnes of water may be filtered each hour. Sexual maturity is reached when they have attained 4–6 m.

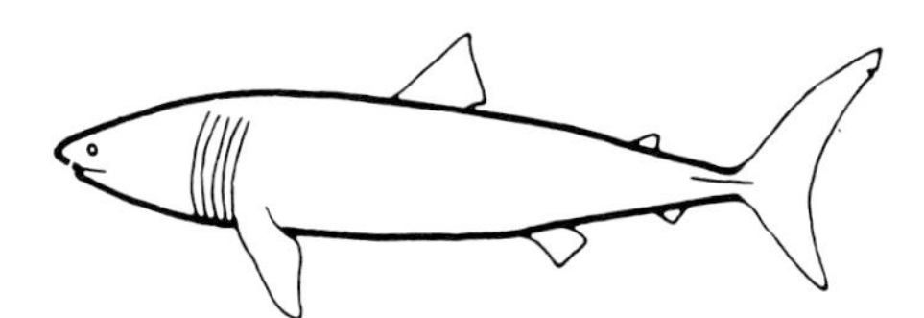

Basking Shark
Cetorhinus maximus

5 very large gill slits;
minute teeth;
young specimens have an elongate snout.
12 m.

Basking Shark, *Cetorhinus maximus* (Brian Pitkin).

Family:

ALOPIIDAE

The very long upper lobe to the tail fin of the **Thresher Shark** is a family characteristic. There are two dorsal fins, the second being the smaller and about the same size as the anal fin. The tail stalk is compressed laterally. There are spiracles and the eyes have nictating membranes. Viviparous.

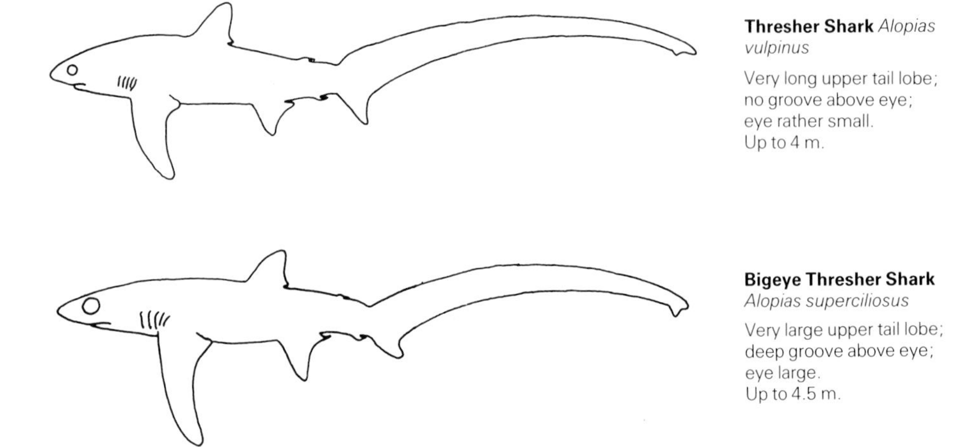

Thresher Shark *Alopias vulpinus*
Very long upper tail lobe; no groove above eye; eye rather small.
Up to 4 m.

Bigeye Thresher Shark *Alopias superciliosus*
Very large upper tail lobe; deep groove above eye; eye large.
Up to 4.5 m.

Thresher Shark
Alopias vulpinus (Bonnaterre)

Mediterranean; Atlantic; all temperate and tropical seas.

Body slender with no keels on the tail stalk. A pit is present on the top of the tail stalk immediately in front of the tail fin; *snout* blunt; *eyes* large; *teeth* small triangular and smooth (see diagram); *gill slits* 5,

very small; *fins* tail fin with upper lobe greatly elongate and measures nearly half the total length of the fish; 2 dorsal fins, the 1st situated between the pectoral and pelvic fins. Pectoral fins long.

Dark blue, grey, brown or black on the back shading to white underneath. The underside of the snout and the pectorals are dark.

Up to 3 m.

Adult fish are found in the open sea, but younger specimens are encountered near shore. Usually solitary but sometimes found in pairs and they have been reported as hunting in pairs. The two sharks will encircle a shoal of small fish and thrash the water with their tails forcing the fish into a smaller and smaller area where they can be attacked more easily. Occasionally fish and sea birds are stunned by the force of the tail. The young are born in summer in small numbers.

Bigeye Thresher Shark
Alopias superciliosus (Lowe)

Warm, temperate Atlantic, Portugal to Morocco, Madeira; Indopacific

Resembles the Thresher Shark (*Alopias vulpinus*) except for larger eyes, more pointed snout, and a deep horizontal groove running from above the eye to above the gill slits.

Up to 4.5 m.

Family:

GINGLYMOSTOMATIDAE

Medium-sized sharks that have a short snout and small mouth. The mouth is connected by a deep groove to the nostril, which has a large sensory barbel. Teeth are small. There are 5 small gill openings. The eyes have no nictating membrane. Bottom-living sharks which feed on invertebrates and small fish. The young are born alive. Found mainly in warm water.

Nurse Shark
Ginglymostoma cirratum (Bonaterre)

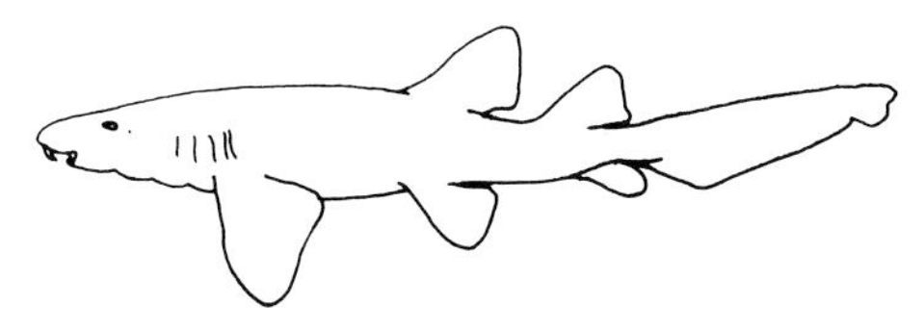

Nurse Shark
Ginglymostoma cirratum
2 similar dosal fins;
snout broad;
long barbel from each nostril.
4 m.

Western Atlantic from South Carolina southwards; eastern Atlantic, Gulf of Biscay; eastern Pacific.

Snout short and broad; *mouth* short and broad joined to nostrils by deep grooves; *nostrils* each have a prominent barbel; *gill* openings are small, the last two very close together.

Yellowish to greyish brown above, light yellowish underneath. There are small dark spots on the young.

Males to 2.6 m, females to 4.3 m.

Live on the bottom inshore, especially near mangrove swamps and sand flats, but also on rocky bottoms. They feed mostly on squids, shrimps, crabs, sea urchins, etc.

Nurse Shark, *Ginglymostoma cirratum* (Dick Clarke).

Family:
SCYLIORHINIDAE

These are the well-known **Dogfish** which are in reality small sharks. They are very widely distributed but always live near, or on, the bottom. Usually two dorsal fins set near the end of the body. The upper lobe of the tail fin is nearly horizontal. There are 5 gill slits, and spiracles are present. The teeth are small and numerous with a variable number of points. The nostrils are connected to the lip edge by a groove and this is a useful characteristic for identification. Oviperous.

Lesser Spotted Dogfish; Rough Hound

Scyliorhinus canicula (L.)
= *Scyllium canicula* Cuvier
Mediterranean; eastern Atlantic north to Norway south to Senegal; North Sea; English Channel.

Body slender and elongate gradually tapering towards the tail; *snout* rounded; *eyes* oval, without nictating membranes but with a thick fold of skin on lower margin; *nostrils* present and connected to the mouth by grooves which form a flap. The simple shape of this flap makes the Lesser Spotted Dogfish (diagram above right) easily distinguished from the Large Spotted Dogfish (diagram below right) (*Scyliorhinus stellaris*); *gill slits* 5, small, the last two situated over the pectoral fin; *skin* rough; *fins* 2 dorsal fins, the 2nd dorsal situated immediately behind the anal fin, the 1st behind the pelvic fin. Tail fin only slightly directed upwards, the lower lobe larger than the upper.

Back brown or reddish-brown, grey or yellow-grey with many small brown and black spots about same diameter as pupil, sometimes with small white spots. Sides lighter and gradually shading to grey-white underneath.

Up to 75 cm.

The Dogfish are nocturnal and during the day are frequently seen 'asleep' on mud or sand. Found at 3–400 m. At night they become active and feed on any small animals they find living on the bottom. At the end of the summer they move into deep water for mating and then return to shallow water for spawning, which occurs at different times according to locality. The eggs have yellowish-brown transparent cases 5–6.5 cm long and 2–3 cm wide with tendrils at the corners by which they are attached to seaweeds and other suitable objects. Day (1880) reports that the female will be found wriggling round and round any suitable object until she has securely attached the tendrils. She then draws the 2 eggs from her body and the remaining tendrils can be attached also. The embryos take about 9 months to develop and can be seen clearly through the cases.

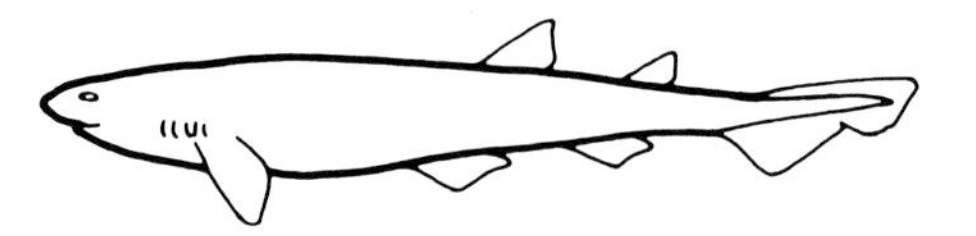

Lesser Spotted Dogfish; Rough Hound
Scyliorhinus canicula
Lower lobe of tail fin larger than upper and not directed upwards; snout flattened; 2nd dorsal fin immediately behind anal fin; type of nasal grooves (see text). Eastern Atlantic; Mediterranean. 75 cm.

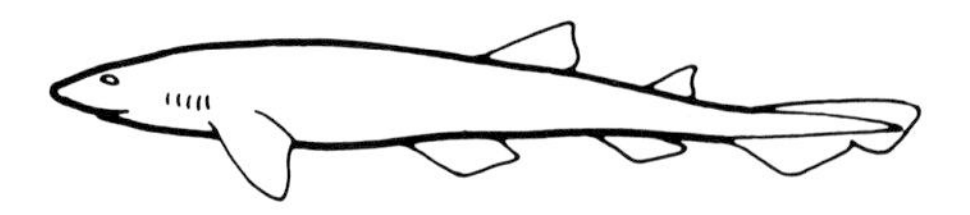

Large Spotted Dogfish; Nurse Hound; Bull Huss
Scyliorhinus stellaris
Lower lobe of tail fin larger than upper and not directed upwards; snout flattened; 2nd dorsal begins above anal fin; type of nasal grooves (see text). Eastern Atlantic; Mediterranean. 1 m.

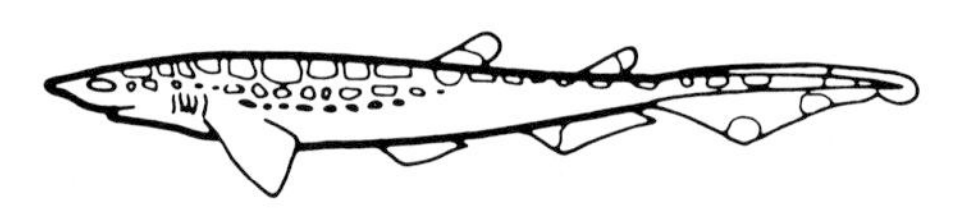

Black-mouthed Dogfish
Galeus melastomus
Lower lobe of tail fin larger than upper and not directed upwards; snout flattened; blotches along back, arranged in rows. Eastern Atlantic; Mediterranean. 90 cm.

Lesser Spotted Dogfish, *Scyliorhinus canicula* (Brian Picton).

Large Spotted Dogfish; Nurse Hound; Bull Huss

Scyliorhinus stellaris (L.)

Mediterranean; eastern Atlantic north to Norway; North Sea; English Channel.

Body slender and elongate gradually tapering towards tail; *snout* slightly more rounded than in the Lesser Spotted Dogfish; *eyes* oval, without nictating membranes but with a thick fold of skin on lower margin; *nostrils* present. Nasal flaps complicated do not connect with the mouth (diagram opposite below) as in the Lesser Spotted Dogfish (*Scyliorhinus canicula*) (diagram opposite above); *gill slits* 5 small, the last two situated over the pectoral fins; *skin* rough; *fins* 2 dorsal fins, the 2nd commences above the middle of the anal fin and the 1st directly behind the pelvic fin. Tail fin only slightly directed upwards, the lower lobe larger than the upper.

Back greyish or brownish gradually becoming lighter on sides and nearly white underneath. The back and sides have circular brown spots larger than pupil with a light centre and many smaller black spots. The colour is not a good guide to these fishes as it can vary with age and locality.

Up to 1 m.

The Large Spotted Dogfish is less common than the Lesser Spotted (*Scyliorhinus canicula*) and is usually found amongst rocks in quiet water. Most common at 20–60 m. They are nocturnal and feed by scavenging or on any animals that live on the bottom. The eggs are laid in cases which are dark brown, 10–13 cm long and 3.5 cm wide with tendrils at each corner. They are usually laid attached to seaweed probably throughout the year.

The skin is very rough and was used for polishing wood, alabaster and copper under the name of 'rubskin'.

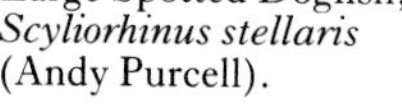

Large Spotted Dogfish, *Scyliorhinus stellaris* (Andy Purcell).

Black-mouthed Dogfish

Galeus melastomus Rafinesque-Schmaltz
= *Pristiurus melanostomus* Müller & Henle

Mediterranean; eastern Atlantic north to Scandanavia south to Madeira.

Body very slender and elongate; *snout* somewhat elongate; *eye* large, oval without nictating membranes; *gill slits* 5, the 5th situated over the pectoral fin; *fins* 2 dorsal fins both small and similar in shape. 1st dorsal situated behind the pelvic fin. Anal fin has an elongate base. Pectoral fin large. Tail fin long, more than ¼ total body length, only slightly directed upwards with the lower lobe larger than the upper. Along the upper edge of the fin is a row of flat spines arranged like a saw.

Back brown to yellow, sides lighter, underneath whitish. Along the back and sides are irregular darker blotches more or less arranged in rows. The inside of the mouth is black.

Up to 90 cm.

Found in deep muddy water, usually between 180–900 m but occasionally as shallow as 55 m. They are carnivores feeding on bottom-living animals but they also hunt fish in mid-water. Eggs are laid throughout the year in the Mediterranean and are distinct as the egg-cases have no tendrils at the corners. They measure about 6 cm long and 3 cm wide.

Family:
CARCHARHINIDAE

Moderate to large **sharks** with 5 gill slits. The upper lobe of the tail is longer than the lower; the tail is therefore not 'half-moon' shaped. There are 2 dorsal fins of which the 1st is the largest. There is an anal fin. The eyes, which are round or slightly oval, have a nictating membrane. The spiracles are very small or absent. The teeth are blade-like with only 1 point and only 1 or 2 rows are functional. Viviparous.

Blue Shark
Prionace glauca (L.)

Mediterranean; all tropical and temperate seas.

Body long and slender with distinct upper and lower pits immediately in front of the tail fin; *snout* long and pointed; *spiracles* not present; *eyes* with nictating membranes; *teeth* (see diagram) pointed; *gill slits* 5, the 5th situated over the pectoral fin; *skin* with very small scales hence a smooth feel; *fins* 2 dorsal fins, 1st much longer than the 2nd and situated between the pectoral and pelvic fins. Tail fin large, upper lobe longer than lower and with a notch on lower surface. Pectoral fins very long and sickle-shaped.

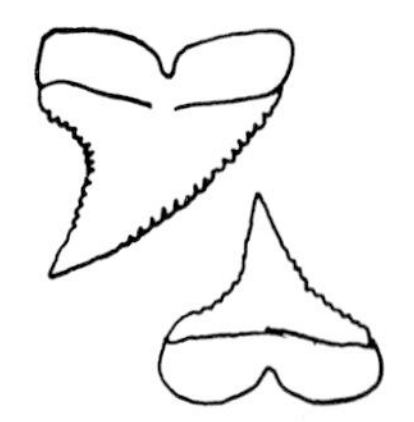

This shark is easily distinguished in life by its uniform dark blue back, lighter sides and white undersurface.

Up to 4 m, occasionally longer.

The Blue Shark is an extremely voracious, mainly nocturnal animal, which pursues and feeds on shoals of fish, other sharks, squids and occasionally man. Although usually found near the surface over deep water, they do, however, approach shore during the summer months, where, in south-west England they are fished from the bottom. They may form more or less unisexual shoals. Young fish are born alive during the summer. They are

Blue Shark, *Prionace glauca* (Marty Snyderman).

about 60 cm long and there may be more than 50 in a litter.

Sandbar Shark

Carcharhinus plumbeus (Nardo)
= *Carcharhinus milberti* (Valenciennes)

Both sides of Atlantic, southern New England to Florida, Portugal to North Africa; Mediterranean.

Body stout with snout short and rounded; *teeth* upper teeth toothed and broadly triangular, lower teeth with narrow erect cusps; *fins* 1st dorsal much higher than 2nd dorsal; pectorals triangular and relatively long.

Grey to bronze above, white below.

2 m.

Usually found on sandy bottoms from estuaries to the edge of the Continental Shelf. Feeds on molluscs, fish and crustacea.

Blacktip Reef Shark

Carcharhinus melanopterus (Quoy & Gaimard)

Mediterranean between Tunisia and Israel; Indopacific.

Body rather slender with snout short and blunt; *fins* rear dorsal fin slightly more than half height of front dorsal fin; pectoral fins rather broad.

Lemon-brown above, pale below. All fins tipped with black especially the front dorsal fin; trailing edge of tail fin has a black rim.

1.6 m.

It has migrated through the Suez Canal from the Indopacific waters of the Red Sea,

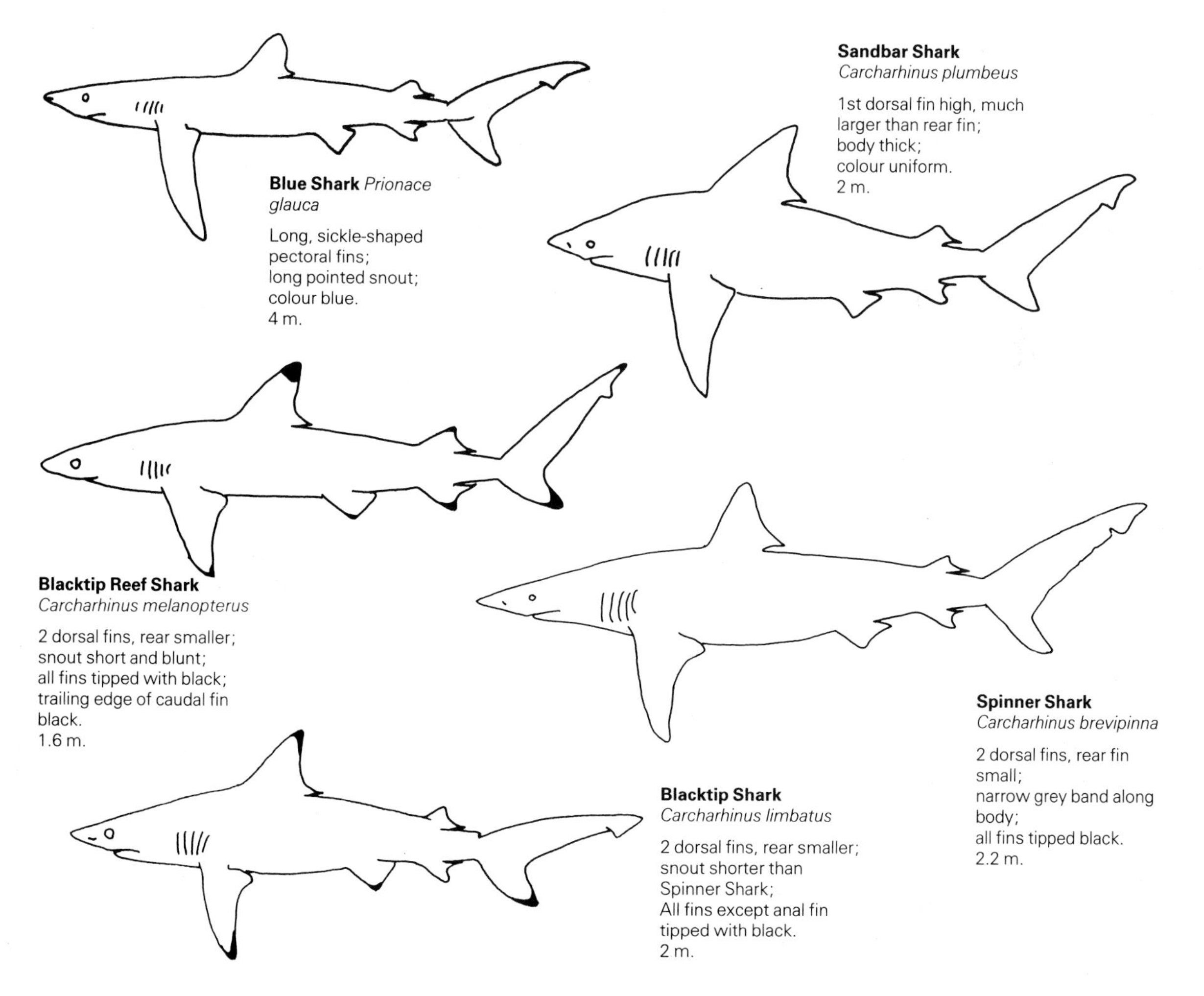

Blue Shark *Prionace glauca*
Long, sickle-shaped pectoral fins;
long pointed snout;
colour blue.
4 m.

Sandbar Shark
Carcharhinus plumbeus
1st dorsal fin high, much larger than rear fin;
body thick;
colour uniform.
2 m.

Blacktip Reef Shark
Carcharhinus melanopterus
2 dorsal fins, rear smaller;
snout short and blunt;
all fins tipped with black;
trailing edge of caudal fin black.
1.6 m.

Spinner Shark
Carcharhinus brevipinna
2 dorsal fins, rear fin small;
narrow grey band along body;
all fins tipped black.
2.2 m.

Blacktip Shark
Carcharhinus limbatus
2 dorsal fins, rear smaller;
snout shorter than Spinner Shark;
All fins except anal fin tipped with black.
2 m.

Blacktip Reef Shark, *Carcharhinus melanopterus* (R.H. Johnson).

and is now established in the southern and eastern Mediterranean.

Spinner Shark
Carcharhinus brevipinna (Muller & Henle)
= *Carcharhinus maculipinnis* (Poey)

Eastern Atlantic south from Spain into both sides of the tropical Atlantic; Mediterranean.

Snout long and conical, length between tip of snout and front of nostril 1.1–1.4 times the distance from the front of the nostrils to the mouth; *teeth* spike-like serrate above, serrate or smooth below, rear teeth with smooth trailing edge.

2.2 m.

Grey to bronze above, pale below. All fins tipped with black except when very young.

A fast-swimming shark of coastal and open water where it feeds mostly on fish. Like the Blacktip Shark (*Carcharhinus limbatus*) it has been observed to leap from the water, rotate more than once and fall back again.

Blacktip Shark
Carcharhinus limbatus (Valenciennes)

West Atlantic from New England to southern Brazil; western basin of Mediterranean; warm, temperate and tropical waters of the Atlantic; Indopacific.

Snout rather long, distance from tip of snout to nostrils equal to, or less than, the distance from the front of the nostril to the mouth. *Fins* rear dorsal fin much smaller than front; *teeth* as for the Spinner Shark (*Carcharhinus brevipinna*) but rear teeth have a notch (see diagram right).

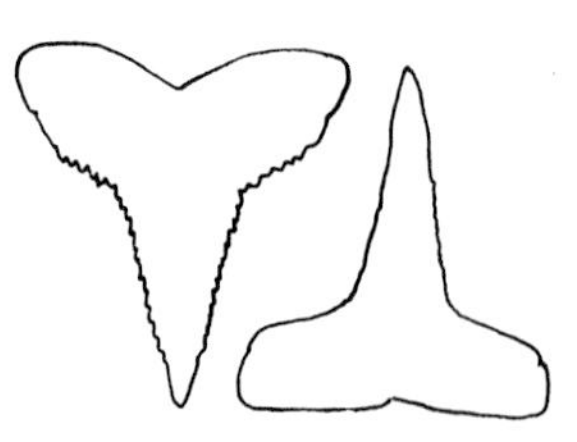

All fins tipped with black except the anal fin.

2 m.

Fast swimmers near the surface, they are often in shoals. Like the Spinner Shark it has been observed to make spiral leaps into the air.

Family:
TRIAKIDAE

The **Smooth Hounds** live in moderately shallow water. They have 2 dorsal fins of which the 1st is in front of the pelvic fins. The tail is not moon-shaped. There in no nictating membrane but the lower eyelid has a longitudinal fold. The teeth are small and rounded, with 3–4 points. Viviparous or ovoviviparous.

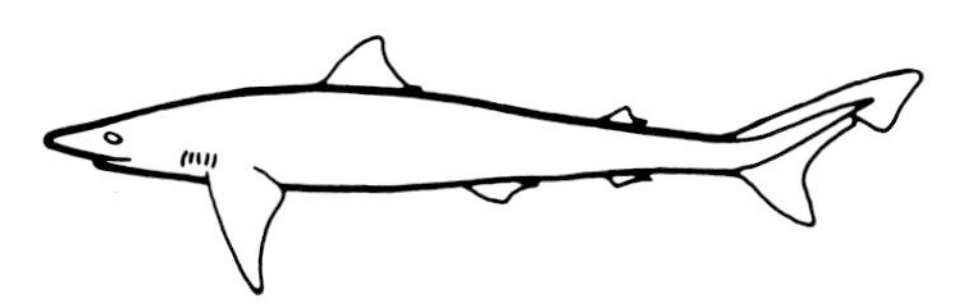

Tope *Galeorhinus galeus*

Lower lobe of tail fin larger than upper and directed upwards; 2nd dorsal fin much smaller than 1st; colour uniform. 2 m.

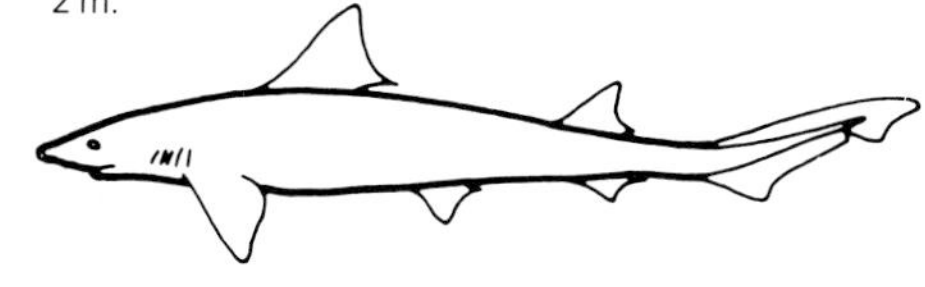

Smooth Hound
Mustelus mustelus

Lower lobe of tail fin larger than upper and directed upwards; 1st dorsal fin in front of pelvic fins; colour uniform. 1.6 m.

Tope

Galeorhinus galeus (L.)
= *Eugaleus galeus* Gill

Mediterranean; eastern Atlantic; North Sea; English Channel; all temperate and tropical seas.

Body slender; *snout* long and pointed; *eyes* oval with nictating membranes; *teeth* (see diagram left) triangular and pointed; *gill slits* 5, the 5th situated over the pectoral fin; *fins* 2 dorsal fins, 1st situated between the pectoral and pelvic fins but nearer the pectoral than the pelvics; 2nd smaller than the 1st and slightly in front of the anal fin. Tail fin with a large notch on the lower edge of the upper lobe. Pectoral fin large.

Uniform dark grey on back, sides lighter, underneath whitish.

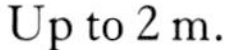

Up to 2 m.

Bottom-living sharks usually found on gravel or sand from shallow water down to 400 m, but they may also swim near the surface. Fairly frequently encountered near coasts in summer either singly or in small shoals. They feed mainly on fish but also on other animals associated with the bottom. Young sharks are born alive during the summer and 52 embryos have been recorded from one female though more usually between 20–40 are born in one litter. The Tope is a common and popular game fish. The fins are used for shark fin soup.

Smooth Hound

Mustelus mustelus (L.)

Mediterranean; eastern Atlantic north to Britain.

Body slender; *snout* rather long and pointed; *eye* oval, without nictating membranes but with a thick fold of skin on lower margin; *teeth* mosaic-like; *gill slits* 5, the 5th situated over the pectoral fin; *skin* rough, the scales oval with ridges which do not extend the full length of the scale; *fins* 2 dorsal fins, 1st situated behind the pectoral fin and the 2nd commences in front of the anal fin. Tail fin angled upwards and has a large notch.

Usually a uniform grey in colour with lighter sides. Occasionally there may be black spots scattered over the surface. Underneath whitish.

Up to 1.6 m.

Usually found on sandy and muddy bottoms from about 5 m down to the Continental shelf. Mainly nocturnal they feed on any bottom-living animals. Viviparous. A maximum of 28 are born in one litter.

Family:
SPHYRNIDAE

The **Hammerhead Sharks** resemble the Carcharhinidae except that the front portion of the head is flattened and is extended into one lobe on each side.

Scalloped Hammerhead Shark
Sphyrna lewini (Griffith & Smith)

Western basin of Mediterranean; warm temperate Atlantic and Indopacific coasts.

Head outline moderately curved with scallops opposite each nostril and on mid-line; *teeth* smooth-edged or slightly rough in old specimens; *fins* rear point of dorsal fin reaches back to at least the origin of the pelvic fin.

Light greyish-brown above, pale below. Pectoral fins have pronounced spot.

15 m.

Feeds on fish, smaller sharks, crabs, etc. They are found from close inshore and in open water, usually swimming at a depth of 10 m.

Scalloped Hammerhead, *Sphyrna lewini* (Christian Petron).

Common Hammerhead Shark
Sphyrna zygaena (L.)
= *Zygaena malleus* Valenciennes

Mediterranean; Atlantic north to Biscay; occasionally English Channel; all temperate and tropical seas.

Head greatly flattened and extended into 2 lobes to give the characteristic 'hammer-head' shape; *snout* smoothly rounded (see diagram); *eyes* situated at extreme ends of the head lobes and with nictating membranes; *teeth* triangular and pointed; *gill slits* 5, the 4th and 5th situated over the pectoral fins; *fins* 2 dorsal fins, the 1st immediately behind the pectoral fin, 2nd much smaller than the 1st and opposite the anal fin. Tail fin with a long upper lobe notched underneath.

Slate-grey, brown-grey or greenish on the back, sides and undersurface lighter. The tips and rear margins of the fins usually darker.

More than 4 m.

Usually found in deep water from the surface down to 400 m, occasionally inshore. They are excellent swimmers and feed mainly on fish but also crustaceans and cephalopods. There are confirmed reports

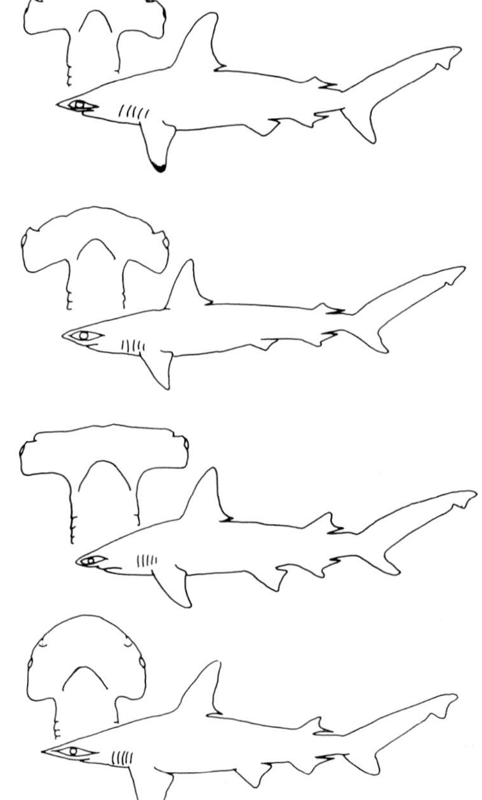

Scalloped Hammerhead Shark *Sphyrna lewini*

Scalloped outline of head; black tip on pectoral fins. 4 m.

Common Hammerhead Shark *Sphyrna zygaena*

Hammer-like shape to head, no indentation at snout;
pectoral fins often shaded darker.
4 m.

Great Hammerhead Shark *Sphyrna mokarran*

Hammer-like shape to head, indentation at snout;
pectoral unmarked.
4 m, can be up to 6 m.

Bonnet Shark; Shovel Head *Sphyrna tiburo*

Shovel-like outline to head;
no fin markings;
western Atlantic.
10 m.

Great Hammerhead, *Sphyrna mokarran* (Doug Perrine).

that they have attacked man. The young are born alive at about 50 cm and with between 37–39 in a litter.

Great Hammerhead Shark
Sphyrna mokarran (Ruppell)

All temperate and tropical oceans.

This shark is very similar to the Common Hammerhead (*Sphyrna zygaena*) except that the centre of the snout is indented and the body is rather heavier.

Bonnet Shark; Shovel Head
Sphyrna tiburo L.

Western Atlantic from Chesapeake Bay south to Brazil; tropical eastern Atlantic.

Head shovel-shaped with rounded outline and not indented at the midline or opposite the nostrils; *teeth* smooth-edged.

Grey or greyish-brown above, paler below. There are no conspicuous markings on the fins.

A very large shark, up to 10 m long.

Often found close inshore in bays and estuaries in shallow water. Rather sluggish. It feeds mainly on crustacea and small fish and has been observed rooting around under coral rocks. Harmless to man.

Bonnet Shark, *Sphyrna tiburo* (Ken Lucas).

Family:
SQUALIDAE

These are small sharks which chiefly live in very deep water. They are distinguished by having no anal fin and in most species there is a spine in front of each of the 2 dorsal fins. There are 5 gill slits, spiracles are present, the eye has no nictating membrane and the nostrils are separate from the mouth. There is a longitudinal fold of skin in the rear part of the body behind the pelvic fins. Viviparous.

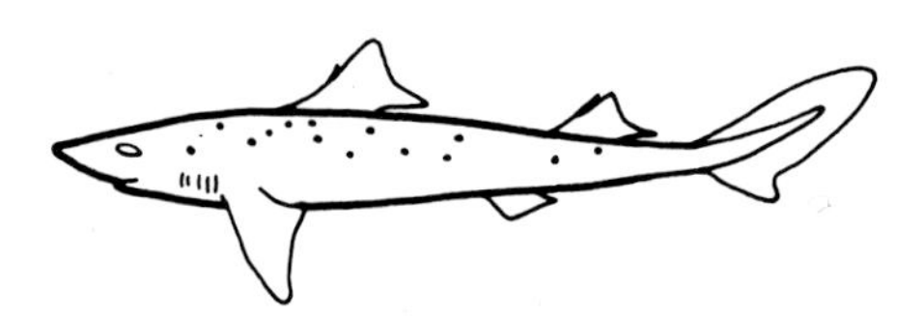

Picked Dogfish; Spurdog; Common Spiny Dogfish *Squalus acanthias*

Spines in front of dorsal fins, 2nd dorsal spine not as long as the fin; white spots on back and sides; no anal fin. 1.2 m.

Picked Dogfish; Spurdog; Common Spiny Dogfish
Squalus acanthias (L.)
= *Acanthias vulgaris* Risso

Western Atlantic from Labrador to Florida; eastern Atlantic from Norway to Morocco; Mediterranean.

Body slender; *snout* rounded; *eye* large and oval; *teeth* 24–28 in upper jaw, 22–24 in lower; *gill slits* 5, all situated in front of the pectoral fins; *scales* small; *fins* 2 dorsal fins with spines in front of them. Both spines are shorter than the fins. No anal fin. Tail fin with large upper lobe and no notches. 1st dorsal fin situated between the pectoral and pelvic fins.

Light grey, dark grey or brownish on the back, lighter on flanks and white underneath. The back and sides have irregular white spots which may disappear with age.

Up to 1.2 m.

These are small common migratory bottom-living sharks which may be found from shallow water down to about 950 m within a temperature range of 6°–15°C. Usually in shoals but may occasionally be found singly. Young fish are born alive after a pregnancy of 18–22 months, the female approaching shore to give birth. They feed mainly on fish but also on crustacea and coelenterates and will attack shoals of fish that have been netted. In 1882 there was considerable concern that some fisheries would have to cease due to the damage caused by these fish. The spines are slightly poisonous. Edible; were once considered excellent smoked.

Blainville's Dogfish
Squalus fernandinus Molina
= *Acanthias blainvillei* Risso

All temperate oceans; Mediterranean.

Body long and slender; *snout* rounded; *eye* oval and large; *gill slits* 5, all situated in front of the pectoral fins; *fins* 2 dorsal fins both with spines in front. The first dorsal spine is shorter than the fin but the 2nd dorsal spine is at least as long as the fin. The 1st dorsal is situated immediately behind the base of the pectoral fin. Tail fin with large upper lobe and no notch. Pectoral fins large. No anal fin.

Uniform grey or brown on the back, sides lighter, white underneath.

Up to 70 cm.

Occasionally seen in groups near the coast but more usually lives from 50 m down to 700 m. Little is known about the habits of this fish except that it is carnivorous.

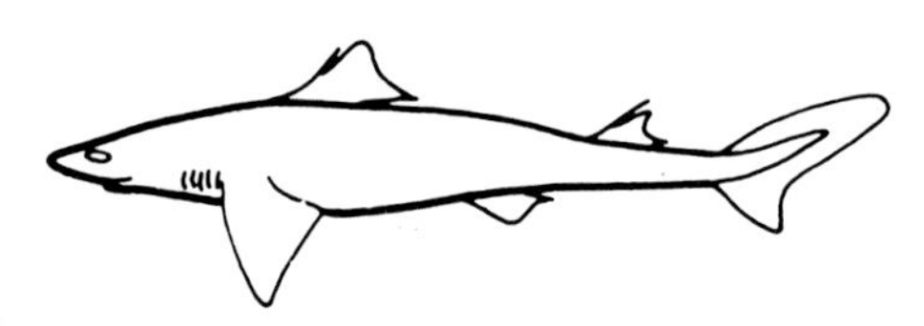

Blainville's Dogfish
Squalus fernandinus

Spines in front of dorsal fins, 2nd dorsal spine at least as long as the fin; no anal fin. 70 cm.

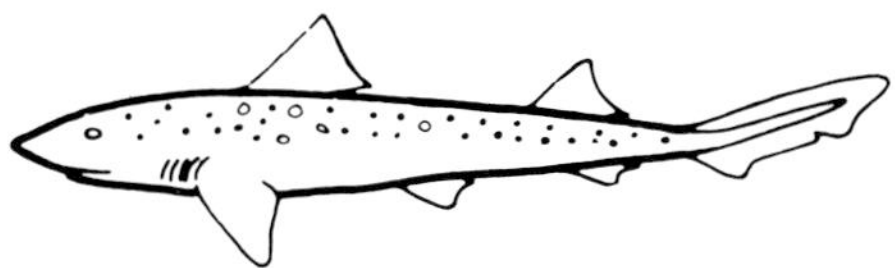

Stellate Smooth Hound *Mustelus asterias*

Lower lobe of tail fin larger than upper and directed upwards; 1st dorsal fin in front of pelvic fins; white spots on back and sides. 2 m.

Stellate Smooth Hound
Mustelus asterias Cloquet

Mediterranean; eastern Atlantic north to Scotland south to Canaries; North Sea; English Channel.

Body slender; *snout* rather pointed; *eye* oval without nictating membranes but with a thick fold of skin on lower margin; *nostrils* completely separate from mouth; *teeth* mosaic-like in 5–7 rows; *gill slits* 5, the 5th situated over the pectoral fins; *skin* rough, the scales rounded with one large central ridge and 2 smaller side ones which extend the full length of the scale; *fins* 2 dorsal fins, 1st starts immediately behind the pectoral fin base, 2nd dorsal starts in front of anal fin. Tail fin angled upwards, a distinct notch on rear edge.

Back light to dark grey; sides lighter, underneath whitish. Irregular white spots are scattered over the back and sides, particularly along the lateral line.

Up to 2 m.

Found on muddy sand at 5–165 m. Predominantly nocturnal, they feed on any bottom-living animals but principally crustaceans and occasionally fishes. Ovoviviparous. Up to 15 may be born in one litter.

Blackspotted Smooth Hound
Mustelus punctulatus Risso
= *Mustelus mediterraneus* Quignard & Capapé

Mediterranean, south Portugal to Morocco.

Body form resembles the Smooth Hound (*Mustelus asterias*), but *scales* are much longer than wide with weak ridges extending along less than half their length.

Grey to brown above with small dark blotches on back and sides, white below. Trailing edges of dorsal fins have dark margins.

60 cm.

Lives on the bottom in coastal water where it feeds on fish, crabs, etc.

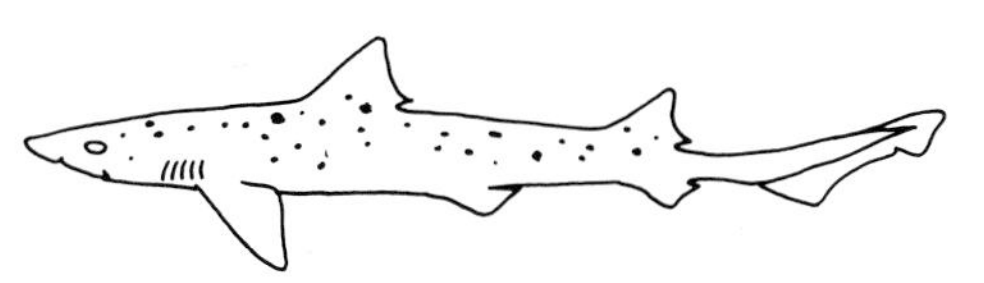

Blackspotted Smooth Hound *Mustelus punctulatus*

Lower lobe of tail fin larger than upper and directed upwards; black spots on back and sides. 60 cm.

Family:
SQUATINIDAE

The **Monkfish** are usually seen on the bottom where their flattened and elongated shape is unmistakeable. The pectoral fins are much enlarged. There are 2 dorsal fins which have no spines and there is no anal fin. There are 5 gill slits, large spiracles, large eyes with no nictating membrane and the nostrils are separate from the mouth. The nostrils have a valve which is generally fringed. The teeth are small and similar in both jaws. Ovoviviperous.

Monkfish; Angel Shark
Squatina squatina (L.)

Mediterranean; eastern Atlantic from Scandinavia to the Canaries; English Channel.

Body flattened with a characteristic outline comprised of head, pectoral and pelvic fins and tail; *head* wide and rounded; *nostrils* with small slightly branched barbels; *eyes* small; *spiracles* large with diameter greater than eye diameter; *scales* small and rough, cover whole of upper surface and most of lower surface; *fins* pectoral fins very large, pelvic fins similar but smaller than the pectorals. No anal fin.

2 dorsal fins situated on tail behind the pelvic fins. Tail fin with lower lobe greater than upper.

Grey, brown or greenish on the back frequently with darker mottlings and occasionally with lines of lighter spots. Undersurface white.

Up to 2 m.

They live at 5–100 m on sand and gravel and are usually seen lying on the bottom or partly buried. They feed on any animals which they find living on the bottom. The young are born alive during summer in Northern Europe and winter in the Mediterranean with 7–25 in a litter.

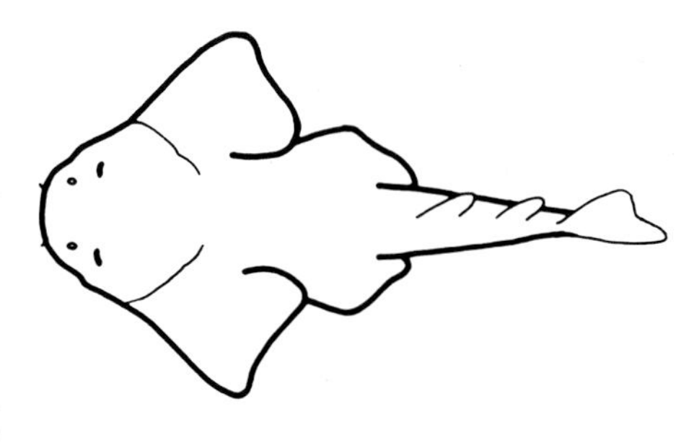

Monkfish; Angel Shark
Squatina squatina

Rounded pectoral fins; nostril barbels small and slightly branched.
Eastern Atlantic; Mediterranean.
2 m.

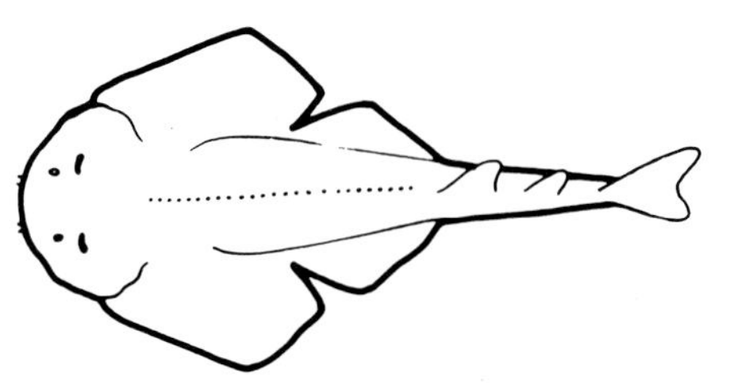

Monkfish *Squatina aculeata*

Row of large spines along centre back.
Eastern Atlantic; Mediterranean; a related species in West Atlantic.
1.5 m.

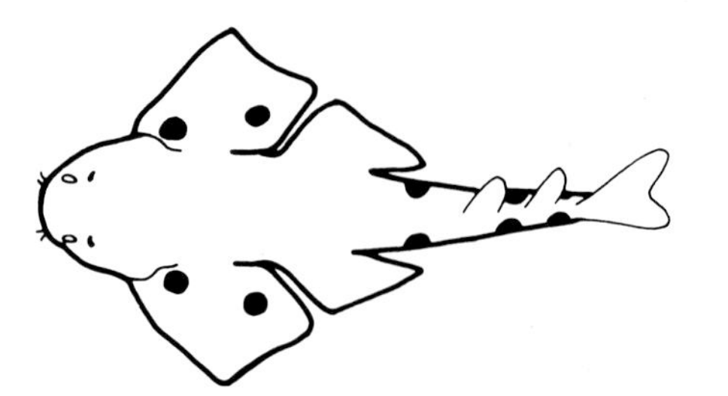

Monkfish *Squatina oculata*

Pectoral and pelvic fins distinctly separate; black spots on pelvic fins; nostril barbels prominent and fringed.
Eastern Atlantic; Mediterranean.
1.5 m.

Monkfish
Squatina aculeata (L.)

Mediterranean and adjacent Atlantic.

Body flattened with a characteristic outline comprised of head, pectoral and pelvic fins and tail; *head* wide and rounded; *nostrils* with small very branched barbels; *eye* width equal to or less than spiracle width; *spiracles* with fringed front edge; *scales* small and rough, cover whole of back but underneath only on the front edge of the pectoral and pelvic fins and along the centre of the tail. There is a distinct row of spines along the centre back and a number of spines between the eyes; *fins* pectoral fins large and rather rectangular. 2 dorsal fins, 1st situated immediately behind the pelvic fins.

Back sandy brown with symmetrically arranged white spots. Undersurface white.

Up to 1.5 m.

Not common, usually found in muddy regions. Very little is known about this fish but its habits are probably similar to the Monkfish (*Squatina squatina*).

A closely related species, *Squatina dumerili*, is found in the western Atlantic from southern New England south to Florida.

Monkfish
Squatina oculata Bonaparte

Mediterranean and adjacent Atlantic.

Body flattened with a characteristic outline comprised of head, pectoral and pelvic fins and tail; *head* wide and rounded; *nostrils* with small very branched barbels; *eyes* equal to or greater than spiracle width; *spiracles* with fringed front edge; *scales* rough and cover whole of upper surface but underneath only on the front edge of the pectoral and pelvic fins and along centre of tail; *fins* pectoral fins large and a distinct gap between them and the pelvics. 2 dorsal fins situated behind the pelvic fins.

Back reddish brown, brownish or grey with small white spots regularly distributed over the surface. There are also larger black spots on the pectoral fins and tail. Undersurface white.

Up to 1.5 m.

Little is known about these fish except that they are usually found at 50–300 m on mud or sand. Living fishes are their usual prey but they may only be able to catch them when the fish swim directly over their heads.

Order:
HYPOTREMATA
Rays, Electric Rays and Guitar Fish.

The front margin of the pectoral fin is continuous with the head. The gill slits are on the lower surface of the head.

Family:
RHINOBATIDAE

The **Guitar Fish** are normally seen resting on the bottom and have a ray-like front half but are more shark-like behind. The pelvic fins are well separated from the pectorals. The two dorsal fins are about equal in length. The tail fin is well developed but there is no well-defined lower lobe. Ovoviviperous.

Guitar Fish
Rhinobatos rhinobatos (L.)

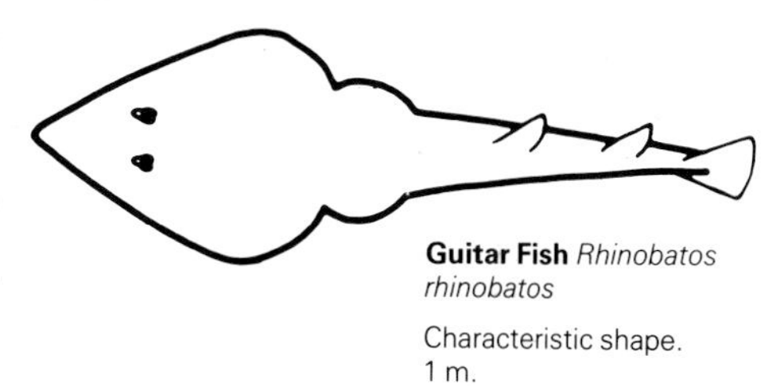

Guitar Fish *Rhinobatos rhinobatos*

Characteristic shape.
1 m.

Mediterranean; eastern Atlantic north to Biscay.

Body front half-flattened and rear half shark-like; *head* is continuous with the body; *snout* elongate and pointed; *eyes* immediately in front of the spiracles; *skin* with a row of spines along the centre back; *fins* pectoral fins continuous with head. 2 dorsal fins situated behind the pelvic fins. Tail fin not lobed.

Back brownish, greyish or yellowish. The dorsal and tail fins usually have whitish edges. Undersurface white.

Up to 1 m.

Found in muddy and sandy areas either resting on the bottom or half buried. The 2 dorsal fins are a characteristic and conspicuous feature of this fish in life. Feeds on any bottom-living animals. Young are born from February to July in the Mediterranean.

A related species, the Blackfin Guitar Fish, (*Rhinobatos cemiculus*) is usually distinguished by a dark blotch on the snout.

Guitar Fish, *Rhinobatos rhinobatos* (Andy Purcell).

Family;
TORPEDINIDAE

The **Electric Rays** are able to deliver a shock of unpleasant intensity to a diver if touched. The head is a distinct disc with a separate rounded tail. There is a tail fin and two dorsal fins. The skin is naked. There is an electric organ on each side of the disc.

Electric Rays are usually encountered almost buried in the sand with only their spiracles and eyes visible. The outline of the disc is just discernible. They eat fishes, which they envelop and stun with an electric discharge.

An electric ray was responsible for the first scientific description of electricity. Aristotle described the shock received when a ray is prodded by a metal rod.

Eyed Electric Ray
Torpedo torpedo (L.)
= *Torpedo ocellata* Rafinesque

Mediterranean; eastern Atlantic north to Biscay south to Angola.

Body disc-shaped and contains 2 large electric organs; *tail* short and fat; *spiracles* wider than eyes with a margin slightly fringed or smooth; *skin* without scales; *fins* pectoral fins large and form outer edge of the disc.

Back light to dark brown with 5 (sometimes 1, 3 or 7) blue spots surrounded by black and light brown rings. Undersurface off-white.

Up to 60 cm.

Found on or half buried in sand and amongst sea-grass beds. From shallow water to 50 m, occasionally down to 200 m. They are solitary and nocturnal and feed on fish and crustacea. The embryos take about 5 months to develop and the number of young in a litter depends on the size of the female.

Marbled Electric Ray
Torpedo marmorata Risso

Mediterranean; eastern Atlantic north to Britany south to South Africa; occasionally English Channel.

Body disc-shaped and contains 2 large electric organs; *tail* short and fleshly; *spiracles* with 6–8 small lobes; *skin* without scales; *fins* pectoral fins large and fleshy and form the outer margin of the disc.

Back light or dark brown with a darker marbling. Undersurface whitish with a dark edge.

Up to 60 cm.

Found in shallow water usually at 2–20 m, occasionally down to 100 m. Usually found on or half-buried in sand and mud. They are solitary and nocturnal and feed on bottom-living animals including fish and crustacea. In the Mediterranean the young are born in September and October with between 5 and 36 in a litter.

In the past the flesh of this fish was recommended to be eaten by epileptics and the electric organs to be applied to their heads for the therapeutic value of the shock. Underwater the shock gives a severe jolt but is not normally dangerous.

Dark Electric Ray
Torpedo nobiliana Bonaparte

Mediterranean; eastern Atlantic north to Orkneys south to North Africa; west Atlantic Nova Scotia south to Chesapeake Bay.

Body disc-shaped and contains 2 large electric organs. Young specimens have 2 dents on the front margin of the disc; *spiracles* with smooth edges; *skin* without scales; *fins* pectoral fins large and form outer margin of the disc.

Back a uniform dark grey or black. Undersurface whitish with dark edges.

Up to 180 cm.

Found on or half buried in sand and mud at 10–350 m. In the Mediterranean only, occasionally above 100 m. They are solitary and nocturnal, feeding mainly on fishes which they catch by stunning with an electric shock. A voltage as high as 220 v has been measured from this fish but underwater the voltage is probably much less, but if touched a diver can receive a sharp shock.

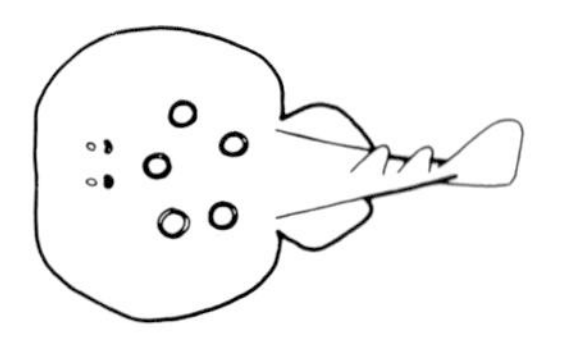

Eyed Electric Ray
Torpedo torpedo
5 large spots on disc; spiracles with no fringes or only slightly fringed. Eastern Atlantic; Mediterranean. 60 cm.

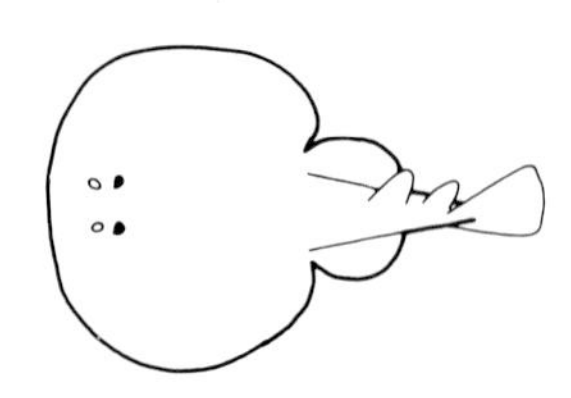

Marbled Electric Ray
Torpedo marmorata
Marbled coloration; 6 to 8 small lobes on spiracle. Eastern Atlantic; Mediterranean. 60 cm.

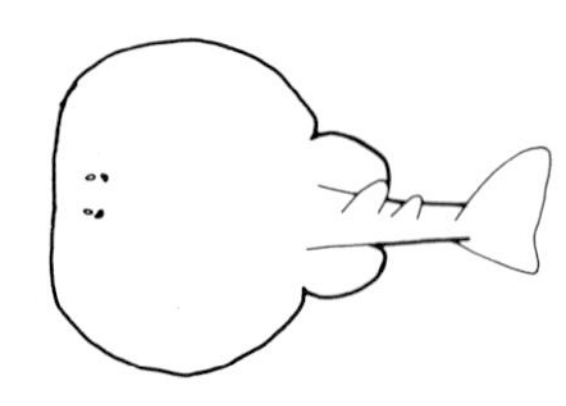

Dark Electric Ray
Torpedo nobiliana
Uniform dark colour; spiracles smooth. Eastern and western Atlantic; Mediterranean. 130 cm.

Marbled Electric Ray, *Torpedo marmorata* (Guido Picchetti).

Dark Electric Ray, *Torpedo nobiliana* (Gilbert van Rijckevorsel).

Family:

RAJIDAE

The disc is diamond-shaped and the tail relatively small. The skin is spiny, the tail especially so. At the rear end of the tail are the 2 dorsal fins. Oviperous.

There is no biological distinction between **Skates** and **Rays** but those with long noses are normally called Skates. Many species are biologically important as they are abundant on the trawling grounds and are good to eat. They are normally seen partly buried in the sand and are easy to overlook. They swim by flapping their pectoral fins in an undulating motion.

The Skates and Rays are very difficult to identify partly because there is great variation within the species and partly because the young differ from the adults and the males from the females.

Starry Ray

Raja asterias Delaroche
= *Raja punctata* Risso

Mediterranean.

Body disc-shaped with a gently undulating front edge; *snout* slightly pronounced but blunt; *skin* rough on upper surface. Row of *spines* (between 60–70) run from immediately behind the eyes to the 1st dorsal fin. Two more rows of spines run along the tail. Adult male fish have patches of spines on the wings and at the sides of the head. There are sometimes spines between the two dorsal fins.

Back brownish to brownish-red, olive-green to yellow with a large number of small black spots and fewer larger yellowish spots surrounded by a ring of small brownish-black dots. The snout is light brown with no patterning.

Up to 70 cm.

Found on or half buried in mud and sandy bottoms usually at 7–40 m, but also down to 100 m. They feed on any bottom living animals including fish and crustacea. The egg capsules are a transparent greenish-brown, rectangular and measure about 45 × 30 mm and have elongated corners. There are a number of rays in the Mediterranean with which this one may be confused including, *Raja brachyura* and *Raja montagui*. *Raja asterias*, however, is far more common than these species.

Starry Ray, *Raja asterias* (Jim Greenfield).

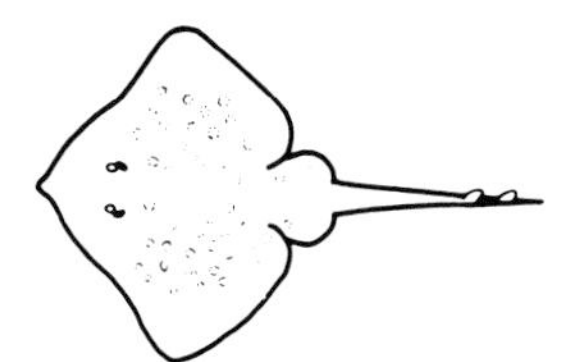

Starry Ray *Raja asterias*

Yellowish spots outlined in dark dots.
Mediterranean.
70 cm.

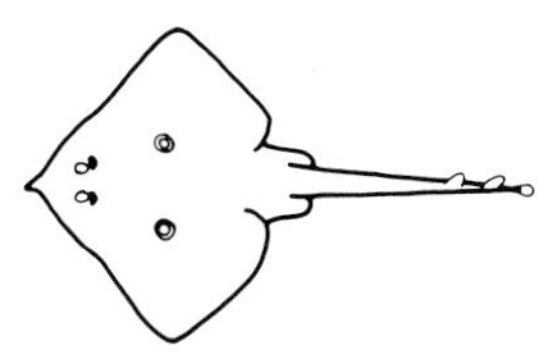

Brown Ray *Raja miraletus*

1 blue spot outlined in black and light brown on each wing;
prominent snout.
Eastern Atlantic;
Mediterranean.
60 cm.

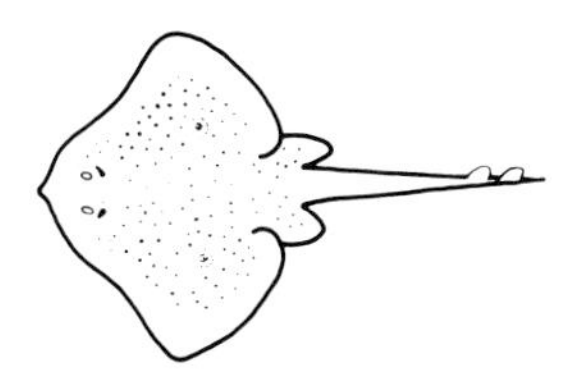

Spotted Ray *Raja montagui*

Small dark spots scattered over surface not extending to wing margins.
Eastern Atlantic;
Mediterranean.
75 cm.

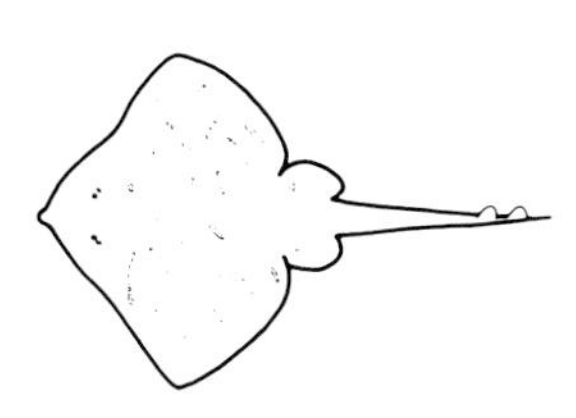

Painted Ray *Raja microcellata*

Greyish or brownish with lighter spots and lines.
Eastern Atlantic.
80 cm.

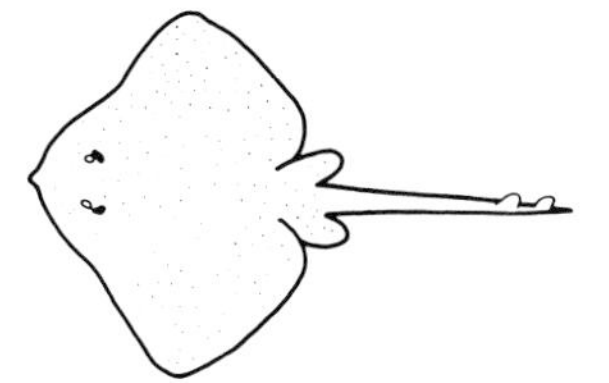

Blonde Ray *Raja brachyura*

Small dark spots scattered over surface and extending to extreme edge of wings.
Eastern Atlantic;
Mediterranean.
115 cm.

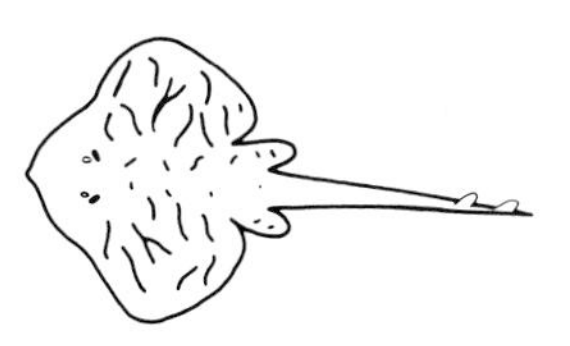

Undulate Ray; Painted Ray *Raja undulata*

Rounded pectoral fins;
wavy dark brown bands.
Eastern Atlantic;
Mediterranean.
120 cm.

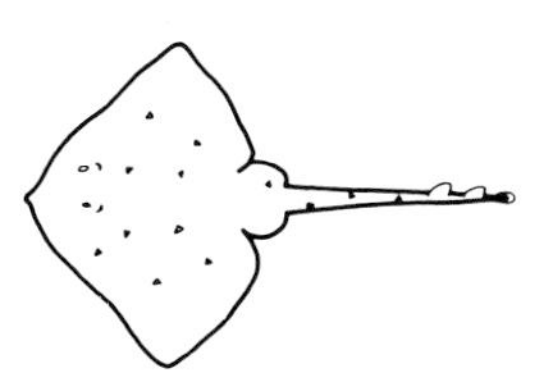

Thornback Ray *Raja clavata*

Large broad-based spines scattered over surface.
Eastern Atlantic;
Mediterranean.
85 cm.

Cuckoo Ray *Raja naevus*

Large black and yellow marbled spot on each wing.
Eastern Atlantic;
Mediterranean.
70 cm.

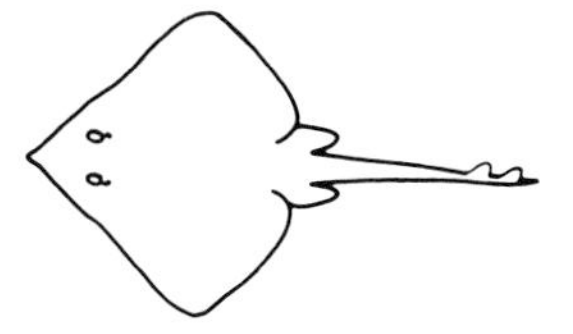

Shagreen Ray; Fullers Ray *Raja fullonica*

Snout slightly elongate;
no distinct markings.
Eastern Atlantic.
110 cm.

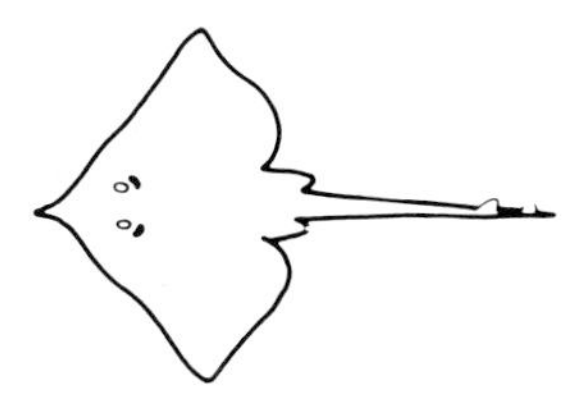

White Skate; Bottle-nosed Skate *Raja alba*

Elongate snout;
sharply angled wings;
underside white with darker edges.
Eastern Atlantic;
Mediterranean.
200 cm.

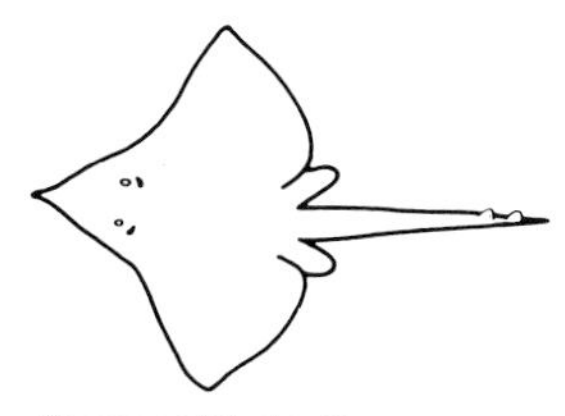

Common Skate; Grey or Blue Skate *Raja batis*

Elongate snout;
front edge of wings slightly concave;
grey underside.
Eastern Atlantic.
235 cm.

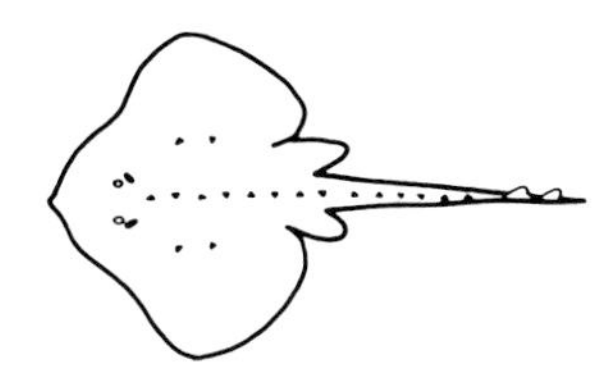

Starry Ray *Raja radiata*

A central row of 12–19 very large spines in the adult;
North eastern and western Atlantic.
90 cm.

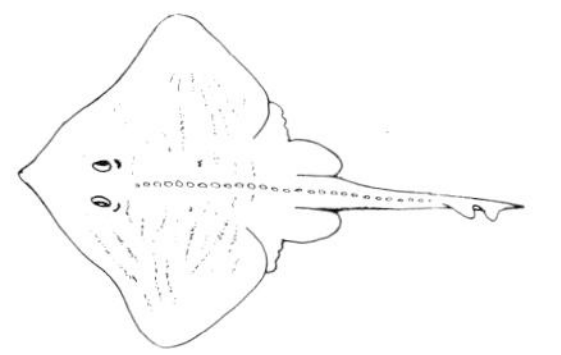

Clearnose Skate *Raja eglanteria*

Single row of thorns along mid-line of back and tail;
dark bars on upper surface.
Western Atlantic.
75 cm.

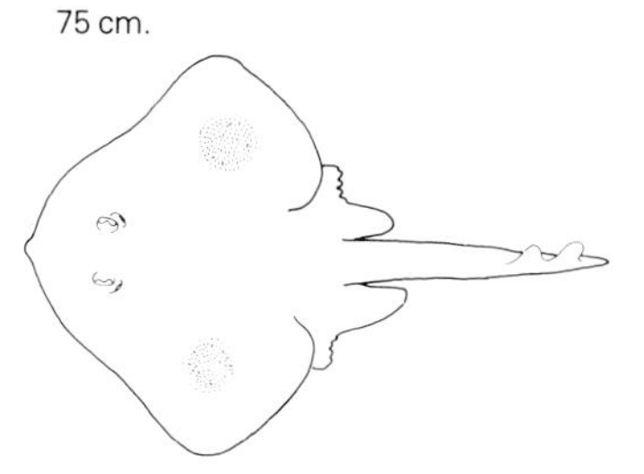

Winter Skate *Raja ocellata*

Large rounded spots (usually);
upper teeth, at least 80 series.
Western Atlantic distribution.
80 cm.

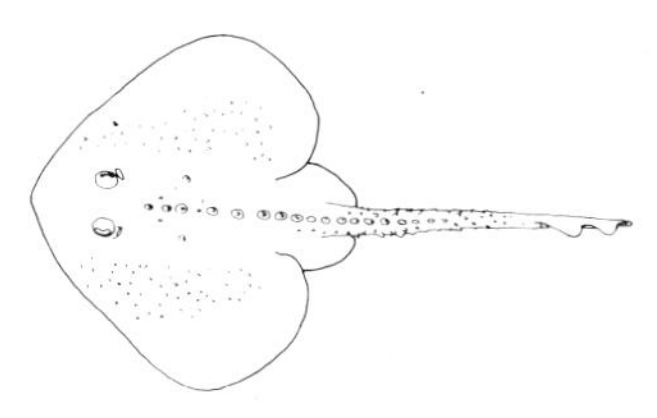

Little Skate *Raja erinacea*

Most of upper surface thorny;
no large dark eye spots;
upper teeth, less than 67 series.
Western Atlantic.
50 cm.

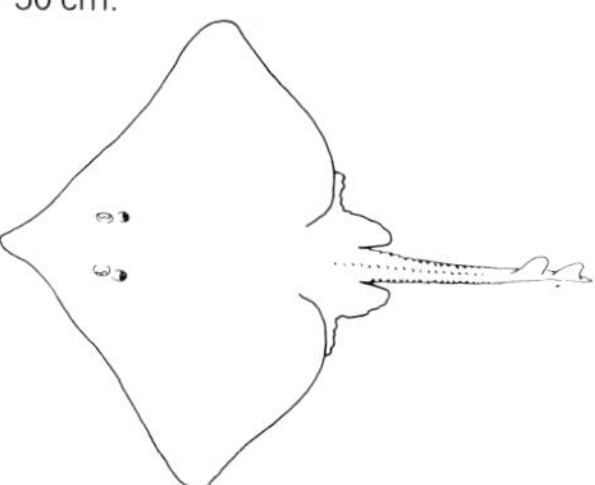

Barndoor Skate; Sharpnose Skate *Raja laevis*

No large thorns on back;
3 rows of thorns on tail;
black pores on underside.
Western Atlantic.
1 m.

Brown Ray
Raja miraletus L.

Mediterranean; eastern Atlantic from Gascony to South Africa.

Body disc-shaped with rather pointed wings; *tail* more than half total length of body; *snout* elongate and pointed; *eyes* larger than the spiracles; *skin* in the adult male spines are present along the front edge of the disc and in small patches on the wings; there is a central row of spines (between 14–18) along the tail which in the male is flanked by 1 row of spines each side, and in the female by 2 rows of spines each side; 2 spines are situated between the dorsal fins and a number of spines around the eyes; *fins* 2 dorsal fins and small tail fin.

Upper surface light brown with darker edges and covered in small black dots. The male fish also has yellowish spots. Snout region lighter. Each wing has a blue spot surrounded by a dark (nearly black) ring and then a yellowish ring.

Up to 60 cm.

Found on or half buried in sand or mud. Usually in water at 90–300 m in depth but in the summer they approach shore and may be found as shallow as 30 m. Egg capsules are laid in the spring in the Mediterranean and in winter in the Atlantic.

Spotted Ray
Raja montagui Fowler

Mediterranean; eastern Atlantic north to Shetlands south to Morocco; English Channel; southern North Sea.

Body disc-shaped with wings angled; *snout* slightly elongate but blunt; *skin* spines only on front part of the disc. In adults they extend behind the eyes. There is one central row of *spines* which runs from behind the eyes along the tail to the 1st dorsal fin (20–30 spines in young, 40–50 in the adult). There may be 1–3 spines between the dorsal fins and adult males have small patches of spines on the wings.

Upper surface yellowish or brownish with small black spots which do not extend to the edge of the disc. Often a yellowish spot surrounded by a ring of black dots is present on each wing. Undersurface white.

Spotted Ray, *Raja montagui* (B. Picton).

Up to 75 cm.

Found on muddy, sandy or sand and rock areas at 25–120 m. They feed on bottom-living animals including fish and crustacea. Egg capsules are laid in spring and early summer in the English Channel and the embryos take about 5 months to develop. The capsules measure 64–77 mm by 37–46 mm and have a short horn at each corner. One side of the capsule is smooth and the other covered with a network of fine fibres.

Painted Ray
Raja microcellata Montagu

Eastern Atlantic north to Ireland and Cornwall south to Morocco.

Body disc-shaped with angled wings; *snout* prominent; *eyes* small; *skin* only front half of surface rough. Central row of small *spines* along back and tail and one similar row each side along the tail.

Upper surface greyish to brownish with small white spots and wavy lines. Undersurface white.

Up to 82 cm.

Found on shallow sandy bottoms. In some localities it may be very common while in other similar areas it may be unknown. Egg capsules are probably laid in summer, they measure 8.7–9.5 cm by 5.4–6.3 cm and have 2 long slender horns at one end and 2 short curved horns at the other.

Blonde Ray
Raja brachyura Lafont
= *Raja blanda* Holt & Calderwood

Eastern Atlantic north to Shetlands, south to Madeira; English Channel; occasionally Mediterranean.

Body disc-shaped with angled wings; *snout* slightly elongate and rounded; *skin* adult fish are rough over the whole upper surface with a central row of small *spines* along the tail. Young fish are smooth and have a central row of spines along body and tail and there are 2 more complete rows of spines along the tail. Adult females have incomplete side rows along the tail. Adult males have patches of spines on the wings and at the sides of the head.

Upper surface light brown with numerous small dark dots which extend to the edge of the disc and larger irregular lighter patches. Undersurface whitish. In young fish the extreme tip of the snout is black.

Up to 115 cm.

Found on muddy and sandy bottoms between about 40–100 m. Young fish are found in shallower water than the adults. They feed on bottom-living animals which include fish, crustacea and molluscs. The egg capsules are large (115–143 mm by 72–90 mm) and are elongated into horns at each corner. Up to 30 may be laid at a time during the spring and early summer. The embryos take about 7 months to develop.

Undulate Ray; Painted Ray
Raja undulata Lacépède

Mediterranean; eastern Atlantic north to Britain; English Channel.

Body disc-shaped with rounded wings; *snout* slightly elongate; *skin* back rough except for the base of the tail and the pelvic fins. A row of irregularly spaced *spines* along the centre back and tail. Adult males have 1 row of spines each side of the central row along the tail while adult females have 2 rows. There are also a number of spines around the eyes.

Upper surface grey-brown, brown or brownish-yellow. Scattered over the surface are darker wavy lines each bordered with white dots. Snout pinkish grey. Undersurface white.

Up to 120 cm.

Found offshore on sand or mud down to 200 m. They probably feed on bottom-living animals, like the other rays. Egg capsules are laid during late summer in the Atlantic and spring in the Mediterranean. The capsules are large and measure 82–90 mm by 45–52 mm, they are reddish brown and have one side covered with fibres; horns at each corner.

Thornback Ray
Raja clavata L.

Mediterranean; eastern Atlantic north to Scandinavia; North Sea; English Channel; West Baltic; Black Sea.

Body disc-shaped with angled wings; *snout* slightly elongate; *skin* very rough on

Undulate Ray (top), *Raja undulata* (R. M. Soames).

Thornback Ray (bottom), *Raja clavata* (B. Picton).

upper and lower surface. The upper surface has large-based *spines* symmetrically scattered over it. In the female these large spines are also present on the undersurface. Young fish and females have a central row of spines down the body and along the tail. The male has only these spines down the tail.

Upper surface greyish or brownish with lighter and darker mottlings. The markings are much clearer in young fish and consist of small black spots and larger yellowish spots surrounded by brown. Undersurface white with a darker edge.

Up to 85 cm.

These rays are very common outside the Mediterranean. They are found on mud, sand, gravel and rocky and sandy areas at 2–60 m, but also down to 500 m. They are nocturnal and feed on bottom-living animals mainly crustacea but also fish, worms, molluscs and echinoderms. The breeding grounds of the Thornback Ray are inshore, the females preceeding the males to the grounds by several weeks. The egg capsules are laid in winter in shallow water and are often seen washed up on beaches, and called 'Mermaids' Purses'. They measure 6–9 cm by 5–7 cm, are greenish-brown and have pointed horns extending from each corner. One side of the capsule is covered with fibres. The embryos take about 5 months to hatch and the young fish are frequently seen close inshore. The males reach maturity at about 7 years and the females at about 9 years.

A large specimen will give an electric shock if lifted into the air by its tail.

Cuckoo Ray
Raja naevus Müller & Henle

Mediterranean; eastern Atlantic north to Scandinavia; North Sea; English Channel.

Body disc-shaped with rounded wings; *snout* not prominent; *skin* upper surface rough except for smooth patches on each wing. In the adult there are 4 rows of *spines* along the tail, the 2 central rows continue onto the disc. Young fish have a central row of tail spines. There may also be patches of spines at the sides of the head, on the wings and around the eyes. The undersurface is smooth except for the front edge; *fins* 2 dorsal fins close together towards the end of the tail.

Upper surface light brown or grey-brown with lighter patches. On each wing is a conspicuous large black and yellow marbled spot. Undersurface white.

Up to 70 cm.

Found at 20–150 m. They feed on bottom-living animals which include crustaceans, worms and fish. The egg capsules are a transparent brown and measure 6–7 cm by 3.5 cm. There are pointed horns at each corner and one pair are at least twice the length of the egg-case. Although little information is available on the habits of this fish it is of considerable commercial importance.

Shagreen Ray; Fullers Ray
Raja fullonica L.

Eastern Atlantic south to Biscay; west Iceland; occasionally Mediterranean.

Body disc-shaped with angled wings; *head* and *snout* rather pointed; *eyes* large, diameter equal to the space between the eyes; *skin* upper surface rough with two rows of *spines* along the tail. There may also be patches of spines at the sides of the head, on the wings and around the eyes. No spines between the dorsal fins. Underneath smooth except for the front edge of the disc, base of tail and tail.

Upper surface a uniform grey or brown with small darker dots. Undersurface white.

Up to 110 cm.

Found at 35–350 m on sandy areas but possibly on rather rocky ground also. They feed on bottom-living animals possibly mainly fish but also crustacea. The amber-coloured egg capsules are laid in deep water and measure 88–99 mm by 46–47 mm with long horns at each corner. One pair of horns are longer than the capsule. They are a commercially important fish caught mainly by long line but also by trawl.

Clearnose Skate
Raja eglanteria Bosc

Western Atlantic from Massachusetts Bay south to Florida.

Disc with a single row of thorns extend-

Eyed Skate, *Raja ocellata* (Gilbert van Rijckevorsel).

ing along the mid-line of the back from between the shoulders to the tail; *snout* rather pointed.

Light and dark brown above, pale below. There are distinctive dark bars on the upper surface.

75 cm.

Common, often comes close inshore but occurs down to at least 120 m. In the northern part of its range it is a summer visitor, but in warmer waters further south it appears to withdraw from shallow water in the hottest season.

Winter Skate
Raja ocellata Mitchill

Western Atlantic, Newfoundland south to North Carolina.

Disc scattering of thorns over back and along the mid-line, the thorns along the mid-line becoming lost with age. The *skin* is smooth between the thorns.

Light brown above with large 3–6 mm spots scattered over the back 8–12 mm apart. 1–4 large eyespots edged with white usually present on each pectoral fin.

80 cm.

Often common on sand or gravel bottoms, sometimes close inshore in very shallow water especially in winter. May move offshore when the sea temperature rises. Known to occur down to at least 100 m. It is known to undertake coastwise migrations of more than 160 km.

White Skate; Bottle-nosed Skate
Raja alba Lacépède
= *Raja marginata* Lacépède

Mediterranean; eastern Atlantic north to Ireland and Cornwall; occasionally English Channel.

Body disc-shaped with pointed wings; *snout* elongate and pointed; *skin* rough in adults except for a smooth patch in the centre of the disc. Young are smooth. There is a central row of curved *spines* along the tail and one lateral row each side of it. Undersurface prickly in the adults except for smooth patches on the wings. Young smooth except on the snout. There is one spine between the dorsal fins.

Upper surface greyish or bluish in the adult and reddish-brown in the young. The whole surface is scattered with irregular lighter spots. Undersurface white with a grey edge to the pectoral fins in the adult and with a black stripe across the rear edge of the pectorals and the pelvics.

Up to 200 cm.

Found on sand and mud at 40–200 m. The younger, smaller fish are usually in the more shallow water. Egg-cases are laid in the spring and the embryos take about 15 months to develop. The capsules measure

16–19 cm by 13–15 cm with horns at each corner and with the larger horns flattened. The White Skate in the seventeeth century was a prized food fish of the French but is now of little commercial value.

Common Skate; Grey or **Blue Skate**
Raja batis L.

Eastern Atlantic north beyond Iceland and Norway; North Sea; English Channel; occasionally Mediterranean.

Body disc-shaped with a slightly concave front edge to the wings, which are sharply angled; *snout* prominent; *skin* rough on whole of upper surface in the male, on the front region in the female but smooth in the young. There is a central row of *spines* along the tail, sometimes other rows are also present.

Upper surface greyish, brownish or greenish with lighter and darker spots scattered over the surface. Undersurface greyish with black dots but becoming lighter with maturity.

Females up to 240 cm, males up to 205 cm.

Common bottom-living fish found on sand and mud at 30–600 m. Young fish may be found from very shallow depths. They feed on bottom-living animals, mainly fish and crustacea, but also they make forays from the sea bottom, when they catch other species of fish and cephalopods. The egg capsules are yellow and measure 143–245 mm by 77–145 mm with horns at each corner. The longer horns have bunches of filaments at their ends. The capsules are laid during the spring and summer.

The Skate is an important commercial fish, but is apparently rarely seen by divers.

Starry Ray
Raja radiata Donovan

North Sea; north of eastern Atlantic; Iceland; also north of western Atlantic.

Body disc-shaped with rounded wings; *snout* only very slightly pronounced; *skin* rough. A central row of 12–19 large broad-based *spines* run across the disc and along the tail. There may be other large spines either side of the central row. Undersurface smooth except on snout.

Upper surface light brown with many small dark spots and fewer faint light cream spots. Undersurface white.

Up to 90 cm.

Found on mud, sand and stones at 20–900 m and usually within a temperature range 1–10°C. They feed on bottom-living animals including fish, crustacea, worms and echinoderms. The egg capsules are small and without horns. These fish are common within their area and an important commercial species.

Little Skate
Raja erinacea Mitchill

Western Atlantic, Nova Scotia south to North Carolina.

Resembles the Eyed Skate (*R. ocellata*) but without eyespots. The only certain distinguishing feature is that the Little Skate has 66 or fewer series of teeth in the upper jaw.

50 cm.

Usually found on sand or gravel bottoms down to 30 m. Often comes close enough inshore to be stranded in rough weather, but sometimes found down to 150 m. Tends to be found in shallower water in summer than in winter. Does not appear to embark on long migrations.

Barndoor Skate, Sharpnose Skate
Raja laevis (Mitchill)

Western Atlantic, Newfoundland south to North Carolina.

Disc no large thorns on the back, but three rows of thorns on the tail; *snout* profile sharp.

Brown upper surface with many scattered brown spots, underside white or blotched grey with many small dart spots and dashes marking the opening of mucous pores.

1 m.

Usually found on sand or gravel bottoms between 10–150 m, but can occur from the tide mark down to 400 m. May occur most often in shallow water in winter and spring. Although it feeds on small, bottom-living invertebrates it feeds more often on small fish than most other skates.

Family:

DASYATIDAE

The disc of the **Sting Rays** is generally more rounded than that of the Rajidae. There are no dorsal fins and none of the fishes in our area have tail fins. There are almost always 1 or 2 spines near the base of the tail with a venom gland at the base and a groove along which the venom is ducted to the tip. The skin is smooth. Viviparous.

The Sting Ray may drive its venomous barb deep into a limb, often causing a wound large enough to need stitching. It causes intense pain quite out of proportion to the size of the wound. People rarely die from the sting but the pain may have a paralysing effect. The wound should be washed in salt water and then plunged into water as hot as can possibly be tolerated. When the pain slackens any piece of barb still in the wound should be removed. Treatment is then the same as for any other wound.

Common Sting Ray

Dasyatis pastinaca (L.)
= *Trygon pastinaca* Cuv.

Mediterranean; Black Sea; eastern Atlantic south to Madeira north to Norway; English Channel; occasionally west Baltic.

Body disc-shaped with rounded angles to the wings and straight front edges. Body width only slightly greater than body length; *snout* fairly pointed; *tail* long, about 1.5 times body length with one (sometimes more) large-toothed *spine* situated about ⅓ the distance along the tail. There are ridge folds along both the upper and lower surface of the tail; *skin* smooth, old fish may have a central row of bony knobs along the mid-line of the body; *eyes* smaller than the spiracles which are directly behind them; *fins* no dorsal fins; *mouth* 5 bulbous papillae on floor of mouth.

Upper surface grey, brown, reddish or olive-green. Young fish may have white spots. Undersurface whitish with dark edges.

Up to 250 cm.

Sting Rays prefer calm shallow water (above 60 m) where they can often be seen on, or half buried in, sand or mud. They can tolerate areas of low salinity and may be found in estuaries. The spine on the tail has a poison gland at its base and is capable of inflicting a very painful wound. Fish may be found with more than one spine as replacement spines develop before the old spine is lost. They are carnivorous and feed

Common Sting Ray, *Dasyatis pastinaca* (Andy Purcell).

on any bottom-living animals, fish, crustacea, molluscs, etc., and can be very destructive in areas where shellfish are cultured. The sting ray is viviparous; the young are born in summer with 6–9 in a litter. A number of legends have grown up around the Sting Ray. In one Circe is reputed to have given her son a spear with a Trygon's spine at the tip with which he later killed Ulysses.

A similar species, *Dasyatis tortonesei* Capapé, has 3 filamentous papillae on the floor of the mouth; Mediterranean.

Dasyatis centroura (Mitchill)

Mediterranean; eastern Atlantic north to Gascony; north western Atlantic, Cape Cod south to Florida.

Body disc-shaped with rounded wings. The front edge of the wings are slightly, undulate. Disc width greater than length; *snout* small but pointed; *eyes* very small; spiracles large; *tail* at least twice as long as disc width with one or more poisonous *spines* situated towards the base. There is a well developed ridge on the underside of the tail; *skin* there are a number of broad-based spines scattered over the surface and in a row along the mid-line of the back. Young fish may not have any spines; *fins* 2 dorsal fins.

Upper surface yellowish to olive-green. Lower surface whitish with a dark margin.

More than 3 m. It is one of the largest Sting Rays in the world.

Not very common, they are found on the bottom at 10–45 m, usually more common near the coasts in summer. Their habits are similar to the other Sting Rays.

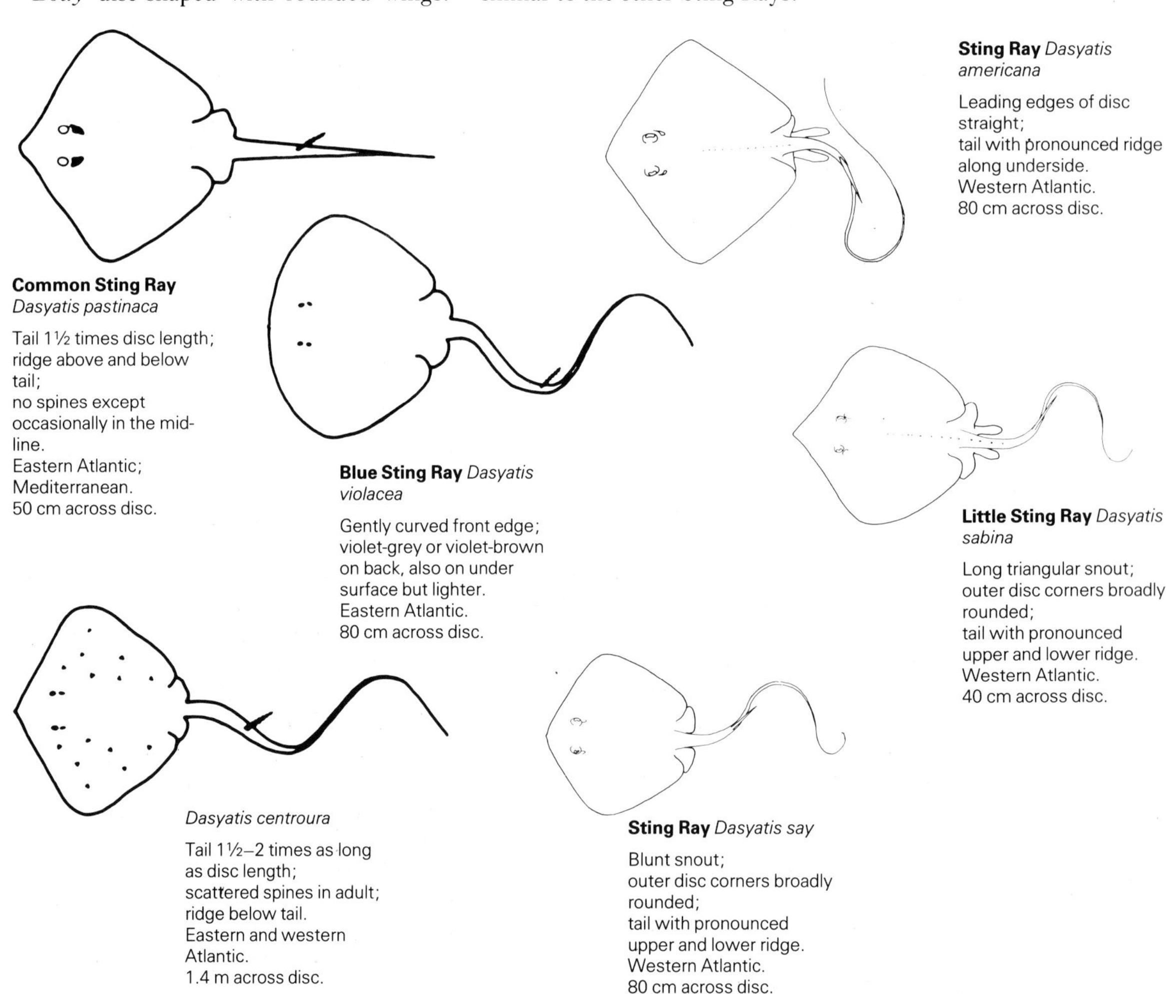

Common Sting Ray *Dasyatis pastinaca*

Tail 1½ times disc length; ridge above and below tail; no spines except occasionally in the mid-line. Eastern Atlantic; Mediterranean. 50 cm across disc.

Sting Ray *Dasyatis americana*

Leading edges of disc straight; tail with pronounced ridge along underside. Western Atlantic. 80 cm across disc.

Blue Sting Ray *Dasyatis violacea*

Gently curved front edge; violet-grey or violet-brown on back, also on under surface but lighter. Eastern Atlantic. 80 cm across disc.

Little Sting Ray *Dasyatis sabina*

Long triangular snout; outer disc corners broadly rounded; tail with pronounced upper and lower ridge. Western Atlantic. 40 cm across disc.

Dasyatis centroura

Tail 1½–2 times as long as disc length; scattered spines in adult; ridge below tail. Eastern and western Atlantic. 1.4 m across disc.

Sting Ray *Dasyatis say*

Blunt snout; outer disc corners broadly rounded; tail with pronounced upper and lower ridge. Western Atlantic. 80 cm across disc.

Sting Ray, *Dasyatis americana* (Dick Clarke).

Blue Sting Ray
Dasyatis violacea (Bonaparte)
= *Trygon violacea* Bonaparte

Mediterranean; warm water of eastern Atlantic.

Body disc-shaped with curved wings. Front edge of disc curved; *snout* small, pointed; *tail* about twice the length of the body and very slender. In young fish the tail may be up to 3 times the body length. One or more poisonous *spines* are present; *skin* there are a large number of small spines scattered along the mid-line of the body; *fins* no dorsal fins.

Upper surface greyish or brownish with a violet tinge. Undersurface lighter.

Up to 1 m.

Found from shallow water to below 100 m. Although found on the bottom this fish leads a more active, free-swimming life than its near relatives. Feeds on small crustacea, fish, molluscs and cephalopods.

Sting ray
Dasyatis americana Hildebrand & Schroeder

Western Atlantic, New Jersey south to Brazil.

Disc diamond-shaped, with straight leading edges meeting at an angle of about 110° at snout; *thorns* small prickles on each side of mid-line; *tail* ridge fold running along the upper side of the tail from the spine to the tip, slight ridge fold only along the upper surface.

Olive-brown above matching the shade of the bottom, whitish underneath with margins edged with grey or brown; tail spine dark.

80 cm across the disc.

A southern visitor only in the northern part of its range. Feeds on bottom-living crustacea, molluscs, small fish, etc. Has been observed to swim in pairs above the bottom.

Little Sting Ray
Dasyatis sabina (Lesueur)

Western north Atlantic, Chesapeake Bay south to Florida and Gulf of Mexico.

Disc with broadly rounded outline to pectoral fins; *snout* pointed and triangular; *tail* very slender towards the tip, there is a pronounced ridge fold along the top and bottom of the tail; *spine* is more long and thin than in other sting rays in the area.

Brown or yellow-brown above, paler below.

40 cm across the disc.

Usually found in depths of 3 m or less. Frequently enters fresh water.

Sting Ray
Dasyatis say (Lesueur)

Western Atlantic, New Jersey south to northern Argentina.

Disc broadly rounded, the upper surface of disc is relatively smooth; *snout* blunt; *tail* pronounced ridge folds along the top and bottom.

Reddish-brown or olive-brown above, sometimes with blueish spots; the end of the tail is dusky or black.

80 cm across disc.

Like other sting rays, it has the habit of lying on the bottom partly covered with sand. Often found in very shallow water following the advancing tide in to feed; when its pectoral fins break water it is often mistaken for a shark. Rarely deeper than 9 m. In the summer it migrates hundreds of kilometres northward along the coast of the USA.

Family:
GYMNURIDAE

The **Butterfly Rays** are closely allied to the sting rays described in the previous section. They differ in the disc, which is much wider than long, the short tail, and in having no papillae in the roof of the mouth. Representatives are found in both eastern and western Atlantic, but they are not often seen and only the Lesser Butterfly Ray from the western Atlantic is included here.

Smooth Butterfly Ray
Gymnura micrura (Bloch & Schneider)

Western Atlantic, New York south to Brazil.

Disc smooth, at least 1.6 times wider than long; *tail* no spine, upper surface of tail with ridge; *spiracle* has no tentacle-like structure.

Changes shade to match background, the back is marked with spots, blotches and vermiculations, the tail has irregular cross-bars, the underside is whitish.

60 cm across disc.

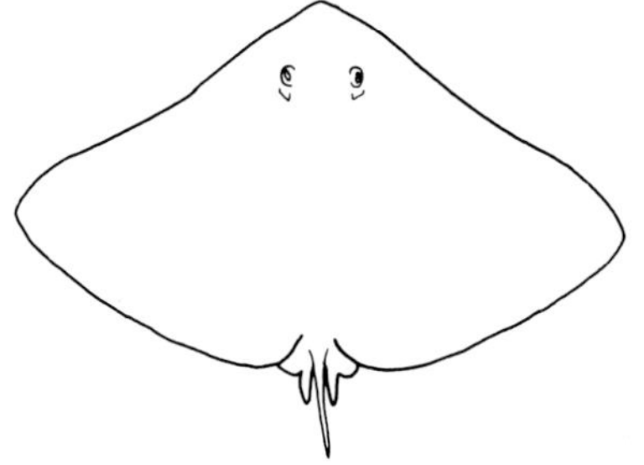

Smooth Butterfly Ray
Gymnura micrura
Disc broader than long;
no tail spine;
tail short.
Western Atlantic.
60 cm.

Prefers sandy bottoms in very shallow water, often moving in with the tide. May enter brackish water but not fresh water.

Family:
MYLIOBATIDAE

The **Eagle Rays** are more often seen actively swimming than are the other rays. The disc is diamond-shaped and is wider than it is long. The tail is long and slender and at its base there are usually one or two venomous spines. The head is distinct forward from the level of the eye. The eye and spiracle are carried on the side of the head.

Eagle Ray
Myliobatis aquila

Mediterranean; eastern Atlantic north to Scotland, occasionally Norway.

Body disc-shaped. Body width nearly twice body length. The angles of the wings are pointed and have convex rear edges; *head* from eye level clearly distinct from the rest of the disc; *snout* rounded; *tail* twice as long as body. One or more spines about ¼ the distance along the tail; *eyes* situated at the sides of the head; *teeth* 7 rows of large mosaic-like teeth; *skin* smooth. Large fish may have small spines along the mid-line of the body; *fins* one small dorsal fin in front of the tail spine.

Upper surface brown, grey, greenish or yellowish with a darker tail. Undersurface whitish with darker edges to the pectoral fins.

Up to 2 m.

Eagle Rays are actively swimming fish which are often encountered near the surface and sometimes even jump out of the water. They are, however, also found on sandy or muddy bottoms down to 300 m. They feed on bottom-living animals, molluscs, crustacea, etc., which they expose by flapping their wings to disturb the sand or by digging with their snouts. 3–7 young are born in each litter.

Bull Ray
Pteromylaeus bovinus (Geoffroy Saint Hilaire)
= *Myliobatis bovina* (Geoffrey Saint Hilaire)

Mediterranean.

Body disc-shaped, very pointed wings with convex rear edges; *tail* very long and

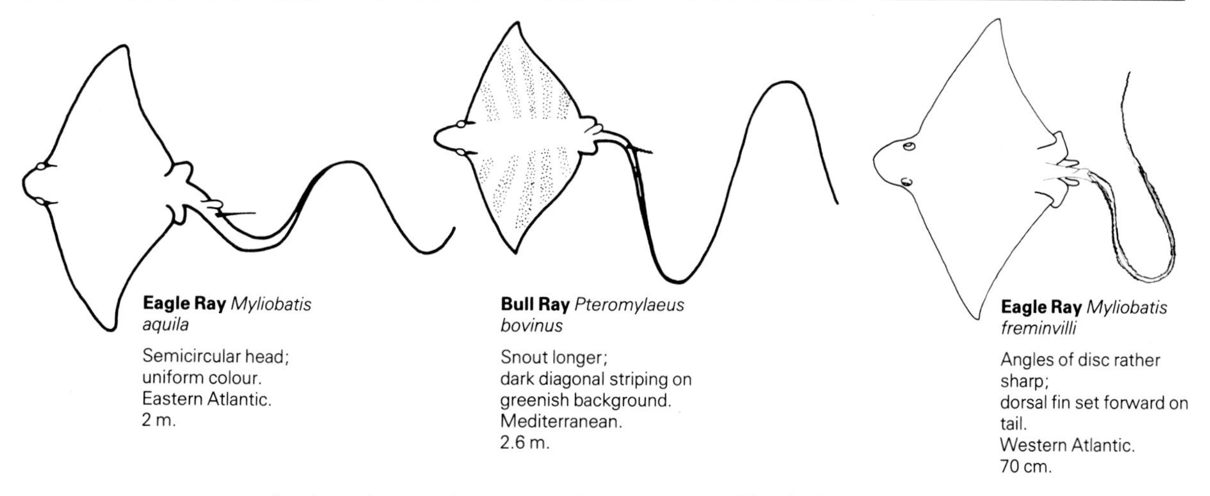

Eagle Ray *Myliobatis aquila*

Semicircular head; uniform colour. Eastern Atlantic. 2 m.

Bull Ray *Pteromylaeus bovinus*

Snout longer; dark diagonal striping on greenish background. Mediterranean. 2.6 m.

Eagle Ray *Myliobatis freminvilli*

Angles of disc rather sharp; dorsal fin set forward on tail. Western Atlantic. 70 cm.

slender, about twice length of the disc. 1 spine situated towards the base of the tail; *head* separate from the body beyond the eyes; *snout* long and rounded; *eyes* situated at sides of the head; *spiracles* elongate and situated almost directly behind the eyes; *teeth* mosaic-like in 7–9 rows; *fins* one small dorsal fin located immediately behind the pelvic fins.

Upper surface greenish-brown with darker diagonal striping. Older fish are uniform in colour. Undersurface white.

Up to 2.6 m.

This ray is found in deep cold or temperate water on mud or sand. In spring they approach shore when they may be seen by divers. They feed on shellfish and fish and are a menace to the shellfish-culture industry. The young are born during late summer and early autumn.

Eagle Ray
Myliobatis freminvilli Lesueur

Western Atlantic, New York south to Brazil.

Disc outer corners abuptly angled, rear corners angled at about 80°; *skin* smooth in young but older specimens have up to 8 low, conical tubercles on the mid-line between the shoulders; *tail* with a relatively large dorsal fin set near the base of the tail.

Grey-reddish or chocolate-brown, the edges of the disc paler. Underside pale.

70 cm across disc.

Feeds on hard-shelled molluscs and crustacea. A summer and early autumn visitor.

Myliobatis goodei, may be present in the more southerly part of the range. This differs in the blunter angles of the disc, and the dorsal fin which is set further back.

Family:
MOBULIDAE

Resemble the Myliobatidae except that the 2 pectoral fins form two 'horns' at the front of the head. There are large gill slits. The food is plankton filtered from the water.

Devil Fish
Mobula mobular

Mediterranean; eastern Atlantic north to Britain.

Body disc-shaped; 2–3 times as wide as long. Wings pointed with convex front edges and concave rear edges; *head* from the level of the eyes the head is distinct from the body. It is wide and has large 'horns' extending at each side of the head; *eyes* situated at each side of the head; *tail* slender with a spine near base; *teeth* minute in 150–160 series; *fins* small pelvic fins; 1 dorsal fin level with the pelvic fins.

Upper surface dark brown or black. Undersurface white.

Up to 6 m.

Large fish which 'fly' through the water by gently flapping their very large pectoral fins. Usually in small groups, occasionally near shore. Although extremely large, these fish are harmless; they feed on small planktonic crustacea and fish which are funnelled into the mouth with the aid of the horns and filtered from the water by the gill apparatus. The young are born alive.

Lesser Devil Ray
Mobula hypostoma (Bancroft)

Western Atlantic, North Carolina south to Brazil.

Similar to the Devil Ray except that it is much smaller, the mouth is on the underside of the head and there are no tail spines.

Disc 1 m wide.

Travel in schools feeding on small plankton and fish.

The **Manta Ray** (*Manta birostris* Walbaum) is similar, but the mouth is terminal. It is much larger, up to 6.7 m across disc. Generally found in warmer water than our area.

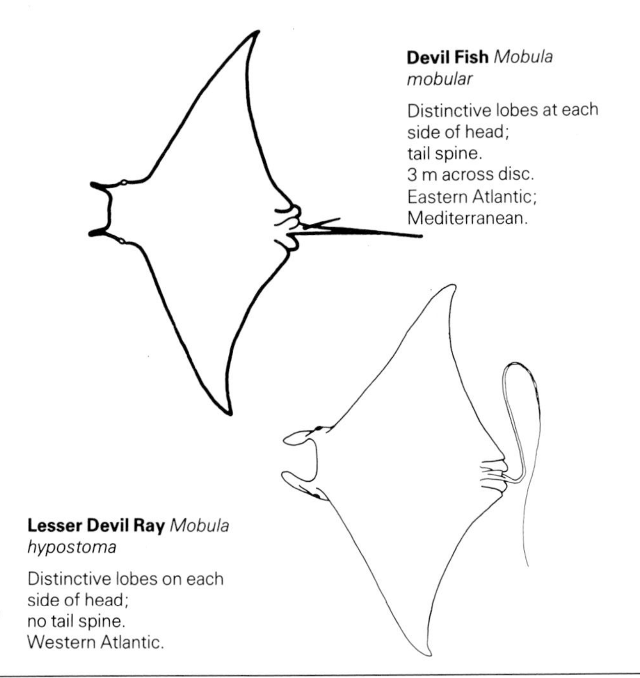

Devil Fish *Mobula mobular*

Distinctive lobes at each side of head;
tail spine.
3 m across disc.
Eastern Atlantic; Mediterranean.

Lesser Devil Ray *Mobula hypostoma*

Distinctive lobes on each side of head;
no tail spine.
Western Atlantic.

Family:

RHINOPTERIDAE

Medium to large rays with the disc much wider than long, and the head elevated above the level of the disc with the eyes and spiracles on the side. The mouth, which is on the underside of the head, has no papillae on its floor. The pectoral fins project forward forming a district lobe which protrudes from below the snout.

Cownose Ray
Rhinoptera bonasus (Mitchill)

Temperate waters of the United States south to Brazil.

Disc much wider than long; *skin* smooth; *head* conspicuously concave set above level of disc; *teeth* 7 rows in each jaw; *fins* pectoral fins form protruding flap under head, dorsal fin set forward on tail.

Brownish above, pale below; there may be narrow radiating lines from the centre of the disc both above and below.

Up to about 1.5 m across disc.

Often in shoals feeding on hard-shelled molluscs, etc., which it exposes by flapping its pectoral fins. Found in shallow water.

Lusitanian Cownose Ray
Rhinoptera marginata (Geoffroy Saint-Hilaire)

Atlantic coasts of Morocco to southern Spain; Mediterranean; not Black Sea.

Disc smooth, outer edges of wings slightly directed backwards; *head* distinctly concave; *teeth* in 9 rows in upper jaw, 7 rows in lower jaw; *fins* pectoral fins form flap under head which is raised above the level of the disc, dorsal fin set forward on the tail; *tail* about the same length as the body.

Uniform greenish-brown to bronze above, pale below.

Disc 2 m wide.

In most warm temperate coastal waters where they form large swimming groups.

Cownose Ray *Rhinoptera bonasus*

Front of head concave;
pectorals form flap under head;
teeth in 7 rows in upper jaw.
Western Atlantic. 1.5 m.

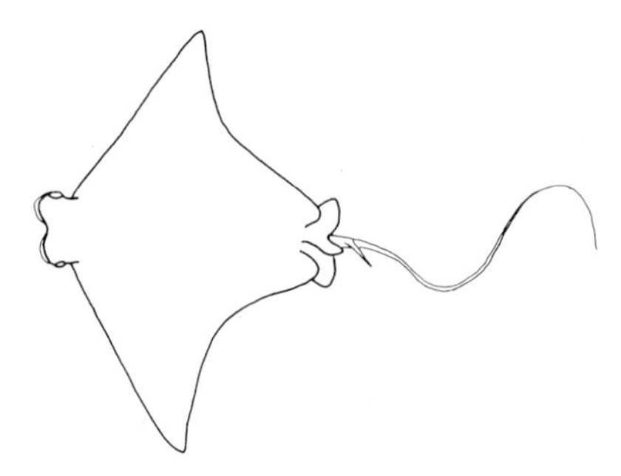

Lusitanian Cownose Ray *Rhinoptera marginata*

Front of head concave;
teeth in 9 rows in upper jaw;
colour uniform.
Eastern Atlantic; Mediterranean.

Class:

OSTEICHTHYES

Sub Class:

ACTINOPTERYGII

Order:

CHONDROSTEI

Family:

ACIPENSERIDAE

The **Sturgeons** have asymmetrical (heterocercal) tails, the body has several rows of bony plaques and the mouth is on the underside of the head.

Sturgeons are valuable fish since their eggs constitute caviar and their flesh is very good to eat.

Sturgeon
Acipenser sturio (L.)

North-eastern Atlantic, especially near the Rivers Gironde and Guadalquivir; Baltic; USSR, especially in lake Ladoga; Mediterranean, especially in the Adriatic.

Snout elongate; *mouth* protrusible and situated on the underside of the head. Mouth opening oval; *barbels* 4 between the snout and mouth; *armour* body covered with 5 rows of bony plaques (the row running down the centre of the flanks has 26–35 plaques), no scales; *fins* the tail is asymmetrical and shark-like, the spinal column being extended into the upper lobe of the tail.

Grey or green above, paler below; the fins are pinkish.

Sometimes attain 4 m but females usually up to 1.5 m, the males smaller.

Generally live over sandy or muddy bottoms in brackish water. They feed on small animals and detritus which they dislodge by rooting with their snouts and sucking them up through their extensible mouths. In spring they penetrate far up rivers to spawn. Males outnumber females and are smaller. The eggs stick to the bottom weed and stones. Caviar consists of the eggs taken from the female before they are laid. Sturgeon do not normally stray far from the rivers where they spawn but occasional specimens are taken which must have travelled hundreds of miles. A related species confined to the Adriatic, *Acipenser maccari* Bonaparte, has 33–42 plaques in the lateral row along the centre of the flanks and a shorter snout.

Atlantic Sturgeon, *Acipenser oxyrhynchus* (Gilbert van Rijckevorsel).

Atlantic Sturgeon
Acipenser oxyrhynchus Mitchill

Western north Atlantic from Labrador to Florida and Gulf of Mexico.

Very closely related to *Acipenser sturio* of the eastern Atlantic and may be the same species. Breeds in fresh water, but has no difficulty adapting to full salinity.

Roots in sand and mud, sucking in large quantities of mud which contains small invertebrates and fish such as sand eels and Ammodytes. Make long coastal migrations, one measured at 1400 km.

Volga Sturgeon
Huso huso (L.)
= *Acipenser huso* L.

Volga Sturgeon, *Huso huso* (Ken Lucas).

Eastern Mediterranean; Adriatic; Black Sea; Sea of Azof and Caspian Sea.

Resembles the sturgeon *Acipenser sturio* except that the snout is less elongate, the bony plaques are smaller and the mouth opening is half-moon shaped.

Very large, may reach 6 m.

Sturgeon *Acipenser sturio*
Body covered with large bony plaques; upper tail lobe longest; mouth opening oval.
North east Atlantic; Mediterranean.
1.5 m.

Volga Sturgeon *Huso huso*
Body covered with small bony plaques; upper tail lobe longest; mouth-opening half-moon shaped.
Mediterranean.
6 m.

Super Order:
TELEOSTEI

This order contains most of the present-day fishes. The vertebrae and skull are well-developed and bony.

Order:
ISOSPONDYLI

1 dorsal fin. Sometimes there is also a flap of tissue in the position of a 2nd dorsal fin (adipose fin). The tail is symmetrical and forked and the swimbladder has an open duct.

Family:
ELOPIDAE

One of the most archaic groups of bony fish now living. The lower jaw has an elongate bony plate between its branches. The mouth is terminal or directed upwards.

Ladyfish
Elops saurus L.

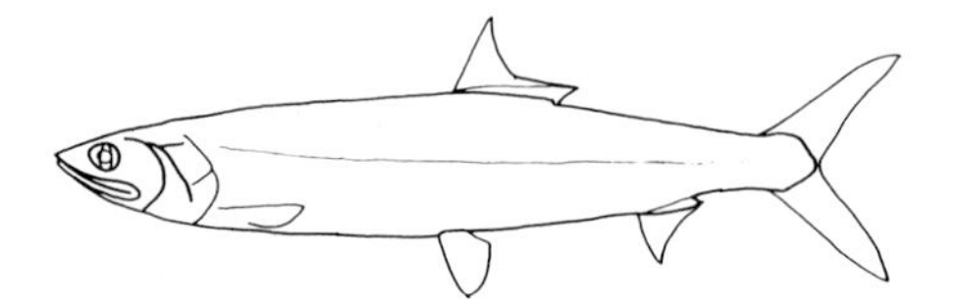

Ladyfish *Elops saurus*
Body slender;
lateral lines straight;
eye set forward and high.
Western Atlantic.
9 cm.

Western Atlantic, Cape Cod south to Brazil.

Body slender, slightly compressed; *scales* small and thin with membrane-like border; *head* flat between the eyes, which are set forward and high on the head; *mouth* nearly horizontal and almost at the front of the head; *fins* tail fin deeply forked with the upper lobe slightly the longer, dorsal fin only has soft rays, there is no long filament-like fin from leading edge as there is in the Tarpon, which scarcely enters our area.

Silvery, with darker upper surface.

9 cm.

An active fish that often swims in schools, frequently skipping over the surface. In inshore brackish water or fully saline seawater, it moves into the sea to spawn. After hatching Ladyfish pass through a leaf-like (leptocephalus) larval stage. An autumn visitor in the northern part of its range.

Family:
CLUPEIDAE

The **Herrings** and their relatives are long-bodied fishes generally living in shoals in open water. The body is covered with large, easily detached scales. There is only one dorsal fin and no lateral line.

These are some of the most important of all food fishes.

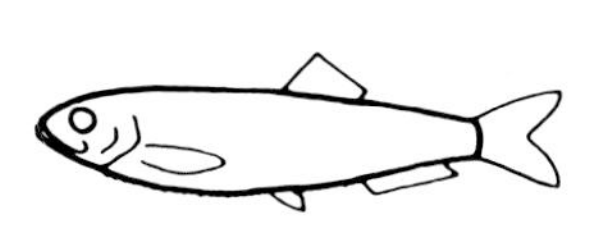

Sprat *Sprattus sprattus*

Single dorsal fin set back;
saw-like keel on belly.
Eastern Atlantic;
Mediterranean.
15 cm.

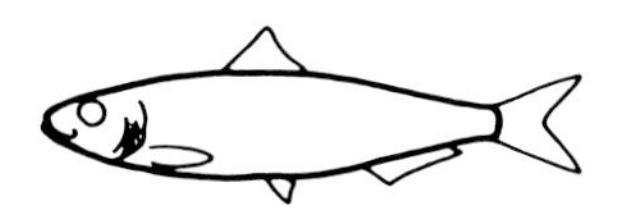

Pilchard (when large)
Sardine (when small)
Sardina pilchardus

Single dorsal fin set
slightly forward;
body not keeled;
gill cover ridged.
Eastern Atlantic;
Mediterranean.
16 cm.

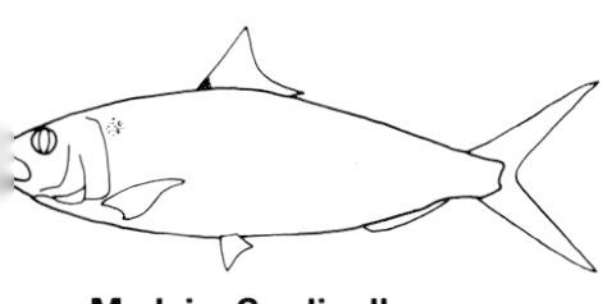

Madeira Sardinella
Sardinella madarensis

1 dorsal fin set forward;
belly keeled;
faint black spot behind gill opening;
black spot at origin of dorsal fin.
Mediterranean and warm eastern Atlantic.
25 cm.

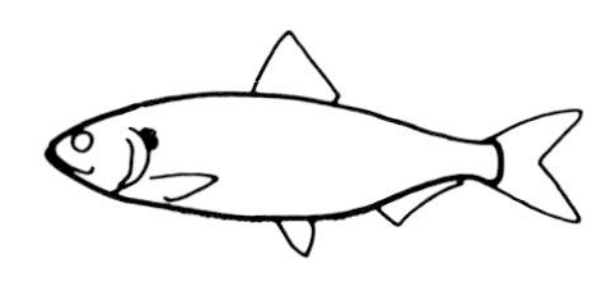

Allis Shad *Alosa alosa*

1 or more spots on flank;
60–80 long thin gill rakers;
upper jaw notched.
Eastern Atlantic;
Mediterranean.
40 cm.

Twait Shad *Alosa fallax*

Usually 5–10 spots along flank;
20–28 stout gill rakers;
upper jaw notched.
Eastern Atlantic;
Mediterranean.
50 cm.

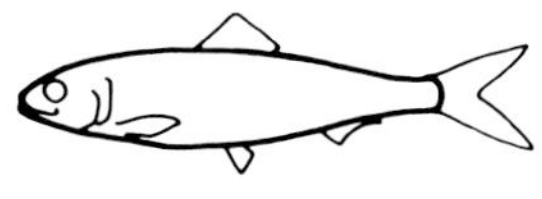

Herring *Clupea harengus*

Single dorsal fin in middle of body;
no saw-like keel along belly;
gill cover smooth.
Eastern and western Atlantic.
40 cm.

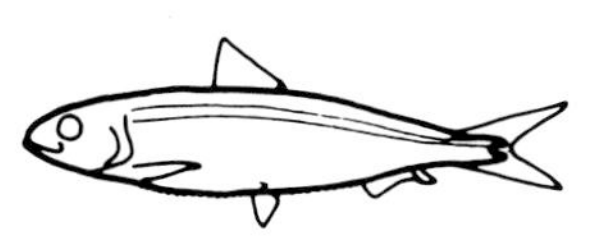

Gilt Sardine *Sardinella aurita*

1 dorsal fin set forward;
belly keeled;
yellow spot behind gill cover;
yellow stripe along flanks.
Eastern and western Atlantic.
23 cm.

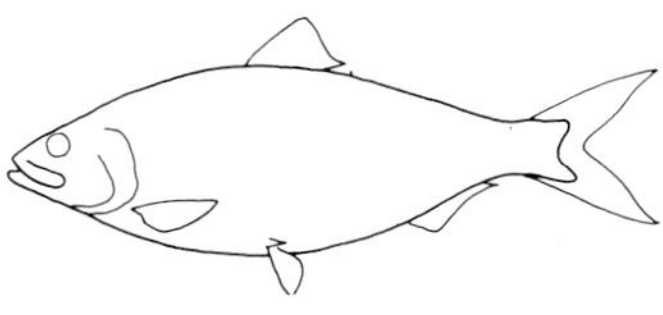

Atlantic Shad; American Shad; Alewife *Alosa sapidissima*

Dark spot on shoulder sometimes followed by smaller dots;
upper jaw notched, lower jaw not strongly projecting;
59–73 gill rakers;
Western Atlantic.
60 cm.

Hickory Shad; Alewife
Pomolobus mediocris

Upper jaw notched, lower jaw strongly projecting;
gill rakers 18–24;
dark spot on shoulder, often followed by several more;
Western Atlantic.
40 cm.

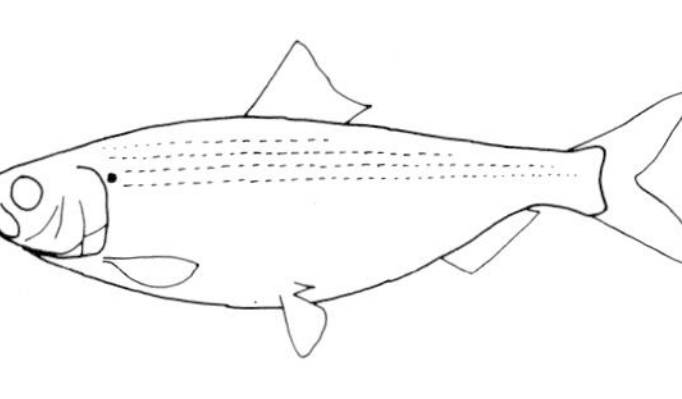

Blueback Herring; Alewife
Pomolobus aestivalis

Eye diameter equals snout;
upper jaw notched, lower jaw not strongly projecting;
gill rakers 38–41; back blue.
Western Atlantic.
30 cm

Grayback Herring; Alewife
Pomolobus pseudoharengus

Eye diameter greater than snout;
upper jaw notched, lower jaw strongly projecting;
gill rakers 38–41;
back grey-green.
Western Atlantic.
30 cm.

Atlantic Menhaden
Brevoortia tyrannus

Body deep, with sharp keel;
scales part exposed, deeper than long;
scale edges toothed;
modified scales in front of dorsal fin.
Western Atlantic.
35 cm.

Anchovy *Engraulis encrasicholus*

1 dorsal fin;
lower jaw much shorter than upper.
Eastern Atlantic; Mediterranean.
20 cm.

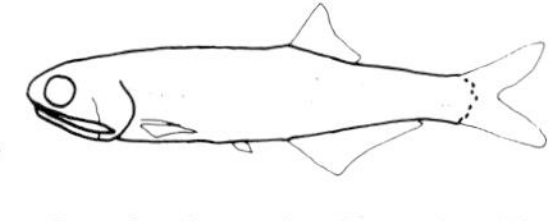

Bay Anchovy *Anchoa mitchilli*

Lower jaw much shorter than upper; single dorsal fin;
dark points at base of anal fin and base of tail fin;
anal fin with 24–30 rays.
West Atlantic.
10 cm.

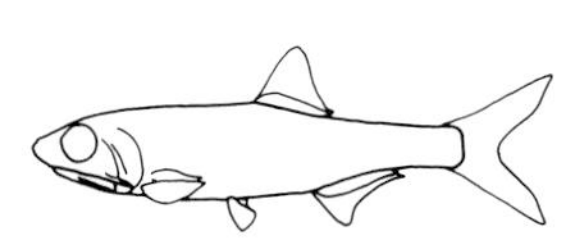

Striped Anchovy *Anchoa hepsetus*

Lower jaw much shorter than upper; single dorsal fin;
silver band along flank;
anal fin with 18–23 rays.
West Atlantic.
15 cm.

Sprat
Sprattus sprattus (L.)

Northern Mediterranean; eastern Atlantic, Norway to Gibraltar; North Sea; English Channel; Baltic.

Body oval in cross section but with a pronounced keel along the belly; *jaws* lower longer than upper; *eyes* the transparent (adipose) eyelids are present but are narrow; *scales* there is a well developed saw-like keel of scales along the belly; *gill cover* ridged; *fins* the dorsal fin begins midway between the hind edge of the eye and the base of the tail fin. The beginning of the pelvic fin is slightly in front of the beginning of the dorsal fin. D 15–19; A 17–23.

Bluish above, silvery on sides and belly. A yellow-bronze band divides the blue back from the silvery flanks.

Up to 15 cm.

It is generally considered that there are 3 sub-species of Sprat, one in the Mediterranean, one in the north-east Atlantic and one in the Baltic. They are able to tolerate water of extremely variable salinity from about 4 parts per thousand to full strength sea water of 36 parts per thousand. Sprats live in a temperature range of 4–36°C. They spawn offshore from late winter to early summer preferring temperatures between 6° and 13°C. Unlike the herring, the eggs float. In the winter large shoals come close inshore and in Norway they are captured in the fjords, canned, and sold as Brisling.

Pilchard (when large)
Sardine (when small)
Sardina pilchardus (Walbaum)

Mediterranean; eastern Atlantic from Canary Islands to southern Ireland; English Channel; rare north to Norway.

Body rather rounded in cross section; *jaws* lower slightly longer than upper; *eyes* there are transparent (adipose) eyelids; *gill cover* ridged; *scales* the belly is smooth with no saw-like scaly keel; *fins* the beginning of the dorsal fin is nearer to the snout than it is to the base of the tail; the pelvic fin begins below the centre of the dorsal fin. The last two rays of the pelvic fin are slightly elongate. D 17–18; A 17–18.

Greenish or bluish above, silvery below. There may be a bluish band along the body.

Up to 16 cm, rarely 25 cm.

A very important food fish, especially in Spain and Portugal. They occur in very large shoals which may enter brackish water. They are found near coasts in the late spring and summer but in the autumn they disappear and probably over-winter in deeper water. Spawning in spring, summer and autumn close offshore but the time depends on the temperature. The eggs are free-floating. The young of this fish are known as Sardines and are canned and exported from Spain and Portugal in large numbers. One of the most important methods of fishing is to attract a shoal with lights and then to surround it with a purse-seine net.

Madeira Sardinella
Sardinella madarensis

Southern and eastern Mediterranean, and from Gibralter south to Angola.

Similar to *Sardinella aurita* (page 52) but distinguished by having more than 70 fine gill rakers on the lower part of the first arch, rather than more than 90 in the case of *S. aurita*.

A faint yellow strip running along the flank originates from a faint black spot behind the gill cover. There is a black spot at the origin of the dorsal fin. The upper pectoral fin rays are white on the outside joined by a black membrane.

25 cm.

Swims in shoals in open coastal waters down to a depth of 50 m. Feeds on plankton. Enters the brackish water of bays and estuaries.

Allis Shad
Alosa alosa (L.)

Mediterranean, Biscay to northern Ireland, rare further north; Baltic.

Jaw lower jaw slightly prominent, mouth oblique; upper jaw notched to receive a knob in the centre of the lower jaw; *eyes* have transparent (adipose) eyelids: *gill cover* has radiating ridges; *fins*

Herring, *Clupea harengus*; Mackerel *Scomber scombrus* (Gilbert van Rijckevorsel).

dorsal fin begins nearer to the snout than to the base of the tail fin, the pelvic fin begins behind the beginning of the dorsal fin; *scales* there is a row of scales along the belly forming a saw-edged keel. Small scales cover the basal two-thirds of the tail fin; *gill rakers* on the lower branch of the outer gill arch there are 60–80 long thin processes (gill rakers). D 18–21; A 20–26.

Bluish above, silvery below. 1, or sometimes a succession of, dark shoulder spots.

40 cm, sometimes up to 60 cm.

Lives in deep water but in spring or early summer large shoals enter the rivers at night to breed. The eggs sink to the bottom. The young may spend up to 2 years in the rivers before going down to the sea, but the adults do not linger in the rivers after spawning and return immediately to the sea.

Twait Shad
Alosa fallax (Lacépède)
= *Alosa finta* (Cuvier)

Mediterranean; eastern Atlantic from Biscay to Ireland; rare in more northern waters; Baltic.

Resembles the Allis Shad (*Alosa alosa*) except that there are 20–28 stout gill rakers along the lower branch of the outer gill arch.

There is usually a large black blotch on the shoulder followed by a line of 5–10 smaller dark spots.

Up to 50 cm.

Very similar in habits to the Allis Shad. They spawn rather later in the year and do not travel so far upstream to breed. Several different races or sub-species are recognized. The Mediterranean race is known as *Alosa fallax nilotica* Geoffroy.

Herring
Clupea harengus (L.)

North-east Atlantic from northern Europe north to the Arctic ocean. A smaller race lives in the Baltic.

Body oval in cross section, belly not keeled; *jaws* lower slightly longer than the upper which is somewhat notched in the centre; *eyes* there are transparent (adipose) eyelids; *gill cover* is smooth; *fins* dorsal fin begins midway between the snout and the base of the tail fin. The pelvic fin begins beneath the centre of the dorsal fin. D 17–20; A 16–18.

Dark blue or green above, silvery below.

Size variable, up to 40 cm.

The Herring is one of the most important food fishes of northern Europe. There are several races of Herring of which three are particularly distinct. The large Norwegian Herring can be 20 years old and becomes sexually mature at 5 to 8 years. Somewhat smaller are the fish from the southern North Sea that live for perhaps 11 years and mature at 3 or 4 years. The smallest race of Herring is the Baltic Herring or Strömling. These have fewer vertebrae than their Atlantic and North Sea cousins and they become mature when 2 or 3 years old.

Herring are not often seen by divers and this may be because they only come near the surface at night and generally keep near the bottom. They are either caught in drift nets at night or in trawls fishing over the bottom.

Gilt Sardine

Sardinella aurita Valencienne

Southern Mediterranean; both sides of the Atlantic, in the west from southern Spain to southern Africa.

Body oval in cross section, rather more compressed than the Pilchard (*Sardina pilchardus*) and with a pronounced saw-edged keel; *jaws* lower jaw slightly longer than the upper; *eyes* there are transparent (adipose) eyelids; *gill cover* without radiating ridges and with a single vertical one; *fins* the beginning of the dorsal fin is nearer to the snout than to the base of the tail; the pelvic fin begins slightly behind the centre of the dorsal fin; the last two rays of the anal fin are slightly elongate; *scales* there is a saw-like scaly keel along the belly; there are two large transparent scales on each side of the tail base, and one pointed scale towards the rear of the dorsal fin base. D 17–20; A 16–18.

Bluish above, silver below with a longitudinal golden stripe along the flanks that quickly fades after death.

Up to 23 cm, rarely up to 38 cm.

A schooling species, it breeds in late summer near the coasts.

Atlantic Shad; American Shad; Alewife

Alosa sapidissima (Wilson)

Western Atlantic, Nova Scotia south to Florida; introduced to the north-west Pacific coast of America.

Resembles the two eastern Atlantic species except that it has more gill rakers than the Twait Shad (*Alosa fallax*) (59–73 in adults).

Similar pattern to the Twait Shad, having one dark spot on the shoulder, often followed by a line of smaller, less well-marked spots. The back is dark metallic bluish or greenish shading to silver on the lower flanks. No dark lines on scales.

60 cm.

The American Shad, which is the largest member of the herring family, spawns in fresh water, sometimes travelling at least 400 km up the larger rivers to do so. The spent fish return to the sea after spawning and do not return to fresh water until it is time to spawn again. They swim in shoals, and appear to have a very varied diet which ranges from parts of plants, invertebrates and fish. They are an important source of food since at least the arrival of the early settlers; most likely they were also an important food to the indigenous Indian population as well.

Hickory Shad; Alewife

Pomolobus mediocris (Mitchill)
= *Alosa mediocris* (Mitchill)

Western Atlantic, Maine south to Florida.

Resembles the Atlantic Shad (*Alosa sapidissima*) except that it has only 18–24 gill rakers on the lower limb of the first arch, and the number of gill rakers does not increase with age.

Greyish-green above, shading gradually to silver below. In adults there are narrow dark lines running along scales on the upper part of sides. There is a dark spot on the shoulder followed by several less clearly marked spots along the sides.

40 cm.

Probably enter fresh water to spawn, otherwise they feed in the sea on small fish, squid, small crabs and various other small crustacea. Not highly prized as a foodfish.

Blueback Herring; Alewife
Pomolobus aestivalis (Mitchill)
= *Alosa aestivalis* (Mitchill)

Western Atlantic, Nova Scotia south to St Johns River, Florida.

Head cheek longer than deep; *jaws* upper notched, lower not strongly projecting but angled upwards; *eye* relatively small, the diameter of the eye more or less equals the length of the snout; *gill rakers* 38–41.

Bluish above, shading rather abruptly into silvery sides; dark lines along the scales of the back in adults. A single dark spot on the shoulder in adults.

30 cm.

Enters brackish or fresh water to spawn, but does not penetrate far into fresh water, mainly in April in Chesapeake Bay, later in more northerly part of its range. Spent fish return to the ocean soon after spawning. The Blueback and the closely related Grayback (*Pomolobus pseudoharengus*) are often caught together, and often classed together as 'Alewives', so their relative abundance is not known. Possibly there are more Graybacks in the northern part of its range.

Grayback Herring; Alewife
Pomolobus pseudoharengus (Wilson)
= *Alosa pseudoharengus* (Wilson)

Western Atlantic, Gulf of St Lawrence, Nova Scotia, south to St Johns River, Florida.

Resembles the Blueback Herring (*Pomolobus aestivalis*) and the young may be very difficult to distinguish. Adult Graybacks differ from Bluebacks in the larger eye, its diameter being greater than the length of the snout.

Back greyish-green merging gradually into the silver of the flanks.

30 cm.

Spawn further upstream in fresh water than the Bluebacks and arrive at the mouth of rivers about a month earlier. Spawning fish apparently return to the stream of their birth. Land-locked populations occur in some freshwater lakes. Alewives can occur in large shoals, and are thus of commercial importance. However, the need to ascend rivers and streams to spawn leaves them vulnerable to obstruction from dams across the rivers and pollution of various kinds.

Atlantic Menhaden
Brevoortia tyrannus (Latrobe)

Continental waters of North America from Nova Scotia to Florida.

Body rather deep and laterally compressed; *snout* rather blunt with a notch on mid-line of upper jaw in larger individuals; *gill cover* with prominent radiating ridges in upper part; *scales* tend to adhere to the body, the exposed part much deeper than long, and the trailing edge finely toothed; there is a line of modified scales running along the top of the back forward from the dorsal fin.

Back dark green or bluish shading to silver on the sides with brassy tints. In fish larger than about 7 cm there is a dark spot, about the same size as the pupil, set high on the flanks and close behind the gill cover, larger individuals develop a variable number of other dark spots scattered over the flanks.

35 cm.

They swim in large well-ordered compact schools near the surface and often break the surface of the water. They have been observed to feed by swimming through the water with their mouths and gill openings wide open, straining out food particles with their long, close-set gill rakers. They appear to be particularly prone to parasites such as large isopods and copepods in the mouth.

Very abundant in some years and might be an important source of food if they were not so bony. However, their bodies contain a lot of oil and large numbers are reduced to fish meal and the oil extracted.

Anchovy
Engraulis encrasicholus Cuvier

Mediterranean; eastern Atlantic, Gibraltar to northern Norway.

Body slender and compressed, thicker along the back than along the belly which is not keeled; *jaws* upper much longer than lower, the mouth cleft extends backwards to the hind edge of the eye; *eye* there are no transparent (adipose) eyelids; *scales* there are two large scales on the base of the tail fin. The scales do not form a saw-like ridge along the belly; *fins* the dorsal fin begins midway between the snout and the base of the tail fin. Dorsal fin begins in front of the pelvic fin. D 15–18; A 20–26.

Greenish above, silvery below. In life a grey-blue band divides the dark upper part and the silvery lower part.

Up to 20 cm.

The habits are very similar to the Pilchard or Sardine (*Sardina pilchardus*). The Anchovy is the subject of an important fishery industry in Portugal, Spain and Italy, where they are canned or made into relishes and pastes. They are not eaten fresh.

Bay Anchovy
Anchoa mitchilli (Cuvier & Valenciennes)

Cape Cod and south to Mexico.

Body long and laterally compressed; *head* with protruding snout; *mouth* very large with lower jaw much shorter than upper; *fins* tail fin with 24–30 rays, single dorsal fin.

Greenish with bluish reflections above, silvery below, with a lateral silver stripe about as wide as the eye extending from the gill cover to the tail.

10 cm.

Occur in schools and is reputed to be a somewhat sluggish swimmer. Mostly found in protected brackish water areas especially where the bottom is muddy, also where there is sea grass and along sandy beaches. Occasionally down to 40 m.

Striped Anchovy
Anchoa hepsetus (L.)

From Chesapeake Bay south to the West Indies.

Body long and laterally compressed; *scales* large, about 40 in a lateral row; *head* with protruding snout; *mouth* very large with lower jaw much shorter than upper; *fins* tail fin with 18–23 rays, single dorsal fin.

Pale grey with some iridescent green and yellow on upper surface, underparts silvery; there is a silver band about as wide as the pupil of the eye stretching from the gill cover to the tail.

15 cm.

Differs from Silversides in only having one dorsal fin. Often abundant in coastal waters where it swims in schools. Regularly taken in seine nets near the shore, but have been caught in 80 m of water.

Atlantic Salmon, *Salmo salar* (Gilbert van Rijckevorsel).

Sub-Order:

SALMONOIDEI

Herring-like fishes which have a second dorsal fin composed of a simple flap of tissue without fin rays. Oviducts either absent or incomplete.

Family:

SALMONIDAE

Powerful small-scaled fishes which breed in fresh water. They are distinguished from related families by the vertebrae which are distinctly upturned at the base of the tail.

Salmon; Atlantic Salmon
Salmo salar (L.)

Eastern and western Atlantic south to Biscay; western Atlantic, Hudson Bay south to New England.

Body the tail stalk is narrower than in the Trout; *jaws* the upper jaw bone extends back to the rear edge of the eye (in older males the jaws become grotesquely hooked in the breeding season); *scales* there are 10–13 rows of scales between the 2nd dorsal fin and the lateral line; *fins* there are two dorsal fins of which the 2nd is a simple flap of tissue without fin rays (adipose fin). The tail fin is shallowly forked. 1D 12–14; 2D none; A 9–10.

Sea-caught Salmon are steel-blue above, silvery below with rounded or x-shaped spots scattered on the upper part of the flanks and head. After the fish has entered fresh water the ground colour becomes

Brown Trout, Sea Trout, *Salmo trutta* (Ken Lucas).

greenish or brownish with streaks and spots of orange.

The young fish (parr) are bluish or brownish above with 8–15 broader dark transverse lobes extending onto the flanks with orange marks between each.

Spawning occurs in late autumn or winter in fresh water on gravel bottoms. After spawning the spent and exhausted females return at once to the sea, but the males may stay in fresh water until the end of the spawning season. The adult fish then spend from 5–18 months feeding in the sea before returning again to spawn. The young fish (smolt and parr) spend between 1 and 4 years in fresh water before entering the sea, where they remain for at least a year. The older male parr may become sexually mature before entering the sea and often fertilize the eggs of the adult females.

The adult fish return to the waters where they were spawned, being able to recognize the correct water by its particular smell. In the sea salmon feed on a variety of animals, especially small fish and crustacea. It is a paradox that the adult salmon will take a bait in fresh water but when their stomachs are opened there is no evidence that they have swallowed food. Young salmon in fresh water take any suitable animal food.

Sea Trout (marine); **Brown Trout** (fresh water)
Salmo trutta

Europe and North Africa, both fresh and salt water. Introduced into North America.

Body the tail stalk is broad; *jaw* the upper jaw bone extends back well behind the rear edge of the eye (in older male fish the lower jaw has a hook at the end but this is less pronounced than in the male Salmon); *scales* small, 13–16 rows between the 2nd dorsal fin and the lateral line; *fins* there are two dorsal fins, the 2nd being simply a flap of tissue without fin rays (adipose fin); the tail fin is square-cut or shallowly forked. 1D 12–14; 2D none; A 10–12.

Sea Trout are dark grey above, silvery below with darker mostly x-shaped spots, especially on the gill cover. River-dwelling brown trout are very variable in colour, greenish or brownish above, silvery below with black and red spots on the flanks and gill covers. There are no spots on the tail as there are in the Rainbow Trout *Salmo gairdneri*. The young fish (parr) are marked similarly to the Salmon but often have a white upper front edge to the dorsal fin.

Up to 140 cm.

The Brown Trout and the Sea Trout are the same species but the latter spends the greater part of its life in the sea. Spawning occurs in fresh water in late autumn and winter on shallow gravel bottoms. The young fish (smolts) may migrate down the rivers into the sea or remain in fresh water. Young trout feed chiefly on aquatic insect larvae and small animals that fall into the water. The adults feed also upon fishes and crustacea and winged insects. Sea Trout take a variety of small fishes and crustacea. It is probably the latter which gives the flesh of the Sea Trout its salmon-pink colour.

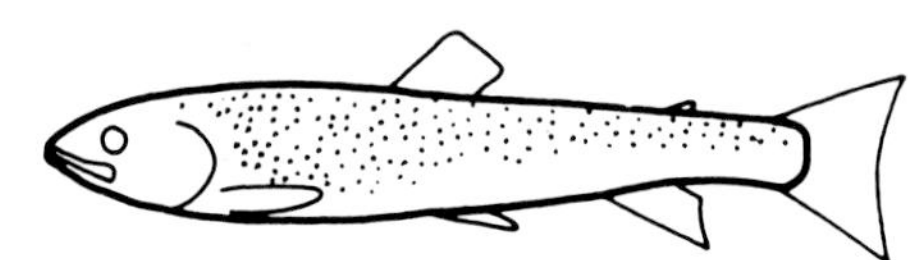

Salmon *Salmo salar*

2nd dorsal fin without fin rays;
upper jawbone scarcely reaches hind margin of eye.

Eastern and western Atlantic.
130 cm.

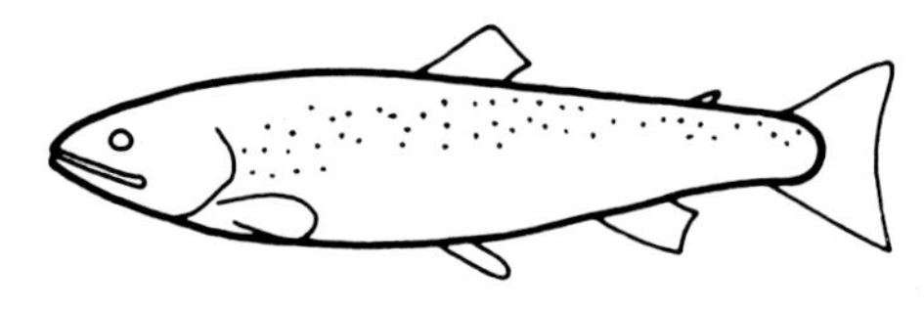

Sea Trout; Brown Trout *Salmo trutta*

2nd dorsal fin without fin rays;
upper jawbone reaches beyond eye.

Eastern Atlantic.
140 cm.

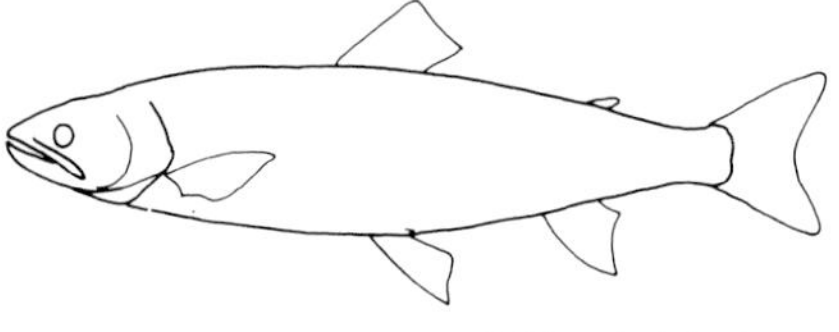

Arctic Char *Salvelinus alpinus*

2nd dorsal fin without rays;
ocean fish plain coloured;
pointed head.

Cold north Atlantic.
60 cm.

Arctic Char, *Salvelinus alpinus* (Gilbert van Rijckevorsel).

Arctic Char

Salvelinus alpinus (L.)

Circumpolar: in the North Atlantic sea-going populations as far south as Newfoundland in the west and Iceland in the east; land-locked populations in deep cold lakes occur further south.

Body trout-like in shape; *head* rather pointed; *scales* very small, about 230 in lateral line; *jaws* almost equal or the lower jaw slightly longer, upper jaw ending slightly behind eye; *teeth* moderate, none on vomer except at its head, teeth on bones at rear of tongue; *fins* tail fin distinctly forked. D 12–16; A 11–15. There is a second (adipose) dorsal fin without rays.

Ocean fish are metallic blue with a silver belly, and may have red, pink or cream spots on flanks; dorsal and anal fins without pigment. At spawning time in fresh water the lower flanks become red and the leading edges of the lower fins edged with white.

60 cm.

Where they are not land-locked, char swim down the rivers to the sea when about 6 years old, but once they have reached the sea do not travel far. Sea-run Arctic Char mature after 6 years, land-locked Arctic Char mature after 2 years. Spawning takes place in September or October in northern waters, two months later in the more southerly parts of its range in Quebec and Newfoundland. Gravelly areas about 1–4 m deep in river pools and lakes are selected for spawning.

An important food fish for the Inuit people.

Family:
OSMERIDAE

Slender rounded bodies, slightly flattened along the sides. There is an adipose fin and the vertebrae are not upturned at the base of the tail.

Smelt; Rainbow Smelt
Osmerus eperlanus (L.)

Eastern Atlantic, northern England south to France; western Atlantic, Labrador south to New Jersey.

Body long and slender; *jaws* large, extending back past the eye, the lower jaw is slightly longer than the upper; *odour* newly-dead fish have a characteristic sweet smell resembling cucumbers, violets or rushes; *teeth* present in both jaws (those in the lower jaw being the largest). There are also teeth on the roof of the mouth and on the tongue; *fins* 2 dorsal fins of which the 2nd is merely a flap of tissue without fin rays (adipose fin). The pelvic fins are inserted below the first rays of the dorsal fin. 1D 9–12; 2D none; A 12–16.

Greenish-grey above, silvery below with a wide silver band running along the flanks. The tail has a dark edge.

Up to 30 cm, sometimes more.

An inshore and estuarine fish which can

survive in completely land-locked situations. They are voracious predators, feeding on small fish and crustacea. They appear to assemble near river estuaries in winter and ascend the rivers to spawn in early spring.

Extremely good to eat when they are sufficiently fresh to retain their cucumber-like smell.

Capelin
Mallotus villosus (Müller)

Arctic waters of the Atlantic; south to Nova Scotia, east and west Greenland, Iceland, White Sea, Spitzbergen, northern Norway.

Body long and slender; *jaws* lower jaw projecting, the line of the jaws extending back to about the level of the pupil; *teeth* minute; *scales* very small, present everywhere on the body and not easily shed; *fins* the second dorsal has no fin rays (adipose fin), which distinguishes it from the Silversides.

Translucent green or olive-green above, silvery below, scales dotted with minute dark specks. Head and back darker at spawning time.

20 cm.

The sexes differ: the base of the anal fin is convex in males, but straight in females; and the pectoral fins are longer in males. In males there is a band of narrow pointed scales running along the flank above the lateral line.

A fish of the open ocean that comes inshore to spawn during June and July. They spawn over coarse sand and gravel bottoms from 50 m or more and right into the surf zone, where they are often stranded. At spawning time they occur in huge numbers, particularly in Newfoundland where they provide a feast for many larger predators including cod, seals, sea birds and whales.

Lizard Fish
Synodus saurus (L)

Mediterranean and temperate waters of the eastern north Atlantic including Canaries, Bermuda and Bahamas.

Body long and rounded, the upper line of the back from eye to tail rather flat; *head* triangular, short upper surface rough; *snout* broader than long; *jaw* directed forward, extending back to behind the eye; *teeth* long and depressable; *fins* two dorsal fins, the second with no fin rays.

Brownish-grey with 8 irregular transverse bars on body which camouflage it against the bottom.

40 cm.

Almost always seen resting on the bottom propped up by its pelvic fins. Often on bottoms with areas of sand and gravel between rocks. Usually at depths of about 10 m. It adopts this position to ambush small fish that swim above it.

Inshore Lizardfish
Synodus foetens (L.)

Western Atlantic, Cape Cod south to Brazil.

Resembles *Synodus saurus* in general body form, except that the tail has no trace of a keel and the snout is not so wide.

Olive or sand-coloured above, paler below; back mottled and upper surface of head distinctly patterned with short wavy lines; underside of head, fins on underside and inside the mouth tinged with yellow; no spot on shoulder, no black tip to snout, little barring on dorsal fin.

30 cm.

This Lizardfish appears to be the only one that regularly penetrates north into our area; further south there are some 7 other species. Found on the bottom in sandy coastal waters. It gets its scientific name *foetens*, which means foeted or stinking, because it soon begins to smell foul when left out in the sun.

Lizard fish, *Synodus saurus* (Andy Purcell).

Smelt; Rainbow Smelt *Osmerus eperlanus*

2nd dorsal fin without fin rays, short; jaws extend beyond eye; smells of cucumbers.
30 cm.

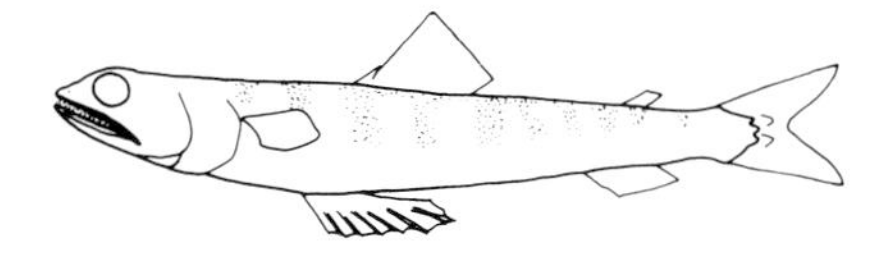

Lizardfish *Synodus saurus*

Long, rounded body; jaws very large, directed forward; rests on bottom supported by its pelvic fins.
Eastern Atlantic.
30 cm.

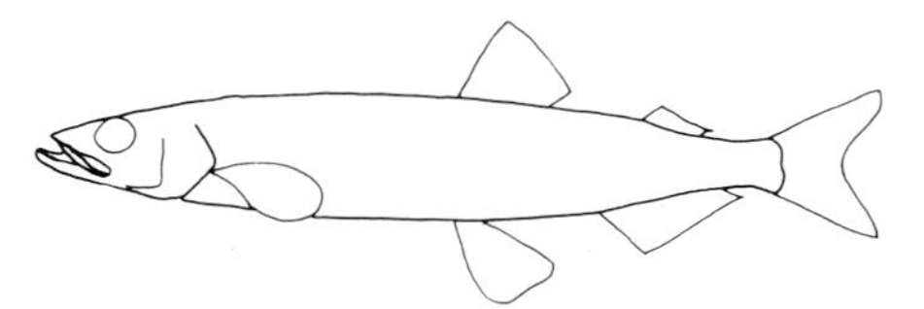

Capelin *Mallotus villosus*

Scales and teeth very small;
2nd dorsal fin without fin rays, longer than tall; projecting lower jaw which extends back to level of pupil.
Cold north-eastern and western Atlantic.
20 cm.

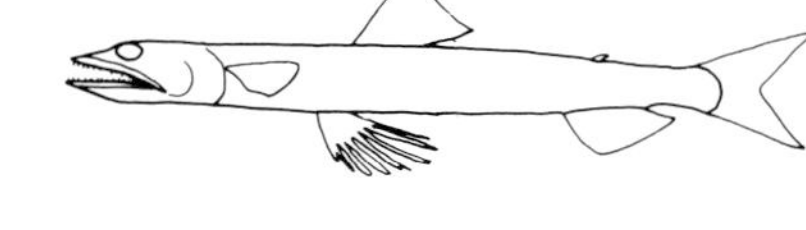

Inshore Lizardfish *Synodus foetens*

Long, rounded body; jaws very large directed forward;
rests on bottom supported by its pelvic fins;
snout not broader than long.
Western Atlantic.
30 cm.

Family:

ARIIDAE **Sea Catfishes**

Hardhead Catfish
Arius felis (L.)

Western Atlantic from Cape Cod south to Texas.

Body elongate tapering back to tail, not flattened, naked; *head* flattened; *barbels* 6 short barbels on chin; *fins* dorsal and anal fins without prolonged first rays, dorsal and anal fins have large serrated spines. D I 7; A 16.

Blue or grey above, white below.

60 cm.

Found in sandy bays, etc., the most abundant of the saltwater catfishes on the United States coast. Not good to eat, and abundant enough to be a considerable nuisance to fishermen.

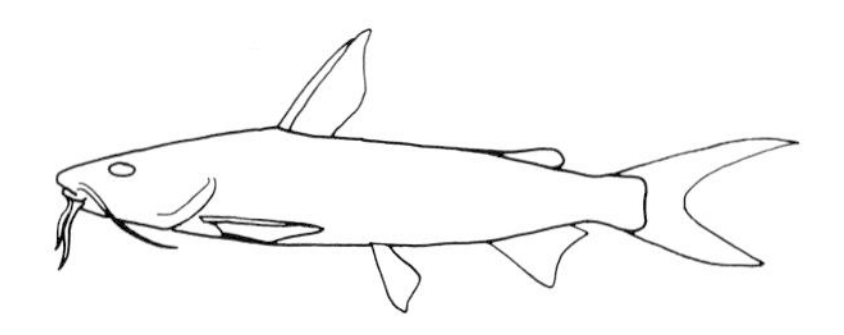

Hardhead Catfish *Arius felis*

Naked skin;
spine on dorsal and pectoral fins;
6 rounded barbles on chin
Western Atlantic.
60 cm.

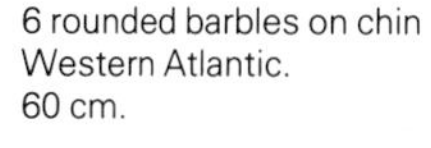

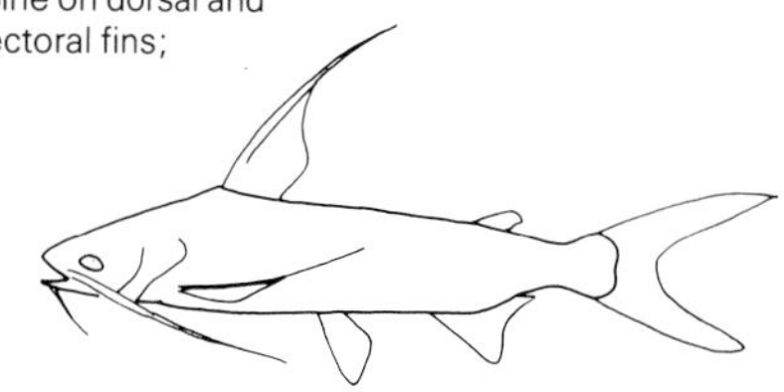

Gafftopsail Catfish *Bagre marinus*

Naked skin;
spine on dorsal and pectoral fins;
4 flattened barbels on chin;
elongate rays on dorsal and pectoral fins.
Western Atlantic.
60 cm.

Gafftopsail Catfish
Bagre marinus (Mitchill)

Western Atlantic from Massachusetts south to Panama.

Body elongate, not flattened, naked; *barbels* 4 long barbels on chin; *fins* first rays of dorsal and anal fins prolonged, first rays, dorsal and anal fins have large serrated spines. D I 7; A 23.

Light blue above, white below.

60 cm.

Found in shallow bays. The larger individuals are a good food fish in contrast to the Hardhead Catfish (*Arius felis*), which is usually not good to eat.

Order:

ANGUILLIFORMES = APODES

The **Morays**, **Conger** and **Freshwater Eels** belong in this family. The body is long and typically eel-like, the skin has no scales or they are minute and embedded in the skin. The dorsal, tail and anal fins form a continuous fringe. No pelvic fins.

The classification of this order is primarily based on skeletal characters.

Family:

ANGUILLIDAE

These **eels** always breed in the sea but may spend their juvenile life in fresh water. Pectoral fins are present.

Eel *Anguilla anguilla*

Lower jaw longer than upper;
pectoral fin round;
eyes round, usually small.
Eastern Atlantic.
140 cm.

Eel
Anguilla anguilla L.

Freshwater and neighbouring coasts of Europe, Asia and North Africa, Atlantic and Sargasso Sea. In North America there is a closely allied species, *Anguilla rostrata*.

Jaws lower jaw longer than the upper; *eyes* round, small in the yellow (immature) phase, much larger in the silver (adult) phase; *scales* not visible but present embedded in the skin; *gill opening* a vertical slit in front of the pectoral fin; *fins* the pectoral fin rounded; when depressed the pectoral fin terminates well in front of the origin of the dorsal fin.

In the immature yellow phase they are olive-brown or greyish-brown above, silver or yellowish-silver below. In the more adult (silver) phase the back is dark grey-green and the flanks and belly are silver. (Colour differences are not, however, always sufficient to separate the two phases.)

Up to 140 cm.

The yellow eel is the form normally seen or caught on hook-and-line. These fish have not yet reached sexual maturity. They are common in lakes and rivers and are often seen lying on the surface of the mud in deep lakes. Sometimes they stand with the upper part of their body upright in the water. Like the Conger and Moray, they apparently spend some time in holes with only the head protruding. Sometimes seen in estuaries and sheltered lagoons where they are often associated with clumps of seaweed.

After a period of 7–20 years, when the females measure between 50 and 100 cm and the males 30–50 cm, a series of remarkable changes associated with the onset of maturity take place. The eye increases in diameter and becomes specialized for vision in deep ocean waters; the gut begins to atrophy and the gonads enlarge. The colour also changes from a yellow-brown to a silver-grey. In autumn the maturing eels set out on a long breeding migration. Usually on moonless nights they travel down the rivers to the sea (sometimes travelling short distances over land). These descending silver eels form the basis of a valuable fishery. Once they have left the coasts there is little further direct knowledge of their movements until their eggs are found in the deep waters of the Sargasso Sea (Western Atlantic). These eggs are

Eel, *Anguilla anguilla* adult (John Lythgoe).

Eel, *Anguilla anguilla* elver (John Lythgoe).

found in spring and early summer; the young tranparent blade-shaped larvae (the leptocephalus) gradually drift back to Europe and North Africa on the prevailing current. The leptocephalus larva changes into the eel-shaped, but still transparent, glass eel during this passive migration. When the coastal waters are reached they assume the normal yellow-brown colour of the young eel (elver). The entire drift across the Atlantic takes 2 or 3 years. When the elvers are about 5–10 cm long they ascend the rivers, or more rarely, settle in coastal waters. The passive migration of the larvae and the young eels from the Sargasso Sea to Europe and North Africa is proved beyond doubt, yet the opposite migration of the maturing eels is supported by no positive evidence. The yellow eel appears to be chiefly nocturnal and feeds upon bottom-living animals including some fish.

American Eel
Anguilla rostrata (Lesueur)

Western Atlantic, Labrador south to Panama.

Differs from the European eel (*Anguilla anguilla*) in having fewer vertebrae (103–111 in the American eel compared to 110–119 from the European eel).

1 m.

Their life cycle appears similar to that of the European eel, although it has been suggested that their spawning grounds may be slightly different.

Family:
MURAENIDAE

The **Morays** have no pectoral fins; the gill opening is small. The teeth are strong.

Moray Eel
Muraena helena (L.)

Moray Eel *Muraena helena*
No pectoral fin;
gill opening small and round;
eyes round.
150 cm.

Mediterranean and warm Atlantic, occasionally reaching as far north as Biscay.

Body long and powerful; *jaws* long, extending back beyond the eye; *teeth* long and sharp; *fins* no pectoral fin; *gill opening* is small and round; *eye* small and round.

Ground colour is usually dark brown mottled or marbled with yellow or whitish spots. The pattern is irregular at the front but tends to become more regular behind. A black spot surrounds the gill opening.

Up to 150 cm.

Occurs from near the surface to deep water. They are normally seen in deep cracks and crevices in the rock with just their head protruding.

Morays are caught rather more frequently during the winter when they approach the coasts to breed; the eggs are pelagic. They are extremely vicious when shot or hooked and they can inflict dangerous tearing wounds which often go septic. The flesh is good to eat, but it has been known since Galen that Moray flesh can be poisonous. The poison is of the 'Ciguatera type' and leads to stomach cramp and nervous disorders. The flesh probably becomes poisonous after the fish has eaten (perhaps indirectly) blue-green algae.

Moray Eel, *Muraena helena* (Andy Purcell).

Family:
CONGRIDAE

The **Congers** have a gill slit, no scales and a pectoral fin.

Conger Eel
Conger conger (L.)

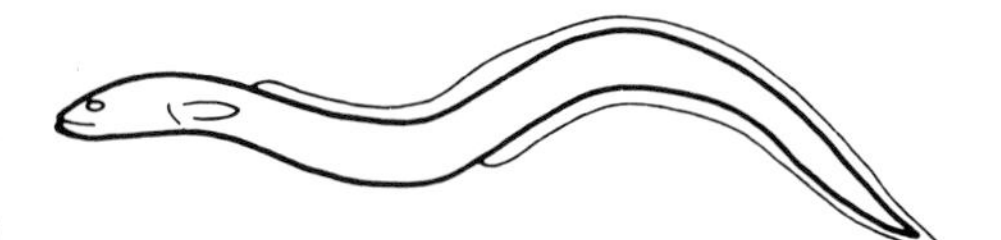

Conger Eel *Conger conger*

Pectoral fin pointed; gill opening a long slit; eyes elliptical. Eastern Atlantic; Mediterranean. 200 cm.

Mediterranean; Atlantic; English Channel; North Sea and occasionally into the Baltic.

Body elongate and powerful; *mouth* large, extending back to the centre of the eye; *jaws* upper jaw slightly longer than the lower; *scales* absent; *gill openings* large, extending low onto the belly; *eye* elliptical; *fins* pectoral fin pointed with 17–20 rays. Dorsal fin origin only slightly behind pectoral.

Uniform grey-brown above, lighter below.

Up to 2 m, occasionally 3 m.

Lives on rocky ground and is particularly abundant in old wrecks. In the daytime the Conger lives in cracks and fissures with the head alone protruding. They may be found from a depth of a few metres to at least 1,000 m. Feed chiefly at night, their diet being mainly bottom-living fishes, crustacea and cephalopods.

Congers probably spawn in mid-water over depths of 3–4,000 m. There is one spawning area between Gibraltar and the Azores and this probably serves the northern European Conger population. There are also spawning areas in the Mediterranean. They have never been caught in the sexually mature state. Spawning occurs in mid-summer and the larvae drift back to the feeding grounds. The change from the larval to the adult form takes about 2 years and the young Congers tend to be found close inshore. The Conger spawns only once, and in the aquarium, at least, the teeth are shed and calcium is lost from the bones so that they become soft and gelatinous. The Conger is good to eat, especially in soups, and once supported an important fishery.

Conger Eel, *Conger conger* (Christian Petron).

Order:

SYNENTOGNATHI=BELONIFORMES

Elongate fish, sometimes extremely so. The lateral line runs low on the belly; dorsal fin set far back, pelvic fin has 6 rays, no spiny rays. Swimbladder closed.

Family:

BELONIDAE

Atlantic Needlefish
Strongylura marina

Western Atlantic from Massachusetts south to Brazil.

Body very long and slim, rounded in cross section; *jaws* very long and almost equal except in juveniles where the lower jaw is much longer; *teeth* pointed; *fins* no spiny rays, pelvic and dorsal fins set well back, no finlets behind. D 15; A 17.

Greenish above, silvery below; the bones are bright bluish-green.

60 cm.

Found in open water near coasts, usually just beneath the surface; have the habit of skittering over the surface. Excellent to eat, but the green bones tend to put people off. May enter brackish or even fresh water.

Garfish; Garpike
Belone belone (L.)

Mediterranean: north-east Atlantic; North Sea.

Body very long; *jaws* elongate, the lower being the longest; *teeth* large and pointed; *fins* no finlets behind either the dorsal or

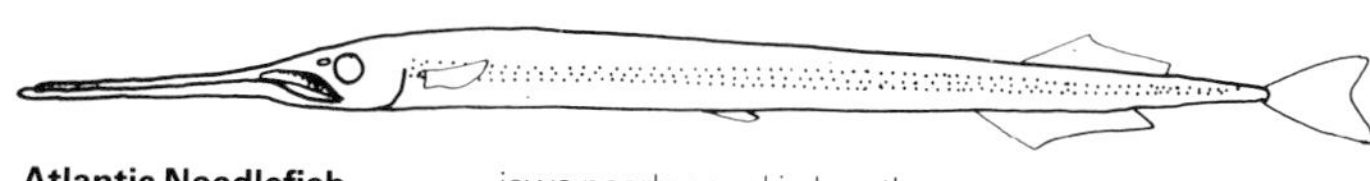

Atlantic Needlefish *Strongylura marina*

Body very long and slim, round in cross section; jaws nearly equal in length; 15 dorsal rays. Western Atlantic. 60 cm.

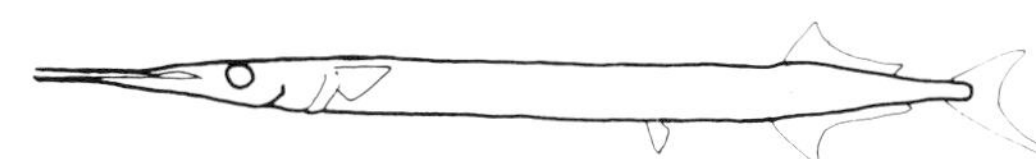

Garfish; Garpike *Belone belone*

Body very long and slim, round in cross section; jaws nearly equal in length; no finlets. Eastern Atlantic; Mediterranean. 80 cm.

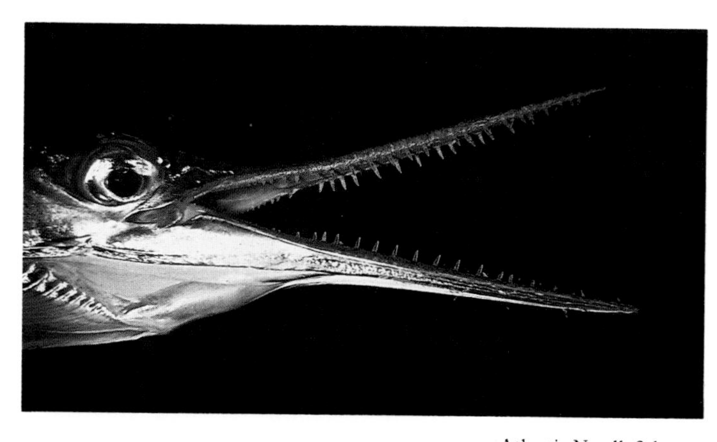

Atlantic Needlefish, *Strongylura marina* (Christian Petron).

Atlantic Needlefish, *Strongylura marina* (James D Watt).

anal fins. The dorsal fin is set slightly behind the anal. D 17–20; A 20–23.

Dark blue or dark green above with silver flanks and a yellowish tinge to the belly. The bones are green.

Up to 80 cm.

Primarily a shoaling fish over deep water, but when the water is warm in summer or early autumn they may come close inshore, especially in the Mediterranean. Typically the swimmer sees them patrolling in small groups just beneath the surface. They are wary fishes and are difficult to approach. Spawning is in coastal waters in summer. The eggs float and have adhesive filaments which become attached to a floating weed, etc. Garfish will take a moving bait and are thought to feed on small fishes (especially atherinids in the Mediterranean), cephalopods, crustacea, etc.

Family:

HEMIRAMPHIDAE

Halfbeak

Hyporhamphus unifasciatus (Ranzani)

Western Atlantic, Maine south to Argentina.

Body very long and slender, oval in cross-section; *jaws* lower jaw long and pointed, upper jaw very short and triangular; *lateral line* runs along lower body for most of its length, but has a branch to the pectoral fin; *fins* dorsal, anal and pelvic fins set far back, no spiny rays, upper and lower lobes of tail fin about equal. D 14–16; A 15–17.

Greenish above, silvery below.

20 cm.

Halfbeaks swim very near the surface, and sometimes schools of them will skitter and leap over the surface of the water, reminiscent of flying fish to which they are closely related. There are several species of Halfbeak, but this is the one most likely to be seen in our area. Usually in open water, but enter bays.

Skipper; Saury Pike

Scomberesox saurus (Walbaum)

Throughout the Atlantic between 30° and 45° N; Mediterranean.

Body long; *jaws* elongate but of equal length except in fishes of 4–15 cm, when the lower jaw is the longest; *teeth* small; *fins* the pectoral fin is directed downwards, finlets behind both the dorsal and anal fins. D 10–12; A 12–14.

Blue or olive above, golden or silvery below. A longitudinal silver band runs along each flank.

Up to 50 cm.

An oceanic species often seen in large shoals just beneath the surface. They appear to approach nearer the coasts in summer and early autumn but probably do not spawn inshore. The eggs are pelagic and are covered with filaments.

They have the habit of leaping out of the water and skipping along the surface to escape nets or predators.

Halfbeak *Hyporhamphus unifasciatus*

Very long and slender; lower jaw very long and pointed, upper jaw very short;
tip of lower jaw red.
Western Atlantic.
20 cm.

Skipper; Saury Pike *Scomberesox saurus*

Elongate body and jaws; finlets behind dorsal and anal fins.
Atlantic; Mediterranean.
50 cm.

Family:
SYNGNATHIDAE

The **Pipefishes** and **Sea Horses** have an external skeleton made up of bony plates. The snout is extended into a long, narrow pipe and the mouth, which is very small, is situated at the tip. All fins except the dorsal fin are very much reduced or absent. The Sea Horses differ from the Pipe Fishes in having prehensile tails and the head set at an angle to the body.

The eggs and larvae are carried by the male until they hatch. In members of the genus *Entelurus* the eggs are attached along the outside of the belly. In *Nerophis* the eggs are held in a narrow groove, whilst males of the genus *Syngnathus* have a brood pouch running along the belly which opens to the outside world through a longitudinal slit. The brood pouch of Sea Horses is almost entirely enclosed by skin and there is only a small opening.

Great Pipefish

Syngnathus acus L.
= *Syngnathus tenuirostris* Rathke

North-east Atlantic, especially west coast of British Isles and Ireland; southern North Sea; England Channel; Biscay; Mediterranean.

Snout long, more than ½ total head length but maximum height less than eight times the maximum length. No ridge along the top margin of the snout; *head* the head behind the eye has a conspicuous bump; *body rings* 17–22 between the base of the pectoral fin and the vent, polygonal in cross section; *fins* tail and pectoral fins present. The dorsal fin extends over 7–9 body rings.

Dark or light grey-brown or greenish above, lighter below. There are darker mottlings usually in the form of cross-bands.

Up to 30 cm, sometimes 50 cm.

Lives on most kinds of bottom, often amongst sea-grass and seaweed, usually shallower than 15 m although it may go deeper. They are able to live in brackish

Great Pipefish, *Syngnathus acus* (B. Picton).

water and are often found in river estuaries. Breeding occurs in summer. The young fish leave the brood pouch after about 5 weeks.

Deep-snouted Pipefish; Broad-snouted Pipefish

Syngnathus typhle (L.)

Mediterranean; north-east Atlantic north to Norway; Baltic.

Snout deep and laterally compressed, usually equalling the head in depth throughout its length; *profile* the top of the head and snout form a continuous line; *body rings* 16–20 (usually 19) rings in front of the vent; *fins* pectoral and tail fins present. The dorsal fin extends over 7–9 body rings and has 29–39 rays.

Light brown or greenish-brown above, lighter below, sometimes with sparse vertical stripes. Head and snout with darker spots.

Up to 35 cm.

Very tolerant of brackish water, it is common in the Baltic, the Dutch Waddenzee and the Black Sea. It assumes an upright stance typically amongst sea-grass but also amongst seaweeds.

Breeds in summer. The young fish are expelled from the brood pouch after about 4 weeks. Feeds on crustacea and small fishes.

Nilsson's Pipefish

Syngnathus rostellatus Nilsson

North Sea; English Channel; west coast of England and Ireland; Biscay.

Resembles *Syngnathus acus* except that the snout is short (less than ½ total head length); *profile* the nape is broadly convex but does not amount to a hump-like crest as in *Syngnathus acus*; *body rings* 13–17 rings in front of the vent.

Up to 17 cm.

Found over sandy bottoms down to about 18 m, especially amongst floating or attached seaweed, sometimes it is seen swimming freely. It can live and breed in brackish water.

Short-snouted Pipefish

Syngnathus abaster Risso

Mediterranean and west coasts of Spain and Portugal.

Snout short and tubular but with a pronounced high membrane-like crest running along its upper margin; *body rings* 14–18 (usually 16) rings in front of the vent; *skin* smooth to the touch when rubbed from tail to head; *fins* tail and pectoral fins present. The dorsal fin extends over 7–11 segments and has 23–34 (usually about 28) rays.

Greyish- or greenish-brown above, paler below. There may be a row of dark or light spots along the flanks which sometimes show darker cross-bands.

Up to 17 cm.

Found on sandy bottoms, often near estuaries.

Syngnathus pelagicus L.

Atlantic, Indian and Pacific oceans; European Atlantic coasts; rare in Mediterranean.

Snout tubular, more than ½ total length of head; *body rings* are smooth, 13–18 in front of the vent, 33–42 behind; *fins* pectoral fin present. The tail fin comes to a point. The dorsal fin has 30–45 rays (typically 40) and extends over 9–11 body rings.

Body a uniform greyish-brown, sometimes speckled or with indistinct cross-bands. There is a dark longitudinal stripe in front of each eye.

Up to 17 cm.

Usually found in association with floating masses of seaweed in the open oceans.

Syngnathus phlegon Risso

European Atlantic coasts; Mediterranean.

Snout tubular and elongate, longer than ½ the total head length; *body rings* 17–19 in front of the vent, 47–50 behind. The ridges along the body are markedly abrasive to the touch when the fish is stroked from the tail to the head; *fins* tail and pectoral fins present. The dorsal fin extends over 12–14 body rings (2–3 in front of the vent) and has 38–46 rays.

Great Pipefish
Syngnathus acus

Hump on nape; snout tubular, more than ½ head length; 17–22 rings in front of vent.
Eastern Atlantic; Mediterranean.
30 cm.

Deep-snouted Pipefish;
Syngnathus typhle

Snout laterally compressed; Eastern Atlantic; Mediterranean.
35 cm.

Nilsson's Pipefish;
Syngnathus rostellatus

Snout short, less than ½ head length; 13–17 rings in front of vent.
Eastern Atlantic.
17 cm.

Short-snouted Pipefish
Syngnathus abaster

Snout with longitudinal crest; skin smooth.
Mediterranean.
17 cm.

Syngnathus pelagicus

Snout long and tubular (more than ½ total head length); tail fin pointed.
Eastern Atlantic.
17 cm.

Syngnathus phlegon

Snout long and tubular (more than ½ total head length); back rough.
Eastern Atlantic; Mediterranean.
20 cm.

Dark-flank Pipefish
Syngnathus taenionotus

Outline of head from snout to nape flat; dark line on flank.
Mediterranean.
20 cm.

Snake Pipefish *Entelurus aequorius*

Pectoral fin absent; tail fin small; dorsal fin set forward.
Eastern Atlantic.
60 cm.

Worm Pipefish *Nerophis lumbriciformis*

No pectoral or tail fins; snout short and up-tilted; back very dark.
Eastern Atlantic.
17 cm.

Straight-nosed Pipefish
Nerophis ophidion

No pectoral or tail fins; snout has longitudinal crest giving a straight profile; dorsal fin extends over 10–12 segments.
Eastern Atlantic; Mediterranean.
30 cm.

Spotted Pipefish
Nerophis maculatus

No pectoral or tail fins; regular spotted pattern; no longitudinal crest on snout; dorsal fin extends over 8–9 segments.
Mediterranean; eastern Atlantic.
30 cm.

Northern Pipefish
Syngnathus fuscus

Body long and slender; body with 54–62 bony rings; long slender mouth; no pelvic fin.
Western Atlantic.
20 cm.

Sea Horse *Hippocampus ramulosus*

Snout more than ⅓ head length.
Eastern Atlantic; Mediterranean.
15 cm.

Sea Horse *Hippocampus hippocampus*

Snout ⅓ or less head length.
Eastern Atlantic; Mediterranean.
15 cm.

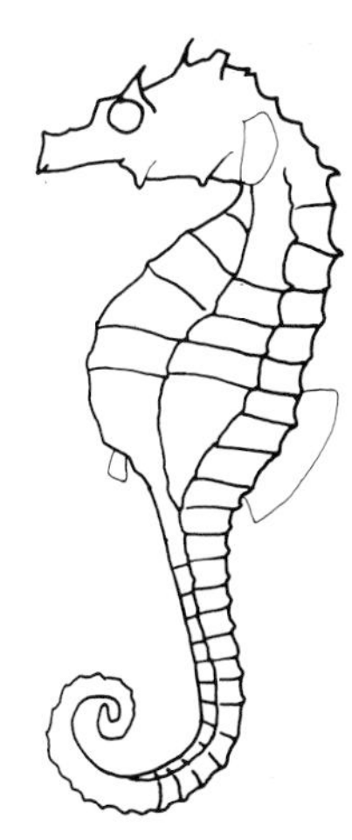

Lined Sea Horse
Hippocampus erectus

dorsal fin rays 19 rays; Western Atlantic;
18 cm.

Either bluish with a row of dark spots along the flanks or silvery yellow with darker vertical stripes.

Up to 20 cm.

An open water species probably found from the surface to 600 m or more.

Dark-flank Pipefish
Syngnathus taenionotus Canestrini

Snout long and tubular; *profile* there is no raised hump on the nape and the line of the head from above the pectoral fin to the snout tip is flat; *fins* the dorsal fin extends over 7–11 body rings. Pectoral and tail fins present.

Grey, often with a dark longitudinal stripe extending along the upper flank from the pectoral fin to just behind the vent.

Up to 20 cm.

Found amongst seaweed in sheltered areas.

Snake Pipefish
Entelurus aequorius (L.)

Eastern Atlantic from Scandanavia to the Azores; rare in the Mediterranean.

Body long and thin and rounded rather than polygonal in cross-section; *body rings* 28–31 in front of the vent and 60–70 behind; *fins* pectoral fin absent, the anal fin is very much reduced with only 4–6 rays. The dorsal fin set forward covering 8 body rings in front of the vent and 3 behind. It has 37–44 rays.

Yellowish- or brownish-grey. On the flanks there are numerous silvery transverse markings often with dark borders. There is a reddish longitudinal band running from the tip of the snout through the eye to the gill cover.

Males reach 40 cm, females 60 cm.

Found amongst marine plants in shallow water down to about 30 m. In summer they are sometimes found swimming near the surface far from the coasts, particularly near floating seaweed. They breed in summer, the young are not fully developed when they leave the brood pouch and live a free-swimming existence in open water.

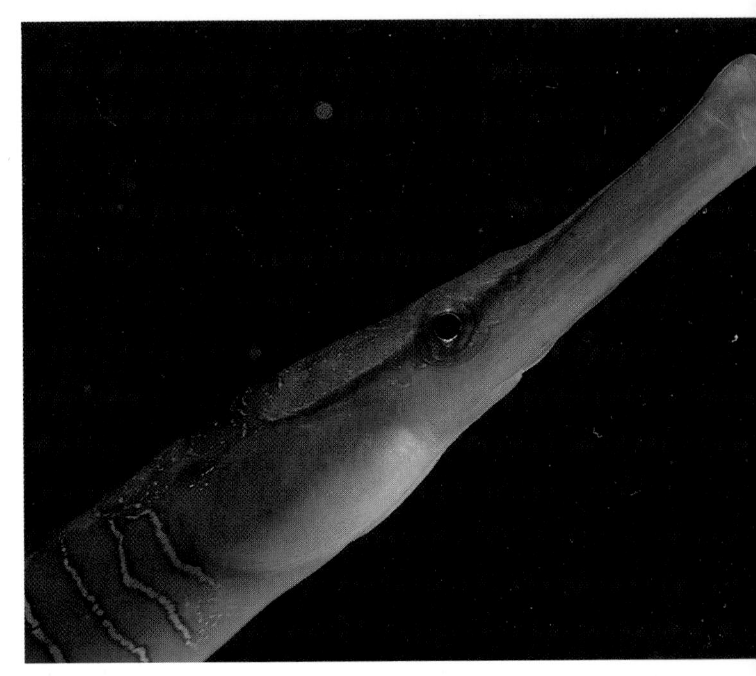

Snake Pipefish, *Entelurus aequoreus* (David Maitland).

Worm Pipefish
Nerophis lumbriciformis (Jenyns)

North-east Atlantic from Scotland to Morocco; English Channel; North Sea.

Body very long and slender and rounded in cross-section; *snout* short and markedly tilted upwards; *body rings* indistinct, 24–28 in front of the vent, 17–18 behind; *fins* there are no pectoral or tail fins. The dorsal fin is set rather far back extending over 2–3 body rings in front of the vent and 4–6 behind.

Very dark greenish-grey or dark brown above, mottled white below.

Up to 17 cm.

The Worm Pipefish is unusual in living in very shallow water, even between tide marks, where they are usually found under stones. They prefer rocky areas where there is a luxuriant algal cover.

Straight-nosed Pipefish
Nerophis ophidion (L.)

North-east Atlantic; Mediterranean.

Body long and slender, rounded rather than polygonal in cross-section; *snout* is half the total head length with a raised longitudinal crest running from between the eyes to the tip of the snout; *body rings* there are more than 25 body rings on the trunk in front of the vent; *fins* no tail or pectoral fins. The dorsal fin extends over about 11 body rings (3–5 in front of the vent, 6–8 behind), it has 33–38 rays.

Colour very variable, dark green, olive, brownish or blackish above with dark vertical lines, pale whitish green, yellow or blue below. Sometimes the front half of the fish has lighter spots arranged in regular vertical rows.

Up to 30 cm.

Lives close to shore amongst seaweeds, especially sea-grass and thong weeds (*Chorda* and *Himanthalia*).

Breeding occurs in summer, the eggs are held in a groove along the belly of the male. The young are released in a relatively undeveloped state and live a free-swimming life for 3–4 months.

Spotted Pipefish
Nerophis maculatus Rafinesque

Mediterranean; eastern Atlantic from Portugal to the Azores.

Body long and slender, rounded rather than polygonal in cross-section; *snout* rounded with no longitudinal crest; *body rings* there are 21–23 body rings on the trunk in front of the vent and 65–74 behind; *fins* no pectoral or tail fin. The dorsal fin extends over 8–9 segments (2–4 in front of the vent, 4–5 behind). It has 24–30 rays.

Greyish, brownish, greenish or greyish-red with rows of yellowish spots with darker edges arranged in vertical bars, sometimes coalescing into vertical bands.

Males up to 20 cm, females up to 30 cm.

Northern Pipefish
Syngnathus fuscus Storer

Western Atlantic, Gulf of St Lawrence south to Florida.

Snout tubular; *body* hexagonal cross-section in front of vent, 4-sided behind; *body rings* 18–21 trunk rings, 34–39 tail rings.

Various shades of brown matching background with irregular crossbars and mottled lower flanks sprinkled with white dots, golden-yellow below.

20 cm.

Found amongst seaweed and eelgrass in coastal water. Enters brackish water and salt marsh, and may travel in association with matts of floating seaweed. Migrates to deeper water in autumn and returns in the spring.

Sea Horse
Hippocampus ramulosus Leach
= *Hippocampus guttulatus guttulatus* Cuvier

Mediterranean; western Atlantic from the English Channel to west Africa; rare in North Sea.

Body bent at the neck, 48–50 body segments often with bony thickenings on the ridges and fleshy tentacles along the back; *tail* prehensile; *snout* long, more than ⅓ head length; *fins* dorsal and pectoral and anal fins, no tail fin. D 18–21; A 4; pectoral fin rays 15–16.

Greenish or brownish, sometimes with white spots on the corners of the body segments.

Up to 15 cm.

Lives in shallow water amongst seaweeds, and also amongst sea-grass (*Posidonia*) meadows where it clings to the vegetation with its tail. Also found in brackish water. Sea Horses are very rarely seen by divers even where they are quite common, presumably because their irregular outline makes them very difficult to see. They swim in an upright position using their dorsal fin as their sole means of propulsion. Spawn in late spring and summer.

Sea Horse
Hippocampus hippocampus (L.)

Mediterranean, French and Spanish coasts, probably not as far north as Britain.

Resembles *H. ramulosus*; the only certain point of difference is the snout, which is short, ⅓ the head length or less; 45–47 body rings; fins D 16–17; A 4; pectoral fin rays 13–15.

This species appears to favour rather muddy bottoms. Spawning is in spring and summer. The male follows the female about presenting his brood pouch to her.

The transfer of eggs normally begins at night, the partners swim side by side and the female transfers a few eggs at a time. The young fish are expelled from the brood pouch fully developed after 4–5 weeks.

Sea Horse, *Hippocampus hippocampus* (Guido Picchetti).

Lined Sea Horse

Hippocampus erectus Perry
= *Hippocampus hudsonius*

Western Atlantic, Nova Scotia south to Uruguay.

Snout about equal to length of head; *head* has weak spines; *body* 32–35 segments; *fins* dorsal fin with 16–21 rays.

Various shades of brown with blotches edged with dark and light margins, no spots.

18 cm.

Sometimes found associated with drifting seaweed.

Family:

GASTEROSTEIDAE

The **Sticklebacks** are marine and freshwater fishes of cool northern waters. They have a series of spines along the back in front of the dorsal fin. The pelvic fin consists of one spine and 1–3 soft fin rays. The body is armoured with bony plates.

Stickleback; Three-spined Stickleback

Gasterosteus aculeatus L.

Fresh waters of Europe, North Africa and North America. Also in brackish and fully marine waters as far south as Northern Ireland and the North Sea.

Body oval and laterally compressed with a slender tail stalk. The flanks are armoured with a series of bony plates which are more numerous in saltwater specimens; *spines* and *fins* 3 spines on the back, the first two large, the third small, situated just in front of the dorsal fin. These spines are usually folded back on the body and are thus not conspicuous. The pelvic fin consists of a single spine and a much shorter fin ray. The dorsal fin is longer than the anal. D III, 8–14; A I, 6–11.

Silver, olive or blue above, silver flanks and white belly. In the breeding season the males become a brilliant red below and the eye assumes an iridescent peacock blue.

Three-spined Stickleback, *Gasterosteus aculeatus* (Andy Purcell).

Up to 7 cm.

The Stickleback can live both in fresh water and in fully saline coastal waters. It lives in most types of fresh water except for stagnant ponds. In the sea it probably prefers rocky areas with some weed. Spawning is chiefly from April to May. The male builds a nest on the bottom from strands of weed, etc., and this he sticks together with a cement secreted from his kidneys. He herds the female into the nest to lay her eggs. These are guarded by the male and oxygenated by fanning with his pectoral fins. He also cleans the nest of unwanted fragments and drives off intruders. The nest is used for successive spawnings. The breeding behaviour of the Stickleback has been extensively studied since it involves complicated nuptial and threat displays. The red of the belly of the male evokes aggressive behaviour in other males. Their food consists of any aquatic animals small enough to eat.

The **Nine-spined Stickleback** *Pungitius pungitius* is similar except for the extra spines along the back (8–10).

Primarily a freshwater species, it occasionally enters brackish water.

Four-spined Stickleback
Apeltes quadracus (Mitchill)

Coastal waters of north-eastern America, Gulf of St Lawrence south to Virginia.

Resembles the Three-spined Stickleback (*Gasterosteus aculeatus*) in general form except that there are 4 rather than 3 spines on the back, the front 2 spines being about twice the height of the rear 2. No bony plates on the flanks.

Olive-green to brownish above with mottled darker markings; in the breeding season the males may be almost black and the pelvic spine and membrane is red.

5 cm.

Primarily a fully marine species, but frequently enters brackish water; there are land-locked populations, especially in the northernmost part of its range. Nest constructed of seaweeds and sea-grass often in the intertidal zone.

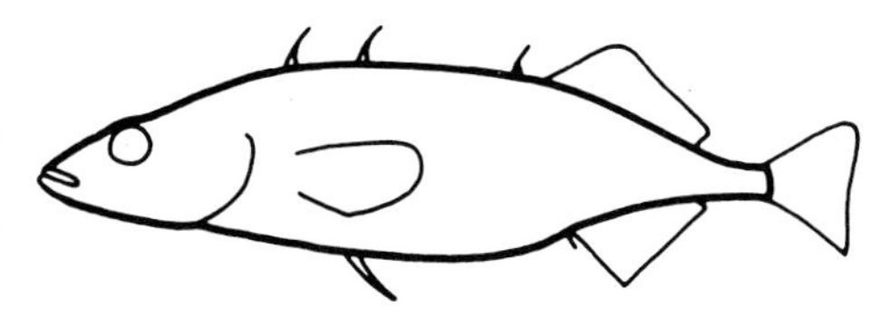

Stickleback; Three-spined Stickleback
Gasterosteus aculeatus

3 spines on back;
bony plates on flanks;
tail stalk with keel.
Eastern and western Atlantic.
7 cm.

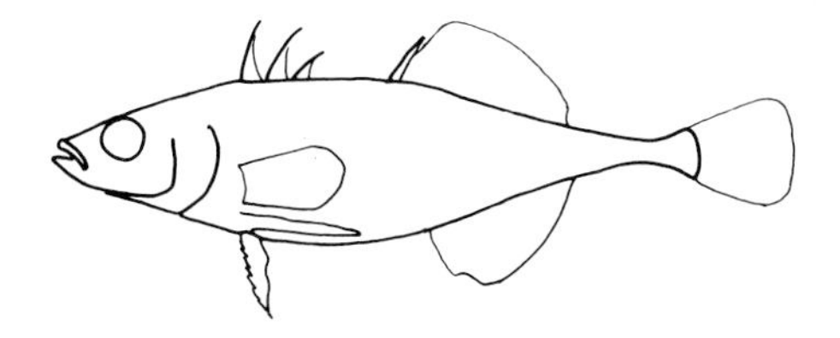

Four-spined Stickleback
Apeltes quadracus

4 spines on back;
no bony plates.
North-west Atlantic.
5 cm.

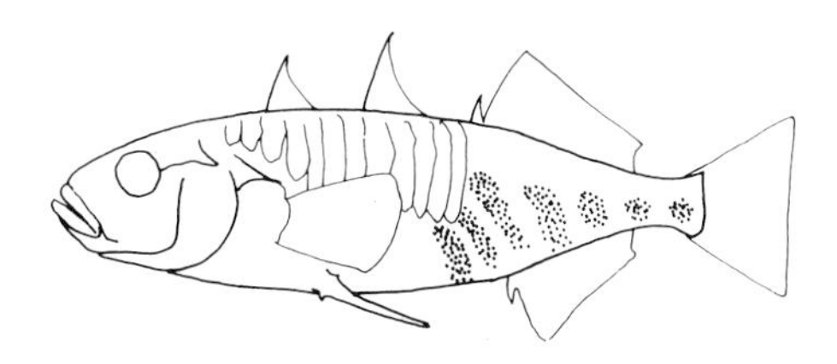

Blackspotted Stickleback
Gasterosteus wheatlandi

3 spines on back;
tail stalk without keel;
bony plates on flanks;
black spots on flanks.
North-west Atlantic.
5 cm.

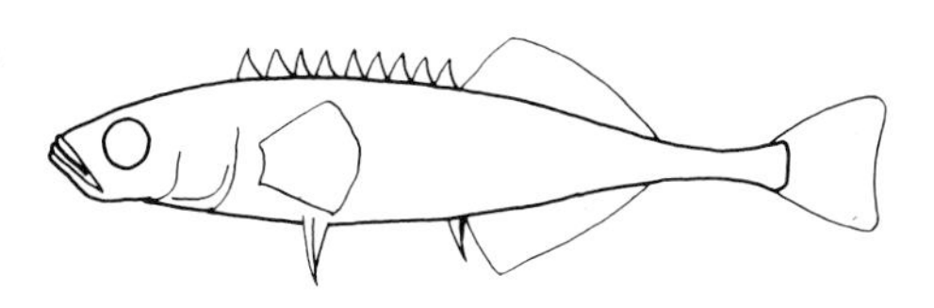

Nine-spined Stickleback
Pungitius pungitius

9 spines on back;
keel on tail stalk;
no large bony plates on flanks.
North-east and western Atlantic.
5 cm.

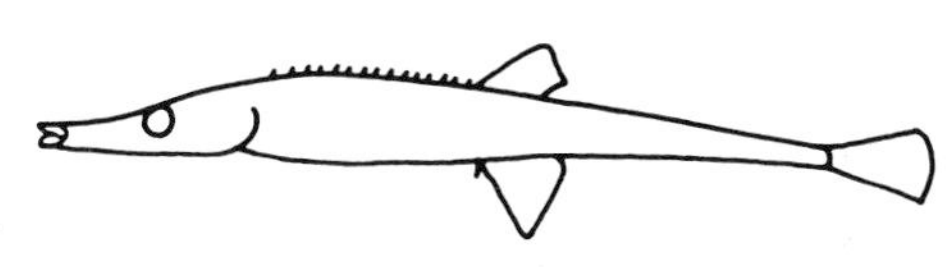

Fifteen-spined Stickleback *Spinachia spinachia*

15 spines on back;
elongate body;
Eastern Atlantic;
20 cm.

Blackspotted Stickleback
Gasterosteus wheatlandi Putnam

North America, Newfoundland south to Long Island.

Resembles the Three-spined Stickleback (*Gasterosteus aculeatus*) in body form. Differs in having fewer bony plates on the flanks (6–11) compared to up to 30, and the tail stalk has no lateral keel.

Yellow or greenish-yellow with black spots on sides. Both sexes become brighter in the breeding season, males become lemon-yellow or golden-yellow with orange pelvic fins.

5 cm.

Primarily a marine species frequently found in water less than 3 m over gravel bottoms, and it can be found in open water associating with floating seaweed. It may come into brackish water to spawn.

Nine-spined Stickleback
Pungitius pungitius (L.)

Circumpolar distribution in the north Atlantic, south to New Jersey and south to Spain.

Similar in body form to the Three-spined Stickleback (*Gasterosteus aculeatus*) from which it differs in having 9 (7–11) dorsal spines and no large bony plates on the flanks.

Olive-green or grey-green above with darker irregular bars, silvery below. In the breeding season reddish tints appear on the head and the belly in males may be black.

5 cm.

Usually found in brackish and tidal pools which it shares with the Three-spined Stickleback, but apparently prefers thicker weed cover.

Fifteen-spined Stickleback
Spinachia spinachia (L.)

North-east Atlantic from Biscay to north Norway; North Sea and Baltic as far north as Åland.

Body very long with a slender tail stalk; there is a raised ridge of about 40 plates in a line down the body; *spines* and *fins* there are 14–16 short spines along the back. The pelvic fin consists of a single spine and a short fin ray. The tail is small and rounded; the dorsal and anal fins are almost opposite and set far back on the body.

The colour is brownish or greenish. In the breeding season the male becomes more blue in colour.

Up to 20 cm.

Typically a coastal marine species, but can tolerate estuarine conditions. It is always found amongst weed sometimes even in rock pools and not deeper than 10 m. Spawning occurs in spring and early summer in very shallow water, sometimes even between tide marks. The nests, which are built amongst seaweed, are woven from plant fibres and are about the size of a fist. It is cemented together with secretions from the kidney. There are 150–200 eggs, which are laid inside the nest and are guarded by the male. The eggs hatch after about 20 days. The fish are said to live for only 2 years, the larger fish being in their 2nd year.

Fifteen-spined Stickleback, *Spinachia spinachia* (Andy Purcell).

Order:

ANACANTHINI

Tail fin formed mainly of dorsal and anal rays; swimbladder; no spiny rays.

Family:

MERLUCCIIDAE

Hake have 2 dorsal and 1 anal fin, the 2nd dorsal and anal fins are notched. There is no barbel. Hake typically live in deep water, moving in to shallower water to spawn. Commerically very important, but only 1 species is likely to be seen in coastal waters in our area.

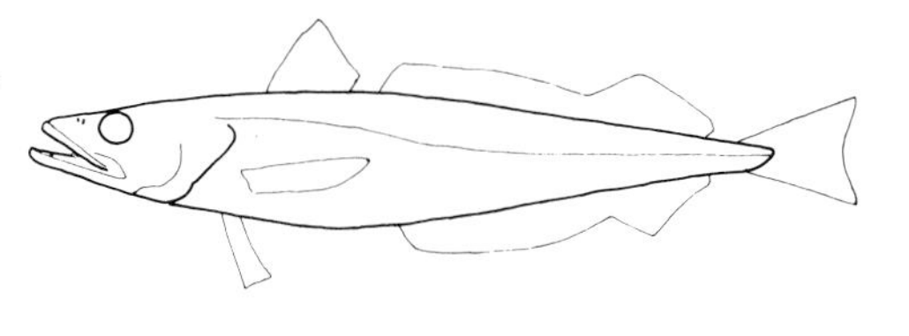

Silver Hake *Merluccius bilinearis*

2 dorsal fins, 1 anal fin; no barbel; pelvic fin normal. Western Atlantic. 75 cm.

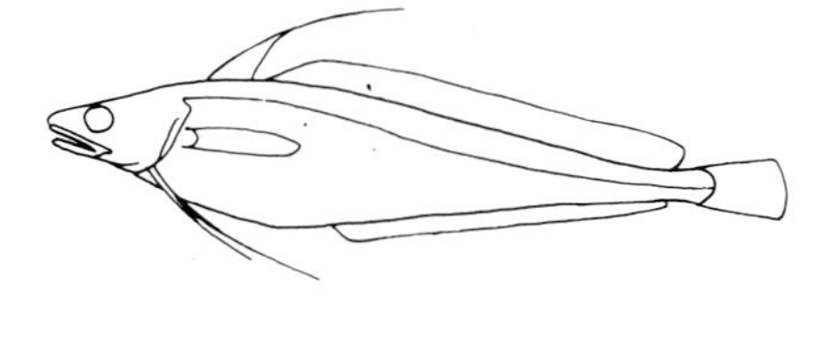

Red Hake; Squirrel Hake *Urophycis chuss*

Pelvic fin consists of 2 long filaments 1st dorsal ray forming filament; back usually reddish. North-western Atlantic. 50 cm.

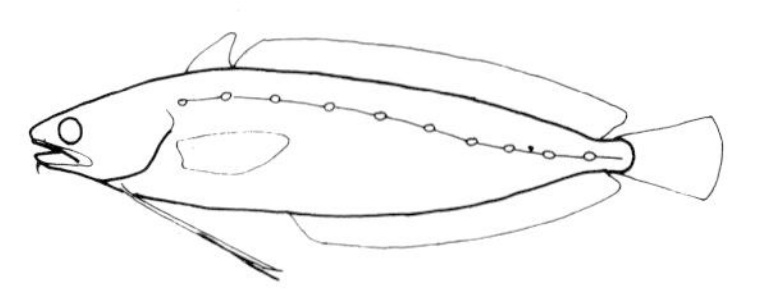

Spotted Hake *Urophycis regia*

Pelvic fin consists of 2 filaments; 1st dorsal ray not forming filament; lateral line interrupted with white spots. North-western Atlantic. 50 cm.

Silver Hake
Merluccius bilinearis (Mitchill)

North American distribution from Newfoundland south to South Carolina.

Body elongate; *jaws* lower longer than upper, with two or more rows of sharp recurved teeth; *lateral line* almost straight; *fins* 2 dorsal fins, 2nd dorsal fin long and almost uniform in height, anal fins similar to dorsal. 1D 11–14; 2D 37–42; A 38–42.

Silvery in life, upper fins black-edged, lower fins white-edged. Axil and edge of pectoral fin blackish. The inside near the throat is dusky blue.

75 cm.

Depth range very varied, from water so shallow that they are sometimes stranded to over 900 m. Appears to prefer water at a temperature of 6–8°C, which is rather warm for most cod species. Moves near the shore in spring to spawn and moves offshore again at the end of the summer. The Silver Hake is a voracious predator especially on other fish including their own species.

Red Hake; Squirrel Hake
Urophycis chuss (Walbaum)

Western Atlantic, southern Newfoundland south to Cape Hatteras.

Body elongate, rounded in front of vent, compressed behind; *head* pointed, upper jaw the longer; *barbel* small barbel on lower jaw; *fins* 1st ray of dorsal fin prolonged into a filament, 2nd dorsal fin lower, long and of uniform height, pelvic fin has 2 filamentous rays, tail fin rather long and rounded. 1D 9–10; 2D 54–57; A 48–50.

Usually reddish or reddish-brown, pale or yellowish below, often with many small black spots.

50 cm.

Adult Red Hake live in deep water over muddy bottoms. They may overwinter in living scallop shells until they are almost 12 cm long moving to warmer water when the temperature drops to 4°C.

Spotted Hake
Urophycis regia (Walbaum)

Western Atlantic, Nova Scotia south to northern Florida.

Body elongate, compressed; *head* pointed, upper jaw the longer; *barbel* small barbel on lower jaw; *fins* first ray of dorsal fin not forming a filament, 2nd dorsal fin lower, long and of uniform height, pelvic fin consists of 2 filamentous rays, tail fin rounded. D 46–51; A 43–49.

Pale brownish, lateral line dark with about 10 white spots evenly spaced along it, 1st dorsal fin dark below with white border.

50 cm.

Found from shallow water down to at least 350 m.

Family:

GADIDAE

An extremely important group of food fishes particularly in northern waters. Barbels are usually present and are a valuable aid in identification. Several cods have three dorsal fins and two anal, which give a characteristic silhouette in the water. Other species are almost eel-like in shape and have one short and one very long dorsal fin.

Whiting

Merlangius merlangus (L.)

Eastern Atlantic from northern Norway to Biscay; North Sea; English Channel; Mediterranean; Faroes and Iceland. There is a subspecies in the Black Sea.

Jaws upper longer than lower; *teeth* sharp and prominent; *barbel* very small or absent; *lateral line* gently curved; *fins* 3 dorsal, 2 anal separated by short interspaces. The 1st anal fin is long and rounded and originates beneath the middle of the 1st dorsal fin. The tail fin is broad and square-cut. 1D 12–15; 2D 18–25; 3D 19–22; 1A 30–35; 2A 21–23.

Body olive, sandy or bluish above, flanks silvery mottled or streaked with gold, belly white. The lateral line is golden bronze. The two anal fins are rimmed with white. There is also an inconspicuous diffuse dark spot at the base of the pectoral fin.

Up to 60 cm, sometimes 70 cm.

The Whiting has considerable commercial importance. It is predominantly a shallow-water species and the bulk of the commercial catch is taken at 25–150 m. They are chiefly caught on sandy or muddy bottoms but venture further above the bottom than Haddock or Cod and are often caught near the surface. They are rarely seen by divers, even in water where they are known to abound. This may be because the neutral colouring of the fish when seen underwater blends in well with the background.

The food is mainly crustacea and fish. The proportions of fish in the diet increases with age. The Whiting appears to feed most actively at dawn and dusk.

Whiting, *Merlangius merlangus* (Andy Purcell).

Bib, *Trisopterus luscus* (Linda Pitkin).

Bib

Trisopterus luscus (L.)

European Atlantic coasts; North Sea from Gothenburg to Biscay; English Channel; western Mediterranean.

Body characteristically deep-bodied; *jaws* upper longer than lower; *barbel* prominant, equalling in length the diameter of the orbit; *lateral line* sharply curved; *fins* 3 dorsal, 2 anal with very short interspaces, the origin of the 1st anal fin is beneath or slightly behind the origin of the 1st dorsal fin. The 1st anal fin is deep and rounded and often joined to the 2nd anal. 1D 11–14; 2D 20–24; 3D 18–20; 1A 30–34; 2A 19–22.

Flanks are pale copper-coloured often with 4–5 broad darker vertical bands, which are sometimes very dark but occasionally very pale or absent. There is a small dark spot at the base of the pectoral fins. The margins of the anal fins are white and the margin of the tail fin is black. The lateral line is golden yellow but shows little contrast against the flanks.

Up to 30 cm, rarely 45 cm.

Of very little commercial importance but is often seen by divers in the English Channel. The young fish are found close inshore, often in large shoals. The larger fish move off into deeper water in depths of 30–100 m and sometimes considerably deeper. They prefer a combination of rock and sand and the young fish, at least, have the reputation of always frequenting one place.

Their food is chiefly molluscs and crustacea but the larger ones take small fish and cephalopods. The main spawning period is March-April but may continue until August. They prefer temperatures of 8–9°C and depths of 50–100 m.

Poor Cod

Trisopterus minutus (L.)

European shores of North Atlantic extending to Trondheim in the north, Kiel in the east; Faroes; round all British coasts; western Mediterranean.

Body rather more slender and delicate than the Bib (*Trisopterus luscus*); *jaws* upper longer than lower; *barbel* long and prominent; *lateral line* less curved than the Bib(*Trisopterus luscus*); *fins* 3 dorsal, 2 anal fins set close together. The 1st anal fin begins between the middle and the end of the 1st dorsal fin. The 1st dorsal fin is rather pointed. 1D 13; 2D 23–26; 3D 22–24; 1A 28–29; 2A 23–25.

Uniform bronze-red above, flanks more coppery and belly a silvery-grey. The

lateral line does not show up well against the flanks. There is a small dark spot at the base of the pectoral fin.

Up to 20 cm, rarely up to 26 cm.

The Poor Cod has no commercial importance. It is found in the same sort of places as the Bib. Small individuals can be very numerous in shallow water, especially off rocky coasts. They appear to be attracted to artificial structures such as cages and framework on the sea bed. The food is chiefly bottom-living crustacea, molluscs and fish. They spawn in March and April, sometimes as late as July, usually in depths of 50–100 m at temperatures higher than 8°C. The males live for 4 years, the females up to 6 years.

Poor Cod (top), *Trisopterus minutus* (Jim Greenfield).

Pollack (bottom), *Pollachius pollachius* (Linda Pitkin).

Pollack

Pollachius pollachius (L.)

Eastern Atlantic from North Cape of Norway to Biscay; south-east Iceland; North Sea; northern parts of western Mediterranean.

Jaws lower longer than upper, especially in larger individuals; *barbels* absent; *lateral line* commences above the gill cover, arches over the pectoral fin and follows the mid-line from the origin of the 2nd dorsal fin to the tail; *fins* 3 dorsal, 2 anal with well defined interspaces. The 1st anal is rather long and rounded and originates below the

Saithe, *Pollachius virens* (Jim Greenfield).

middle of the 1st dorsal fin. Tail fin broad and notched. 1D 12; 2D 19–20; 3D 17–19; 1A 29; 2A 17–20.

Coloration variable, the back is brown or olive, the flanks are paler, there are often dark yellow or orange spots or stripes scattered about the upper part of the body. The lateral line is greenish-brown appearing dark against the flanks.

Up to 80 cm, may reach 120 cm at an age of 15 years or more.

A very popular sport fish, often seen by divers, but with very little commercial importance. The young fish are often seen close inshore amongst rocks and weed where they may be confused with young Saithe (*Pollachius virens*). As the fish become older they move into deeper water, but it is not uncommon to see large specimens in shallow water either individually or in small groups.

The Pollack breeds in shallower water than the Saithe (usually shallower than 100 m) and it prefers the slightly higher temperature of 8–10°C. The eggs and larvae are pelagic.

Saithe; Pollock; Coley
Pollachius virens (L.)

Both sides of North Atlantic, In west: Chesapeake Bay to Hudson Straight. In east: Greenland, Iceland, Barent Sea, British coast and south to Biscay.

Body streamlined and looks very symmetrical; *jaws* in young individuals the jaws are equal in length, in large fish the lower jaw may be longer; *barbel* so small as to be insignificant; *lateral line* almost straight; *fins* 3 dorsal fins, 2 anal with clear interspaces. Beginning of 1st anal fin directly below the beginning of the 2nd dorsal fin. Tail fin broad with notch in rear margin. 1D 13–14; 2D 20–22; 3D 20–24; 1A 25–28; 2A 19–25.

Dark greenish-brown or blackish – older fish are very dark. The belly is silvery white. The white lateral line shows up very clearly against the dusky flanks.

Up to 100 cm, sometimes up to 120 cm.

The Saithe is an important commercial species which is frequently seen by divers. It may be confused with the Pollack

(*Pollachius pollachius*), but its shape and colour together with the colour of the lateral line may allow a positive identification.

The young fish are found in shallow water close inshore amongst rocks and weeds. They are common in harbours. Older individuals move off into deep water where it is believed they undertake considerable migrations. Whilst in inshore waters the young feed on small fish and crustacea. The large, deep-water animals feed almost entirely on fish.

Tend to form schools of uniformly sized fish. The eggs and larvae are pelagic, the young fish being carried inshore by surface currents. The optimum temperature for spawning appears to be 8–9°C but may occur at 3–10°C.

Cod *large fish*
Codling *smaller fish*
Gadus morhua L.

Both sides of North Atlantic; from Cape Hatteras to Nova Scotia; from Biscay to northern Norway; North Sea. There is also a race in the Baltic.

Jaws upper longer than lower; *barbel* long, almost equalling the diameter of the eye; *lateral line* is strongly curved to beneath the leading edge of the 3rd dorsal fin, and from thence runs straight to the tail fin; *fins* 3 dorsal, 2 anal, all fins are rounded with only short interspaces. The 1st anal fin originates below or slightly behind the origin of the 2nd dorsal fin. The tail fin is square cut. 1D 14–15; 2D 18–22; 3D 17–20; 1A 19–23; 2A 17–19.

Colour is very varied. The body is usually greenish-grey spotted with brown or grey. Some specimens taken from weedy, rocky surroundings may be golden or even deep red in colour. The belly is usually a dirty greyish-white. Young fish (5–10 cm) have a distinct 'checkerboard' pattern on upper surfaces and flanks. The lateral line shows up conspicuously light against the ground colour of the flanks and in the sea the lateral line with its characteristic curve may be the only visible part of the fish.

Up to 100 cm; one Cod of 169 cm has been recorded.

Cod (Juvenile) (top), *Gadus morhua* (Jon D. Witman).

Cod (bottom), *Gadus morhua* (Jon D. Witman).

The Cod is the most important commercial fish in its family. The young are often seen by divers either resting or browsing on the bottom. These young fish may be found in very shallow water close inshore but adults are found at all depths down to 500–600 m. Prefer water of 4–7°C.

The Cod is basically a bottom-feeder taking worms, crustacea and molluscs. However, considerable quantities of fish are taken, especially by the larger fish.

Haddock

Melanogrammus aeglefinus (L.)

Both sides of North Atlantic from Cape Hatteras to Cape Cod in the west, Spitzbergen to southern Ireland in the east; also Faroe and Iceland; throughout the North Sea but infrequent in the southern part; sometimes in the eastern English Channel.

Jaws upper longer than lower; *mouth* small; *barbel* short; *lateral line* gently curved; *fins* 3 dorsal, 2 anal. The 1st dorsal is high and sharply pointed; all the fins are separated by distinct interspaces. The origin of the 1st anal fin is beneath the origin of the 1st dorsal fin. The tail fin is slightly notched. 1D 15–16; 2D 19–21; 3D 19–22; 1A 23–24; 2A 22–23.

Dark purple to charcoal on the upper surface, sometimes olive. The flanks are dark silver and the belly is pale. The Haddock is easily distinguished by the large black 'thumbprint' behind the pectoral fin and by the black lateral line. In young fish the thumbprint may be ringed with white and is very conspicuous underwater.

Up to 80 cm, occasionally 100 cm.

A very important commercial species which is eaten either fresh or smoked. Although abundant in many areas they are seldom seen by divers and it is possible that they are frightened by the sound of conventional diving equipment.

The Haddock usually lives very near the bottom but large shoals are occasionally found in mid-water. They chiefly feed on worms, molluscs, echinoderms, but at times many gorge themselves on sand eels, Capelin and Herring spawn. Feeds at depths of 20–300 m.

Haddock, *Melanogrammus aeglefinus* (Bill Hemmings).

Greenland Cod

Gadus ogac Richardson

Cold northern Atlantic, north from Gulf of St Lawrence, west Greenland, northern Norway; also Alaska.

Closely related to the Cod (*Gadus morhua*) but the eye is larger, and the inside of the body wall is dark. The most obvious difference is the lateral line, which is the same colour as the rest of the body rather than light as in the Cod.

This is the common codfish of northern inshore waters, harbours and fjords, tolerant of brackish water. Not considered as good to eat as the Cod.

Atlantic Tomcod

Microgadus tomcod (Walbaum)

North-west Atlantic, southern Labrador south to North Carolina.

Body resembles the Cod (*Gadus morhua*) in general shape, but has a shorter barbel which is only slightly longer than the diameter of the pupil, and the lateral line is less strongly curved. D1 11–15; D2 15–19; D3 16–21; A1 12–21; A2 16–20.

Brown or olive with yellow and greenish tints; back, sides and median fins mottled and marbled. The lateral line is white.

35 cm.

An inshore species that often enters brackish or even fresh water. Some populations are permanently land-locked. The Tomcod is a predator, feeding on almost any animal food that is available. It supports a small commercial fishery. Caught by anglers using hook and line through the ice in winter. The estuarine populations are considered particularly vulnerable to domestic and industrial pollution during the winter when their inshore waters are covered with ice.

Arctic Cod
Boreogadus saida (Lepechin)

Arctic Russia; Arctic North-west Atlantic south to Newfoundland.

Body rather slender, tapering from above gill cover to tail; *jaw* lower jaw projecting; *barbel* slender barbel on chin; *lateral line* inconspicuous and not complete; *fins* 3 dorsal fins with spaces between, tail fin forked. D 11–15; 12–17; 17–23; A 14–20; A2 18–24.

Dark brown to black above, paler below; small black spots on back, all fins dark with paler margins.

25 cm.

One of the most northern fishes, often found in shallow water around floating ice, but may penetrate down to at least 400 m. May occur in very large numbers at depths of 100–200 m. It is a plankton feeder and is itself a very important source of food for larger predators.

Five-bearded Rockling
Ciliata mustela (L.)

North-east Atlantic south to Portugal.

Body elongate; *mouth* rather small, reaches back to the hind edge of the orbit; *barbels* 5, 1 on each of the front nostrils, 2 on the upper lip and 1 on the lower lip. There are no small fringing barbels on the upper lip; *fins* 1D I + ∞; 2D 50–55; A 40–41.

Colour uniform, ranging from reddish-brown to dark brown.

May reach 45 cm in length but usually no longer than 25 cm.

Live mostly between the tides although the larger fish may live beneath the low-tide mark. They tend to prefer soft bottoms near rocks. They spawn offshore in the winter. The surface living 'Mackerel Midge' stages last until early summer, when the silvery colour is lost to the drab brown of the bottom-living form. They feed on small fish, crustacea and molluscs.

Five-bearded Rockling (above), *Ciliata mustela* (Andy Purcell).

Three-bearded Rockling (opposite top), *Gaidropsarus vulgaris* (Jim Greenfield).

Greater Forkbeard, *Phycis blennoides* (Christian Petron).

Northern Rockling
Ciliata septentrionalis (Collett)

North Sea; North-east Atlantic south to Cornwall.

Resembles the Five-bearded Rockling (*Ciliata mustela*) except that the mouth is large, extending well behind the orbit and the upper jaw is fringed with supplementary barbels.

Apparently rare, it is caught at 10–100 m. It is possible that this fish has been confused with the Five-bearded Rockling.

Shore Rockling
Gaidropsarus mediterraneus (L.)

Mediterranean; Atlantic coasts of Europe, west and south England, south Ireland.

Body elongate; *mouth* small, extending to rear edge of eye; *barbels* 3; *fins* 1st dorsal is composed of very fine rays set in a groove on the back; 2nd dorsal and anal fins are long and of uniform height extending back to the tail fin. Pectoral fin has 17 rays. 1D I + ∞; 2D 53–63; A 43–53.

Colour varies with habitat, usually brown or dark brown. Never mottled except occasionally on the lower part of head.

Young silver with bluish or greenish back.

Up to 50 cm.

Lives amongst rocks, often in tide pools and not deeper than 30 m. They feed upon small bottom-living fishes, molluscs, crabs, etc. Spawn close inshore in summer. The young (Mackerel Midges) live just beneath the surface and by the autumn reach about 4 cm in length. They then settle on the bottom in shallow water.

Three-bearded Rockling
Gaidropsarus vulgaris (Cloquet)

Mediterranean; north-east Atlantic north to Faroes.

Differs from the Shore Rockling (*Gaidropsarus mediterraneus*) in the following characteristics: *mouth* large, extends past the eye; *fins* 20–22 rays in pectoral fin. 1D I + ∞; 2D 56–64; A 46–52.

Very characteristic leopard-like pattern of bold dark spots upon a brick-red ground. Fish shorter than 10 cm are not spotted.

Up to 55 cm.

The smaller specimens always occur amongst rocks although larger specimens may be found on soft bottoms nearby. They live from just beneath low-tide level down to about 120 m.

They spawn in winter, the eggs float and the young swim in the surface water. During this so-called 'Mackerel Midge' stage they have a green back and silvery flanks and are often found swarming in great numbers. Once they reach about 3 cm the young take up a bottom-living existence. Here they feed upon bottom-living molluscs, crustacea and fishes.

Greater Forkbeard
Phycis blennoides (Brünnich)

Adriatic; western Mediterranean; eastern Atlantic north to Scotland; rare in North Sea and north to Norway and Greenland.

Barbel on chin; *fins* 2 dorsal and 1 anal. 3rd ray of 1st dorsal is elongate being at least the height of the 2nd dorsal. The 2nd dorsal is long and of uniform height. Pelvic fins consist of a simple branched ray reaching back well beyond the vent. 1D 8–9; 2D 60–64.

Brown or grey, the dorsal, anal and tail fins yellowish with black edges. The young have 2 or 3 black blotches on the back.

Up to 45 cm, occasionally 75 cm.

Usually found on sandy or muddy bottoms from 150–300 m, but may be caught as shallow as 10 m. They feed chiefly on crustacea but fish are sometimes eaten. They spawn from January to May in the Mediterranean and later in northern waters.

Ling
Molva molva (L.)

North-east Atlantic south to Biscay, rarely in cold western Atlantic.

Body elongate; *jaws* upper longer than lower; *barbel* conspicuous barbel on chin and one small barbel near each nostril; *fins* 2 dorsal fins and 1 anal. Tip of pelvic fin does not extend beyond tip of pectoral. 1D 14–15; 2D 62–65; A 58–61.

Ling, *Molva molva* (Jim Greenfield).

Mottled brown or green with a bronze sheen. There is a dark blotch at the rear end of both the 1st and 2nd dorsal fins. Both 2nd dorsal and anal fins are outlined in white. The young have a yellowish-olive ground colour broken up with lilac iridescent lines.

Up to 100 cm, rarely up to 220 cm.

A fairly important commercial fish, it chiefly lives in water of 300–400 m, but immature fish are quite often seen as shallow as 10 m where they tend to favour crevices between rocks. These immature fish live in shallow water for some 2 years (when they may reach 50 cm) and then move into deeper water.

The Ling's diet is not known in detail but it feeds extensively on small fishes, together with some crustaceans and echinoderns.

Mediterranean Ling
Molva elongata (Otto)

Mediterranean and Atlantic north to South-west Ireland.

Differs from Molva molva in the following: *body* longer and more eel-like; *jaws* lower longer than upper; *barbel* forked; *fins* tip of pelvic reaches past tip of pectoral; *eye* small; vertebrae 82–84.

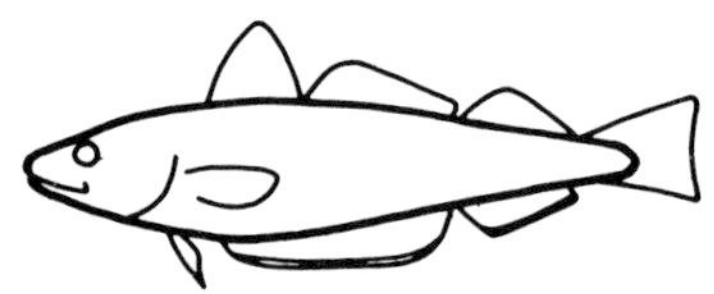

Whiting *Merlangius merlangus*

Long upper jaw; barbel minute or absent.
Eastern Atlantic; Mediterranean.
60 cm.

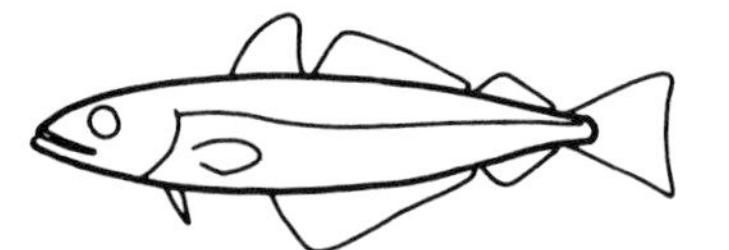

Pollack *Pollachius pollachius*

Lower jaw longer than upper;
spaces between dorsal fins.
Lateral line dark.
Eastern Atlantic.
80 cm.

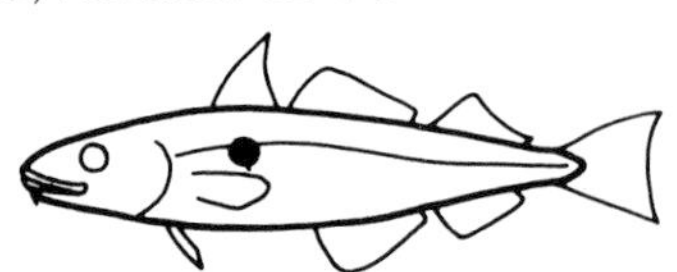

Haddock *Melanogrammus aeglefinus*

Black 'thumbprint'; black lateral line.
Eastern and western Atlantic.
80 cm.

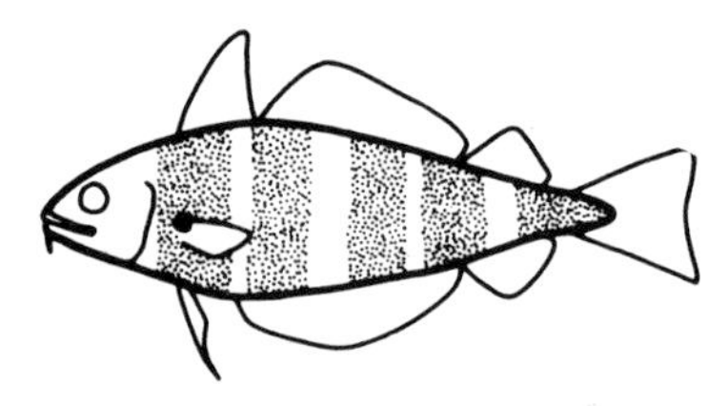

Bib; *Trisopterus luscus*

Usually vertical bands on body;
long barbel.
Eastern Atlantic.
30 cm.

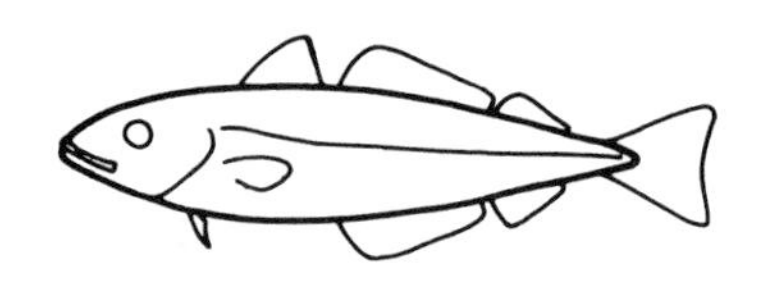

Saithe; Coalfish; Coley *Pollachius virens*

Jaws almost equal in length;
space between dorsal fins.
Lateral line pale.
Eastern and western Atlantic.
100 cm.

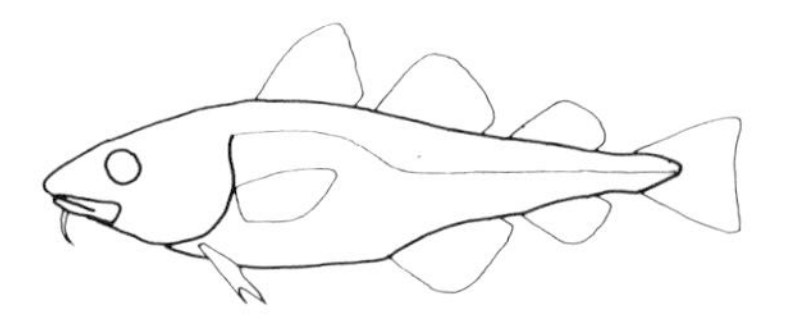

Greenland Cod *Gadus ogac*

Mottled body; barbel long;
arched lateral line, not pale.
Cold north Atlantic.
70 cm.

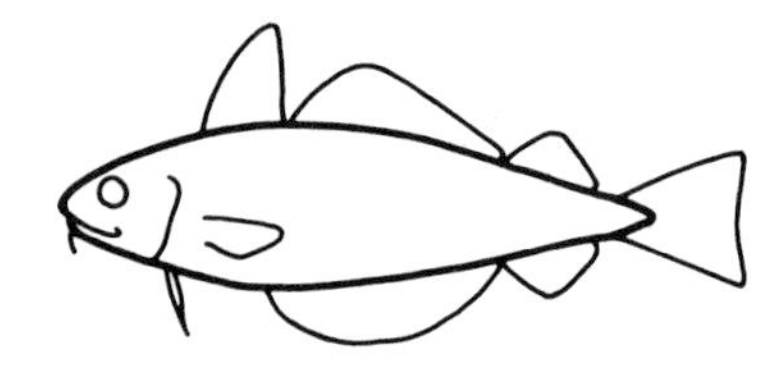

Poor Cod *Trisopterus minutus*

Uniform bronze colour; barbel present.
Eastern Atlantic.
20 cm.

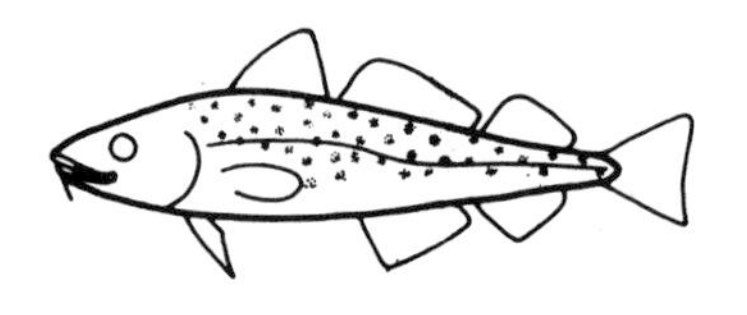

Cod; Codling *Gadus morhua*

Light-coloured arched lateral line;
mottled body.
Eastern and western Atlantic.
100 cm.

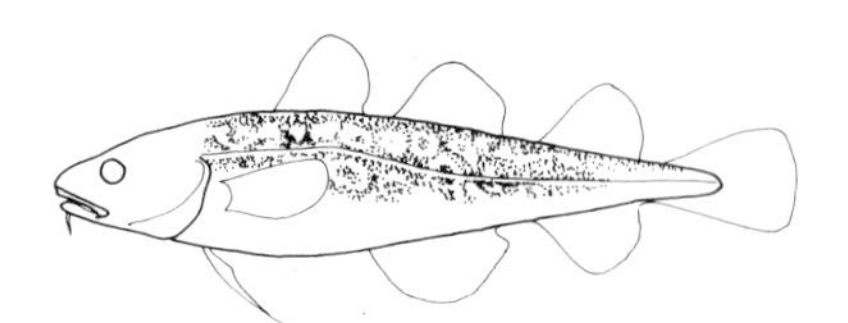

Atlantic Tomcod *Microgadus tomcod*

Barbel short;
slightly arched white lateral line.
Cold north Atlantic.
35 cm.

Grey-brown above, silvery below. Dorsal, anal and tail fins have violet edges. The pelvic fins are blue.

Up to 90 cm.

Found on sandy mud between 200 and 1000 m.

Tadpole Fish

Raniceps raninus (L.)

North Sea and all coasts of England and Ireland.

Body tadpole-shaped. The breadth of the head nearly equals its length and occupies ⅓ of the total length of the body; *barbel* there is a single short barbel on the lower jaw; *fins* 1st dorsal has only 3 rays, 2nd dorsal and anal are long, and of uniform height. 2nd ray of the pelvic fin is elongate, extending back at least as far as the vent. 1D 3; 2D 63–64; A 57–60.

Dark brown with lilac highlights and whitish beneath. The dorsal, tail and anal fins are dark with a light fringe. The lips are light in colour and are probably very conspicuous underwater.

Up to 30 cm.

Solitary fish which live from immediately beneath the low-tide mark to about 100 m. They feed on bottom-living animals such as molluscs, crustacea, annelids and echinoderms. The smell of the fresh fish is unpleasant. They spawn in summer in 50–75 m at temperatures of 10–12°C.

Tadpole Fish, *Raniceps raninus* (John Neuschwander).

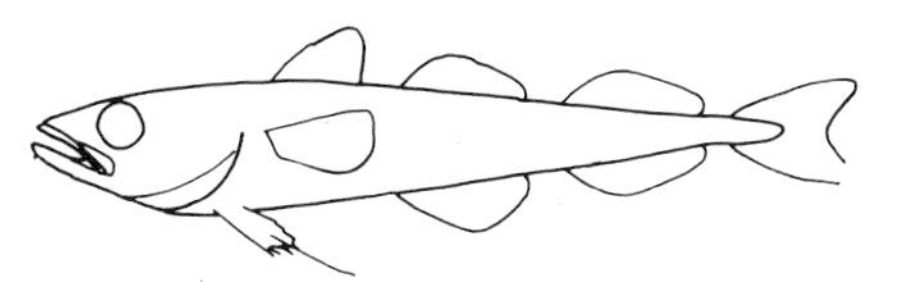

Arctic Cod *Boreogadus saida*

Long body; forked tail;
spaces between dorsal fins.
Arctic.
25 cm.

Shore rockling
Gaidropsarus mediterraneus

Three barbels; not patterned.
Eastern Atlantic; Mediterranean.
50 cm.

Ling *Molva molva*

Elongate body;
unforked barbel on chin;
upper jaw longer than lower.
Eastern Atlantic.
100 cm.

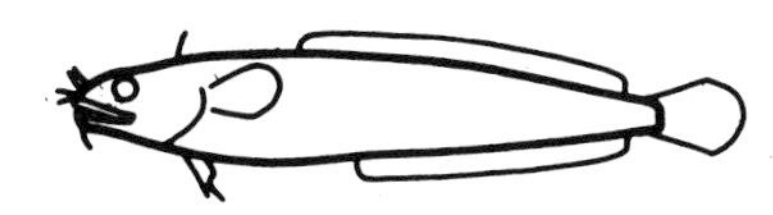

Five-bearded rockling *Ciliata mustela*

5 barbels; upper jaw not fringed;
shallow water.
Eastern Atlantic.
25 cm.

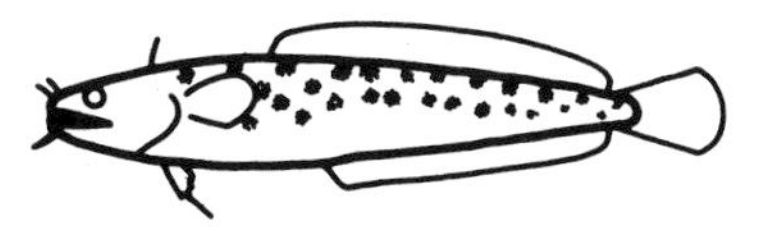

Three-bearded Rockling
Gaidropsarus vulgaris

3 barbels; brick red with leopard spots.
Eastern Atlantic; Mediterranean.
55 cm.

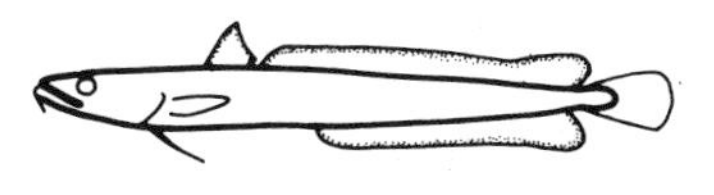

Mediterranean Ling *Molva elongata*

Elongate body;
forked barbel on chin;
lower jaw longer than upper;
dorsal, anal and tail fins with violet edges.
Mediterranean.
90 cm.

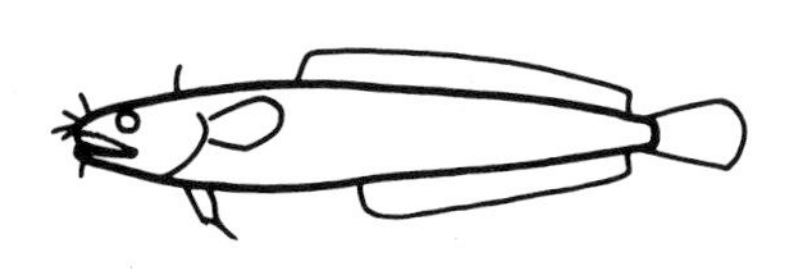

Northern Rockling *Ciliata septentrionalis*

5 barbels; upper jaw fringed.
Eastern Atlantic.
60 cm.

Greater Fork Beard *Phycis blennoides*

Pelvic fin a single branched elongate ray;
elongate 3rd ray of 1st dorsal fin.
Eastern Atlantic; Mediterranean.
45 cm.

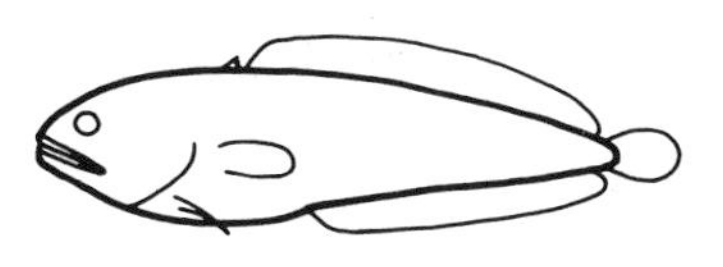

Tadpole Fish *Raniceps raninus*

Tadpole-shaped;
1st dorsal fin has only 3 rays.
Eastern Atlantic.
25 cm.

Order:
ZEIFORMES

Deep, laterally flattened bodies with tall dorsal fins, large eyes and protrusible mouths.

Family:
ZEIDAE

John Dory
Zeus faber (L.)

John Dory *Zeus faber*
Unmistakeable shape; round spot on side of body.
40 cm.

Mediterranean and eastern Atlantic from Scotland to West Africa; rare in North Sea.

Body round and very flattened laterally; *mouth* protractile; *scales* on each side of the base of the dorsal and anal fins there is a row of armoured scales with sharp spines; *fins* there is a single dorsal fin. The spiny rays are long and the tissue of the fin membrane is prolonged into a long filament. The anal fin has 4 strong spiny rays. D IX-X, 21–25; A III-IV, 20–23.

Grey or yellowish with indistinct mottling. There is a dark central spot with a lighter margin.

Up to 40 cm, rarely up to 50 cm.

A solitary fish which ranges in depth from a few metres to 200 m. It is often trawled up from sandy ground but in the summer the young fish are sometimes seen by divers amongst the weed growing on rocky bottoms. The John Dory feeds almost entirely on fishes which it approaches slowly (it is not a strong swimmer), and it then accelerates with a jerky motion, swinging its jaws forward to catch its prey.

They spawn in summer in water shallower than 100 m, but not further north than the Irish sea or further east than the western English Channel. In the warmer waters of its range they spawn in spring. The older fishes longer than 20 cm remain near their spawning grounds, but the younger fishes travel further afield.

Very good to eat although its grotesque appearance may be disconcerting. This is one of the fish that is supposed to taste better after it has been dead for two or three days.

John Dory, *Zeus faber* (C. McTernan).

Order:

PERCIFORMES

A very large order of perch-like fishes. 1 or 2 dorsal fins. The first dorsal fin is usually spined of if there is only one dorsal fin the front part is spined

Family:

SERRANIDAE

This family includes the well known **Groupers** and **Combers**. They are chiefly found close to the coasts and are carnivorous. Many species may change from one sex to the other as they get older and at least one species, the Comber (*Serranus cabrilla*), is able to fertilize its own eggs and is thus a true hermaphrodite.

The gill cover has three spines. There is generally a single dorsal fin but in a few species it is divided. The anal fin has 3 spines and 7 soft rays. 1 spine and 5 soft rays in the pelvic fin.

Marine Goldfish
Anthias anthias (L.)

Mediterranean; eastern Atlantic north to Biscay.

Body oval but very flattened laterally; *profile* steep; *snout* short and rounded; *front gill cover* with a toothed edge; *gill cover* with 2 or 3 spines; *scales* large and continue onto head; *fins* 1 dorsal fin with the 3rd spine greatly elongated. Pelvic fins long; tail fin deeply forked with the lower lobe longer than the upper. D XI, 15; A III, 7.

Back and sides red, shading to pink underneath. There are three yellow stripes along the sides of the head which extend onto the front part of the body. Underwater these fish appear grey with long flowing fins.

Usually 15–24 cm, rarely longer.

Commonly found below 30 m amongst rocks where groups, sometimes large, hover just inside the entrances of caves and crannies. They feed on small fish and crustacea and spawn in the spring.

Marine Goldfish, *Anthias anthias* (Andy Purcell).

Parrot Seaperch
Callanthias ruber (Rafinesque)

English Channel south to Mauretania and Azores; Mediterranean.

Body slender; *lateral line* incomplete, ending a little behind dorsal fin; *fins* soft rays of dorsal fin longer than spiny rays, there are no elongate spiny rays on the dorsal fin, tail fin deeply forked, the two tips extended into long filaments.

Red or pale red, fins yellow.

20 cm.

Judging by the lack of photographs of this species, it is rarely seen. Lives near rocks on muddy bottoms and in submarine caves 50–500 m. It would not be surprising if individuals frequent caves at somewhat lesser depths.

Golden Grouper

Epinephelus alexandrinus Valenciennes

Mediterranean, and neighbouring Atlantic coast as far south as the Cape Verde islands.

Body oval, but more slender than *Epinephelus guaza*; *jaws* the lower is longer than the upper and very prominent; *front gill cover* has a toothed rear edge; *gill cover* has 3 spines; *fins* 1 dorsal fin. Tail fin either square cut or slightly concave. D XI, 16; A III, 8.

Uniform brown in colour with sides lighter and yellowish underneath. There may be lighter blotches along the sides and back but they are much less distinct than the markings of *Epinephelus guaza*. Young fish have 4–5 distinct dark longitudinal stripes along the body and across the gill covers. These stripes are present in the adult but are much less conspicuous.

Up to 80 cm.

Like *Epinephelus guaza* these fish are common amongst rocks but have a larger and less distinct home territory. They frequently swim in pairs or small groups. Another point of difference with *E. guaza* is that they are stronger swimmers and move greater distances from the rocks into midwater, where they hunt. They are rarely seen hovering around holes. They chiefly feed on fishes. They spawn in summer, possibly at the full moon.

Dogtooth Grouper

Epinephelus caninus (Valenciennes)

Eastern Atlantic, from Gibralter south to Senegal; southern Mediterranean.

Resembles the Dusky Grouper, *Epinephelus guaza*, in general form. Differs in the teeth, which have more developed canine-like teeth; *fins* tail fin square-cut or slightly concave. D IX, 13–14; AIII, 8.

Purplish-grey. There are 3 dark lines radiating from the eye to the edge of the gill cover.

150 cm.

Has often been confused with the Dusky Grouper, and its exact range is not known. Found on muddy and sandy bottoms.

White Grouper

Epinephelus aeneus (Geoffroy Saint-Hilaire)

Eastern Atlantic from British Islands south to Angola; southern Mediterranean.

Body rather elongate; *scales* small, embedded in the skin; *fins* D X-XI, 14–16; A III 8–9, tail fin rounded.

Golden Grouper, *Epinephelus alexandrinus* (Andy Purcell).

Greenish-grey often with indistinct darker oblique bands on the body. There are 2–3 lighter oblique bands on the head.
110 cm.
Found on sandy and muddy bottoms at 2–300 m.

Grouper; Dusky Grouper
Epinephelus guaza (L.)

Mediterranean; occasionally eastern Atlantic, rare north of Biscay.
Body oval; *jaws* the lower slightly in front of the upper; *teeth* both jaws contain long, strong sharp and moveable teeth; *front gill cover* has a toothed rear edge; *gill cover* has three short spines; *scales* small and continue onto head, lower jaw scaled; *fins* tail fin rounded. D XI, 13–16; A III, 8–9.
Back green-brown with lighter sides and yellowish underneath. There is a lighter green mottling over the head, back and sides. On the head the spots tend to radiate from the eye and on the body they tend to form vertical bands. The dark fins usually have a lighter edge except the dorsal fin which has an orange edge.
Up to 1.4 m.
These fish are common from 10–400 m amongst rocks where there are plenty of holes and caves. They are usually solitary and have well defined home territories which contain a number of refuge holes, often with one or two primary holes that may be occupied by more than one fish. They are not very strong swimmers and spend a considerable amount of time hovering just outside their holes. They are frequently encountered by divers, both in holes where they sit slowly moving their fins, and in open water where, if not frightened, they frequently turn to face the diver. They spawn during summer possibly at the full moon. They are carnivorous feeding on molluscs, crustacea and, to a lesser extent, fish.

Comb Grouper
Mycteroperca rubra (Bloch)

Eastern Atlantic, Bay of Biscay south to Angola; southern Mediterranean; also western Atlantic off Bermuda, Caribbean and Brazil.
Body rather robust; *front gill cover* with marked notch and lobe; *scales* rough to the touch; *fins* tail fin square-cut. D XI, 15–17; A III, 11–12.
Reddish-brown with irregular dark lines and blotches. In the young there is a black saddle on the tail stalk.
80 cm.

Dusky Grouper, *Epinephelus guaza* (Leo Collier).

Wreck Fish
Polyprion americanum Bloch & Schneider

Mediterranean; throughout Atlantic but only occasionally north of Ireland; English Channel.
Body deep and heavily built; *head* large; *profile* fairly steep with a slight concavity above the eyes with one group of lumps between the eyes and another on the forehead; *jaws* the lower jaw is clearly longer than the upper; *teeth* are small, strong, sharp and conical; *front gill cover*

Snowy Grouper, *Epinephelus niveatus* (Guido Picchetti).

Wreck Fish, *Polyprion americanum* (Andy Purcell).

with a toothed rear edge; *gill cover* has a distinct bone running lengthways across it; *scales* are small and cover the head and body and extend onto the base of all the fins but to a greater extent onto the dorsal, anal and tail fins; *fins* 1 dorsal fin with very strong spines; tail fin rounded or straight. D XI, 11–12; A III, 8–10.

The adult fish has a dark brown back, lighter sides and yellowish belly. Sometimes with faint lighter blotches on the back and sides. Young fish may have scattered irregular large white and smaller black blotches. The edge of the tail fin is white.

Up to 2 m.

These fish live in deep open water but are often encountered around floating wreckage or seaweed. They usually live between 100–200 m, but occasionally as deep as 1000 m. They may either be solitary or in large numbers particular when encountered in association with floating debris. Usually these schools of fish are young specimens; the adults lead a more nomadic, bottom-living existence. Their habit of following floating material does sometimes bring the fish close inshore. However, in the Mediterranean there is a general movement inshore during the summer and autumn. They spawn in the Mediterranean between January and April and feed on fish, crustacea and molluscs.

Snowy Grouper
Epinephelus niveatus (Valenciennes)

Western Atlantic from Massachusetts south to Brazil; eastern Pacific.

Teeth a few canines present; *nostril* rear nostril about 4 times larger than front nostril; *fins* D XI, 13–14; A III, 9.

The young and adults differ in coloration. The young have about 8 irregular vertical rows of white spots on the body, the tail stalk has a black saddle, the tail fin is yellow. Adults are grey with irregular white blotches and have no saddle on the tail stalk.

90 cm.

The young frequent shallow water near reefs, adults live deeper in 240–485 m.

Rock Hind
Epinephelus adscensionis (Osbeck)

Western Atlantic, Massachusetts south to Brazil.

Nostril rear equal or slightly larger than front; *teeth* canine-like teeth present; *fins* tail fin has rounded margin. D XI, 16–18; A III, 9–10.

Ground colour yellowish-brown with head, body and fins spotted with dark orange or orange-brown; 3 dark blotches along the base of the dorsal fin; dark saddle mark on tail stalk.

60 cm.

Found around jetties, oil rigs, and reefs.

Red Grouper
Epinephelus morio (Valenciennes)

Western Atlantic, Massachusetts south to Brazil; west Africa.

Teeth two moderate canines in the front of each jaw; *fins* tail fin square-cut or the outer margins extended, 2nd dorsal spine the longest. D XI, 16–17; A III, 9.

Reddish with paler blotches, becoming less distinct with age; there is no dark saddle on the tail stalk. Dorsal, anal and tail fins dark with pale margins.

125 cm.

Usually found on offshore banks.

Warsaw Grouper; Black Jewfish
Epinephelus nigritus (Holbrook)

Western Atlantic, Massachusetts south to Brazil.

Bass, *Dicentrarchus labrax* (David George).

Body very robust and sometimes very large; *teeth* strong canines present; *fins* tail fin rounded, 2nd dorsal spine longest. D X, 13–15; A III, 8.

A uniform dark blackish-brown or chocolate-brown.

150 cm.

Scamp
Mycteroperca phenax Jordan & Swain

Western Atlantic, Massachusetts south to Venezuela.

Head front gill cover angular, rear nostril larger than front; *teeth* in inner jaw hinged; *fins* in adults the spines of the tail fin extend beyond the edges of the membrane giving a brush-like appearance, tail fin has 12 soft rays and is shallowly forked. D XI, 15–18; A III, 10–12.

Front part of the dorsal fin and the pectoral fins have broad, dull yellow edges, body covered with dark spots that tend to group into irregular horizontal lines.

60 cm.

Found on offshore banks.

Gag
Mycteroperca microlepis (Goode & Bean)

Western Atlantic, Massachusetts south to Brazil.

Head front and rear nostrils about equal; *front gill cover* angular; *fins* no extended rays on dorsal, anal or tail fins. D XI, 16–19; A III, 11.

Rather variable, usually grey or brownish-grey with darker, double 'kiss' marks on the flanks, usually a dark 'moustache' mark above the upper jaw, fins generally darker with paler edges. Younger specimens from sea-grass beds may be coloured green.

90 cm.

Young found on sea-grass beds, adults on offshore banks and reefs.

Bass
Dicentrarchus labrax (L.)
= *Morone labrax* (L.)

Mediterranean; eastern Atlantic north to Norway south to Senegal; Baltic; southern North Sea; English Channel.

Body long and oval; *head* fairly long, 5–6

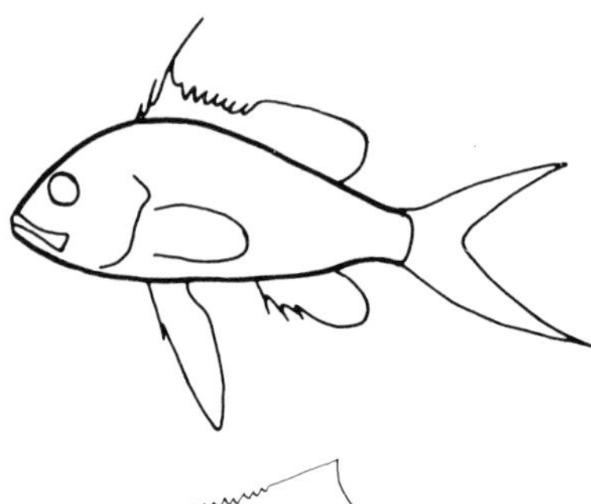

Marine Goldfish *Anthias anthias*

Deeply forked tail fin;
elongate 3rd dorsal spine.
24 cm.

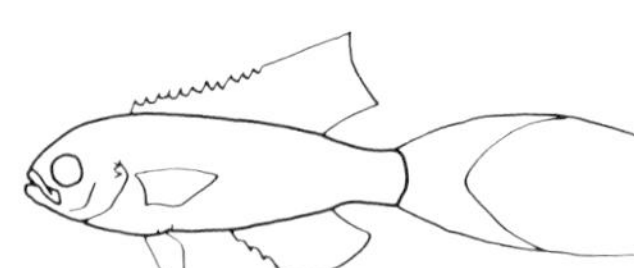

Parrot Seaperch
Callanthias ruber

Deeply forked tail ending in filaments;
no dorsal spine, soft rays tall,
Eastern Atlantic;
Mediterranean.
20 cm.

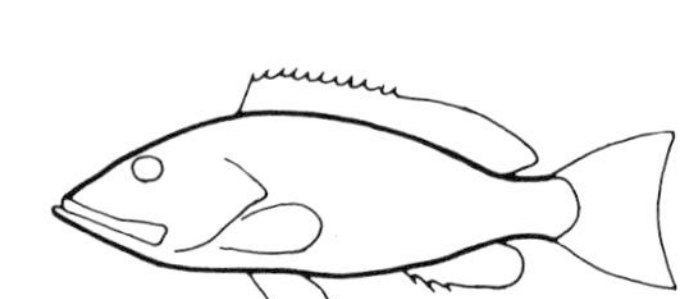

Golden Grouper
Epinephelus alexandrinus

Tail fin square or concave;
16–18 spiny dorsal rays;
longitudinal lines.
Eastern Atlantic;
Mediterranean.
1.4 m.

Dogtooth Grouper
Epinephelus caninus

Tail fin square or concave;
13–14 spiny dorsal rays;
oblique lines on head.
Eastern Atlantic;
Mediterranean.
1.5 m.

White Grouper
Epinephelus aeneus

Tail fin rounded;
head with light oblique lines, greenish.
Eastern Atlantic;
Mediterranean.
1.1 m.

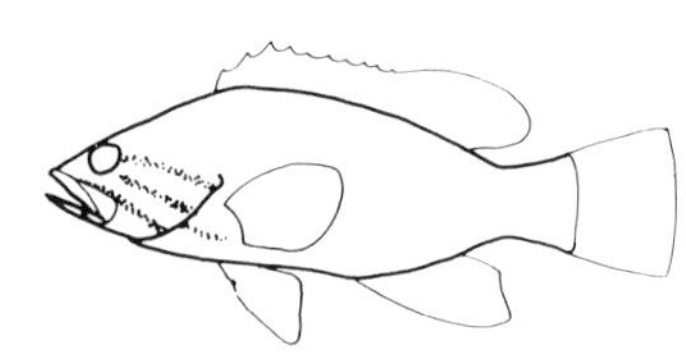

Grouper; Dusky Grouper *Epinephelus guaza*

Tail fin rounded;
no lines on head;
brownish, mottled.
Eastern Atlantic;
Mediterranean.
1.5 m.

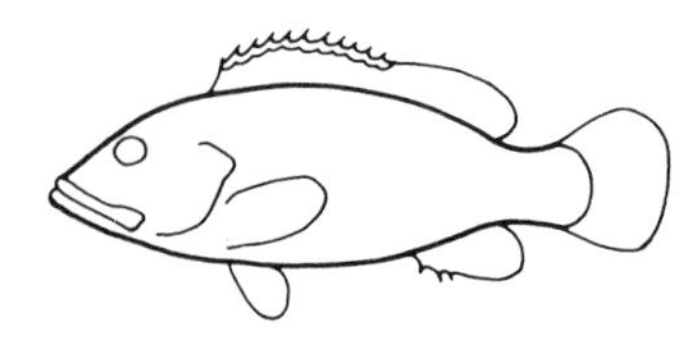

Comb Grouper
Mycteroperca rubra

Tail fin square-cut or concave;
11–12 soft anal fin rays;
wavy dark lines and blotches.
Eastern Atlantic;
Mediterranean.
80 cm.

Wreckfish *Polyprion americanum*

Tail fin rounded;
body deep;
bony lumps on head.
Western and eastern Atlantic; Mediterranean.
2 m.

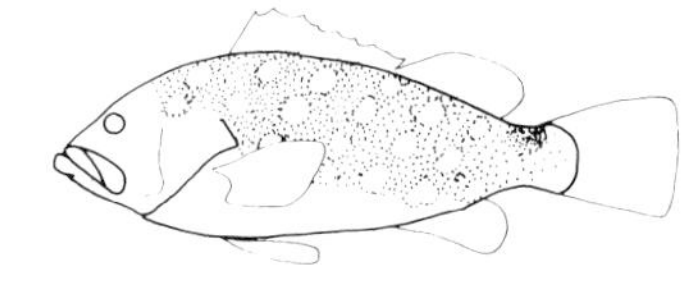

Snowy Grouper
Epinephelus niveatus

Young: irregular rows of white spots;
adults: irregular white blotches.
Western Atlantic.
90 cm.

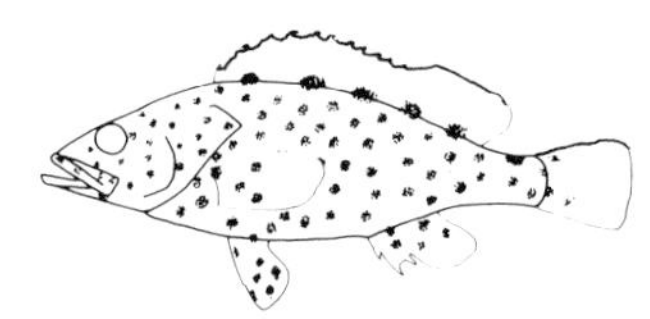

Rock Hind *Epinephelus adscensionis*

Spotted dark orange;
dorsal fin with 5 dark blotches.
Western Atlantic.
60 cm.

Red Grouper
Epinephelus morio

Tail fin square-cut or with elongate margins;
red-brown blotches paler;
dorsal, anal and tail fins dark with pale margins.
Western Atlantic.
125 cm.

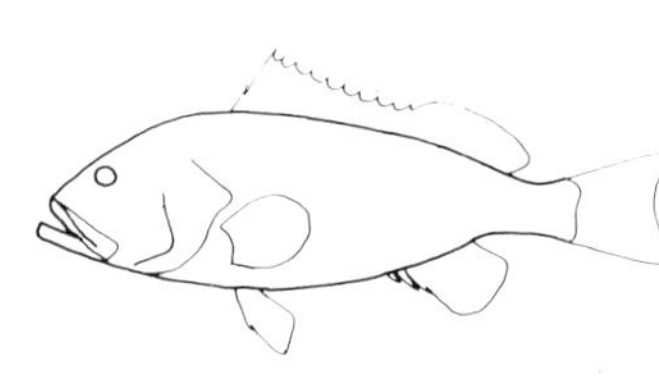

Warsaw Grouper; Black Jewfish *Epinephelus nigritus*

10 (not 11) dorsal spines;
uniform dark brown.
Western Atlantic.
150 cm.

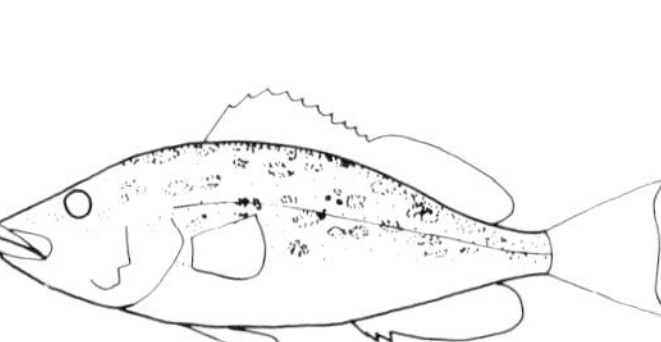

Scamp *Mycteroperca phenax*

Tail fin square-cut or concave;
adults: tail fin rays extended;
fins edged with yellow.
Western Atlantic.
100 cm.

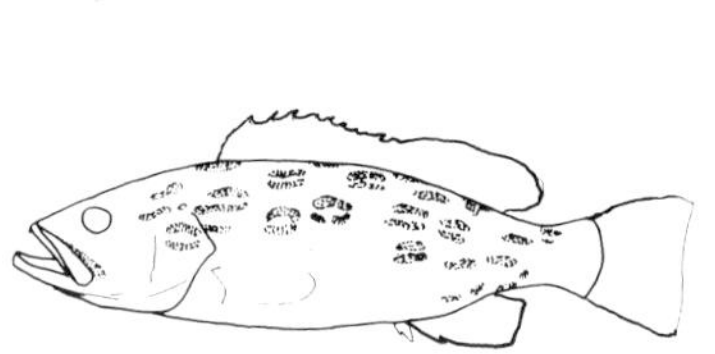

Gag *Mycteroperca microlepis*

No extended rays;
usually brown-grey with 'kiss' marks on sides;
usually dark moustache.
Western Atlantic.
90 cm.

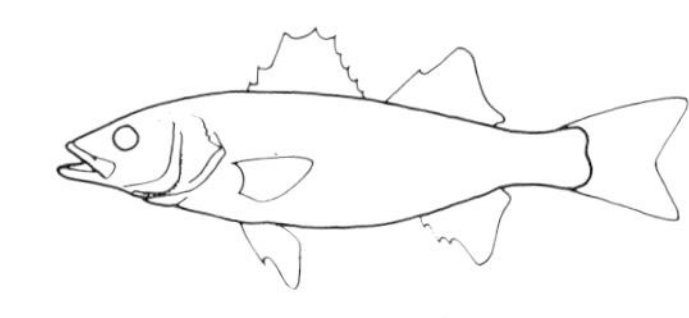

Bass *Dicentrarchus labrax*

Dorsal fins separate;
no spots.
Eastern Atlantic;
Mediterranean.
100 cm.

times the diameter of the eye; *front gill cover* has a toothed rear edge; *gill cover* with 2 spines; *scales* small, 65–80 along the lateral line; *fins* there are 2 dorsal fins, the 1st dorsal is entirely separate from the 2nd. Tail fin slightly forked. 1D VIII-IX; 2D I, 12–13; A III, 10–11.

A silvery fish, grey on the back, lighter on the flanks and white on the belly. Fish up to 10 cm, may have black spots on the sides and back.

Up to 100 cm.

Found at all depths from near the surface to below 100 m. The adults are usually solitary but the young fish do form schools. During the summer they migrate to inshore water where they are frequently found in estuaries and may penetrate far up rivers. They are found over bottoms ranging from rocks and sand to shingle and mud. They appear to occupy the same localities for many months and observations by divers indicate that they may occupy well-defined feeding territories even between the tides. They are carnivorous, the young fish feeding upon small fish and crustacea and the adults to a greater extent upon fish although they may also eat worms, cephalopods and crustacea. In the Mediterranean they spawn in January to March and in the English Channel from March to June. They are edible and a favourite angler's fish.

Spotted Bass

Dicentrarchus punctatus (Bloch)

Mediterraenan; Eastern Atlantic between Biscay and Senegal, common around Gibraltar.

Body long and oval similar to the Bass (*Dicentrarchus labrax*) but with a slightly shorter head and snout; *snout* about 4½ times the diameter of the eye; *teeth* the vomer teeth are T-shaped; *front gill cover* has a toothed rear edge; *scales* 57–68 along the lateral line; *fins* 2 dorsal fins, the 1st fin entirely separate from the second. Tail fin

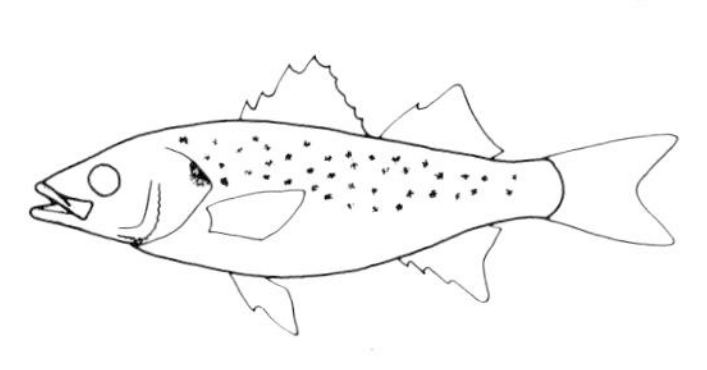

Spotted Bass
Dicentrarchus punctatus

Dorsal fins separate;
dark spots on body;
black spot on gill cover.
Eastern Atlantic;
Mediterranean.
60 cm.

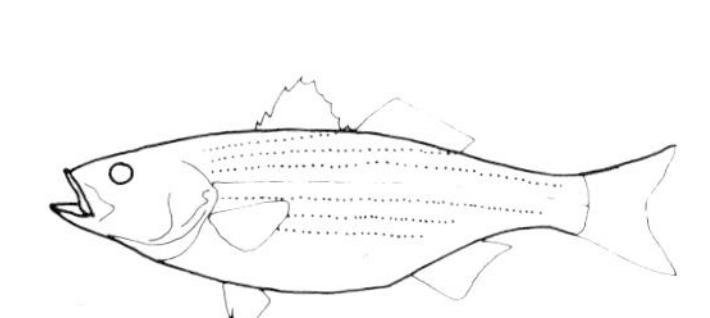

Striped Bass; Rockfish
Morone saxatilis

Dorsal fins separate;
long body;
stripes on flanks.
Western Atlantic.
80 cm.

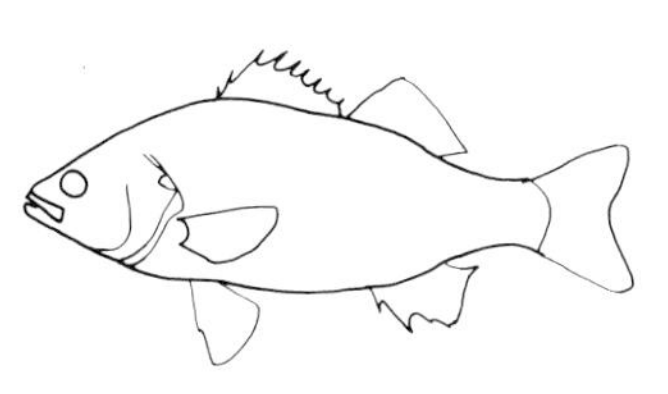

White Perch *Morone americana*

Body short and deep;
no space between dorsal fins;
no stripes on flanks.
Western Atlantic.
35 cm.

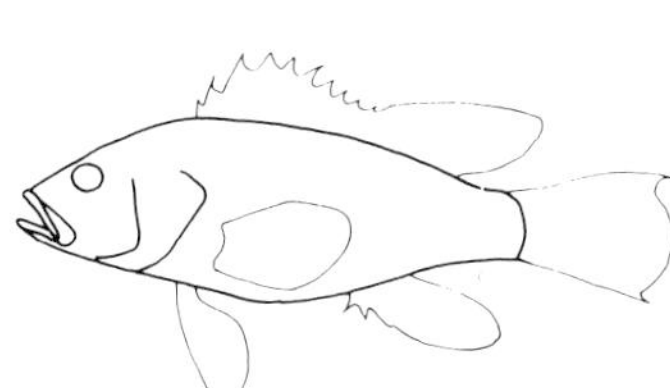

Black Seabass
Centropristis striata

Tail rounded, sometimes with margins prolonged;
dark colour, light spots on scales.
Western Atlantic.
60 cm.

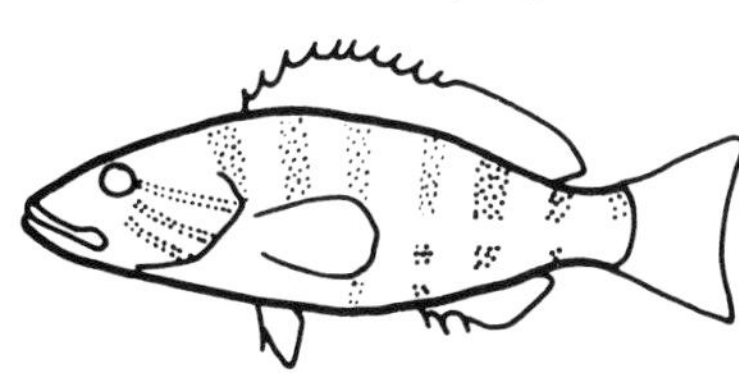

Comber *Serranus cabrilla*

7–9 dark vertical bands crossing;
2–3 longitudinal pale blue stripes.
Eastern Atlantic;
Mediterranean.
20 cm.

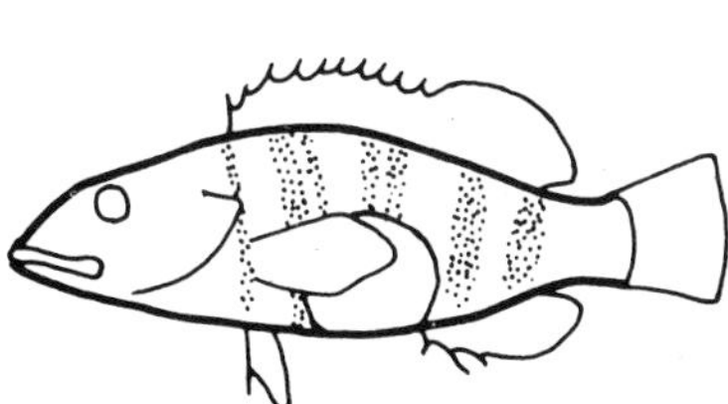

Painted Comber
Serranus scriba

Large blue spot on flank.
Mediterranean.
25 cm.

Brown Comber *Serranus hepatus*

Black spot on dorsal fin.
Mediterranean.
13 cm.

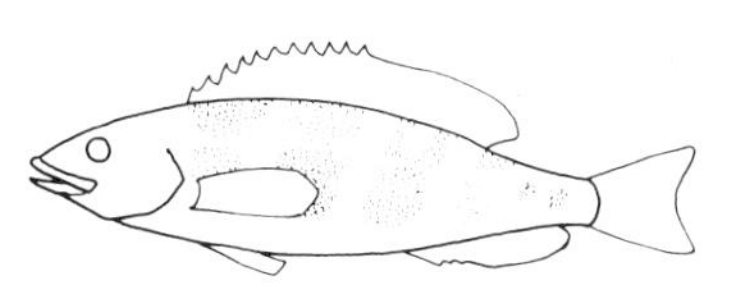

Black-tailed Comber
Serranus atricauda

Brown vertical blotches on sides;
Eastern Atlantic.
35 cm.

slightly forked. 1D VIII-IX; 2D I, 11–12; A III, 10–12.

Silvery but with a dark blue back. Adults with black spots scattered irregularly over the back and sides. There is a conspicuous black spot on the gill cover.

Up to 60 cm.

The habits of this fish are very similar to those of *Dicentrarchus labrax*. They may be found over sand, or mud with sand and rocks, and seem to prefer brackish water. They are carnivorous, feeding upon fish, molluscs and crustacea.

Striped Bass; Rockfish
Morone saxatilis (Walbaum)

Western Atlantic, St Lawrence River south to Florida.

Body elongate; *head* snout pointed, lower jaws projecting; *teeth* sharp, numerous and small; *eyes* large; *scales* 58–64 in lateral line; *fins* 2 well-separated dorsal fins, ID, VII-XII; 2DI, 12. 1 spine precedes pelvic fin. Tail fin distinctly forked.

Dark olive or bluish-black above, white below, often with brassy reflexions; sides with 7–8 horizontal dark stripes. 3 stripes above lateral line extending back to the tail, 1 stripe on the lateral line and 3 below. Stripes are interrupted or absent in young fish.

80 cm.

An important game and food fish. Its numbers appear to have declined recently, perhaps because it enters estuaries and completely fresh water to spawn in spring leaving it exposed to pollution. Some populations are land-locked.

White Perch
Morone americana (Gmelin)
= *Roccus americanus*

Western Atlantic, St Lawrence south to northern Florida.

Body rather deeper than in other members of the genus in our area, tail stalk robust; *head* rather pointed, lower jaw projects very slightly, edges of gill covers rough; *teeth* unequal and small; *scales* 46–51 in lateral line; *fins* 1st dorsal and anal have very robust spines, 1 spine precedes pelvic fin. ID VII-XII; 2DI, 11–12, tail fin rather stout with a concave margin.

Greenish-brown, greenish-grey shading to paler below; land-locked freshwater forms are darker; fins generally dark. Excellent sport and game fish, less important commercially.

35 cm.

Generally restricted to freshwater lakes and rivers, but does occur in the sea especially in brackish water.

Black Seabass
Centropristis striata (L.)

Western Atlantic, southern Massachusetts south to north-east Florida.

Body a high back and steep profile with flat top to head, 1 sharp spine near apex of gill cover; *fins* the tail fin is rounded, but in larger individuals the margins may be prolonged.

Bluish-black to dark brown, each scale has a lighter patch at the base giving the appearance of many narrow longitudinal stripes. The males are all black except for white areas on the head and edges of fins.

60 cm.

Prefer hard bottoms at 5–30 m depth. Come inshore in summer and move offshore again in late October or November.

Comber
Serranus cabrilla (L.)

Mediterranean; eastern Atlantic north to south-west England; western English

Black Seabass, *Centropristis striata* (Ed Brothers).

Comber, *Serranus cabrilla* (B. Picton).

Painted Comber, *Serranus scriba* (Andy Purcell).

Channel; Red Sea.

Body rather long; *teeth* small and pointed and not moveable; *front gill cover* has toothed rear edge; *gill cover* has two spines; *scales* small and continue onto head and lower jaw. Up to 90 scales along the lateral line; *fins* 1 dorsal fin. Tail fin slightly forked. D IX-X, 13–15; A III, 7–8.

Back and sides are reddish-brown with 7–9 vertical darker bands. There are 2–3 longitudinal bluish or bluish-green stripes along the head and body. There are also longitudinal yellow stripes. The colour, however, can vary with age, season and depth. Fish that live in the deeper ranges tend to be more red.

Up to 15–20 cm, rarely 40 cm.

Usually found between 20–55 m but also down to 500 m over sandy rocky areas and also over sea-grass (*Posidonia*) beds. They feed on fish, crustacea, worms and cephalopods. Spawn from April to July in the Mediterranean and Red Sea and are one of the fishes which are true hermaphrodites, both ovaries and testes mature simultaneously in the same individual.

Painted Comber
Serranus scriba (L.)

Mediterranean Black Sea; eastern Atlantic north to Biscay.

Body oval but less elongate than the Comber (*Serranus cabrilla*); *front gill cover* toothed on rear edge; *gill cover* with 2 spines; *scales* rather larger than those of *Serranus cabrilla; fins* 1 dorsal fin. D X, 14–16; A III, 7–8.

The back and sides are reddish- or yellowish-brown with 4–7 dark vertical bands. There is a large pale blue spot on

Brown Comber, *Serranus hepatus* (Andrea Ghisotti).

Black-tailed Comber, *Serranus atricauda* (Werner Frei).

the flanks which prevents confusion with any other species; and a characteristic intricate blue and reddish patterning on the snout and head which resembles arabic characters and gives the fish the name *scriba*.

Up to 25 cm.

Found in shallow water usually above 30 m amongst rocks and areas of sand and sea-grass. They are solitary fish and like *Serranus cabrilla* are true hermaphrodites spawning in early summer in the Mediterranean. They are carnivorous and feed upon fish, crustacea and molluscs.

Brown Comber
Serranus hepatus (L.)

Mediterranean; eastern Atlantic north to Portugal, south to Senegal.

Body oval but rather deeper than either the Comber (*Serranus cabrilla*) or the Painted Comber (*Serranus scriba*); *eye* large, nearly equal to the length of the snout; *front gill cover* toothed along entire edge; *gill cover* with 2 spines; *scales* 50–60 along the lateral line; *fins* 1 dorsal fin; tail fin founded. D X, 11–12; A III, 7.

Reddish- or yellowish-brown with 3–5 vertical dark bands on sides. White on belly and chin. There is a characteristic black spot on the dorsal fin behind the last spine. The pelvic fin is black.

Up to 13 cm.

Found near sand, mud, rocks, and sea-grass beds from about 5–100 m. They are carnivorous, feeding on small fish, crustacea and molluscs. Like related species this fish is a hermaphrodite spawning in the spring and early summer.

Black-tailed Comber
Serranus atricauda Gunther

Eastern Atlantic from Spain south to South Africa; eastern Mediterranean.

Body elongate; *head* jaws large; *scales* small, none between eyes; *fins* single X, 15–16; A III, 8; tail fin square-cut.

Large vertical reddish-brown blotches on sides; blue dots on dorsal, anal and tail fins.

35 cm.

Found over hard rocky bottoms down to about 90 m.

Family:
APOGONIDAE

Small fishes with 2 short, well-separated dorsal fins. 2 spiny rays in the anal fin. The mouth and eye are generally large.

Cardinal Fish
Apogon imberbis (L.)

Cardinal Fish *Apogon imberbis*

Red;
fins not elongate.
Eastern Atlantic,
Mediterranean
15 cm.

Mediterranean; eastern Atlantic around Madeira, Azores and the Canaries.

Body short and deep; *head* large; *eyes* very large; *front gill cover* toothed; *mouth* large; *teeth* small and pointed, similar in both jaws; *fins* 2 dorsal fins, the 1st smaller than the 2nd, which is opposite and similar to the anal fin. 1D VI; 2D I, 9; A II, 8.

Bright orange-red with small black dots irregularly scattered over the surface. There is a darker area on the tail stalk which is sometimes present as 2 or 3 spots. The tips of the dorsal and anal fins are darker. Undersurface lighter. The eye has two light horizontal bands.

10–15 cm in length.

These fish are found from 10–200 m. They are commonly seen in small groups in caves or crevices in rock falls, where they hover near the entrances. At night they venture further afield. Breeding occurs in the summer, which mating being accompanied by quivering movements. After spawning the eggs are bound together into one mass with threads which are present at one end of each egg. The male fish then takes the egg mass into his mouth where it remains until just before the eggs are hatched, when he spits them out. Feed on small crustacea, larvae and eggs.

Cardinal Fish, *Apogon imberbis* (Guido Picchetti).

Family:
POMATOMIDAE

Ferocious fishes hunting in schools in all the warm waters of the world. There are 2 dorsal fins of which the 1st is lower than the second.

Blue Fish
Pomatomus saltator (L.)

Mediterranean and in all warm seas.

Body rather long and laterally flattered; *mouth* large, extending back behind the eye; *jaws* lower jaw longer than upper; *scales* very small, 95–106 along the lateral line; *fins* 2 dorsal fins, the 2nd being much taller than the first. The anal fin is preceded by 2 short spines. 1D VII-VIII; 2D I, 25–28; A II + I, 25–27.

Blue-green above with silver flanks and a whitish belly. There is a black spot at the base of the pectoral fin.

Up to 60 cm, occasionally up to 100 cm.

A ferocious and swift predator living in large shoals in open water. They come into coastal water and will attack fishes almost as long as themselves. A hunting school will leave a trail of blood and maimed fishes behind it.

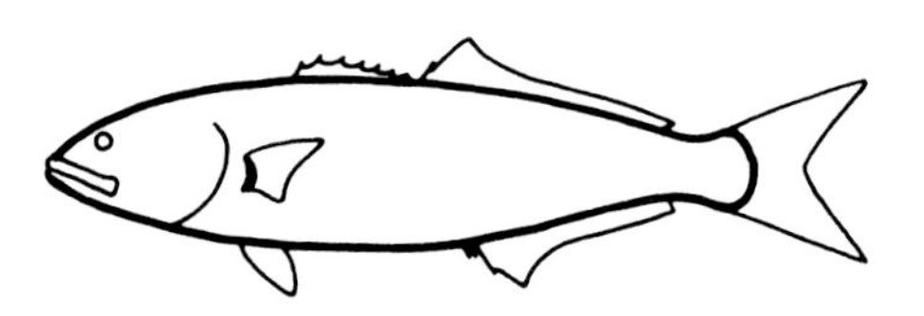

Blue Fish *Pomatomus saltator*

2nd dorsal fin taller than 1st dorsal;
2 spines before anal fin.
60 cm.

Family:
RACHYCENTRIDAE

Cobea
Rachycentron canadum (L.)

Western Atlantic from New Jersey south to Brazil; worldwide in warm seas; rare off Europe and Mediterranean.

Body long and streamlined and head flat; *mouth* large; *jaws* lower jaw longest; *teeth* conical teeth in both jaws; *fins* tail fin rounded in young; shallowly forked in adults the upper lobe being slightly longer. 1st dorsal fin of isolated low spines which can be depressed, 2nd dorsal and anal long, the front rays the highest. D VII-VIII, 30–31; A III, 20–23.

Dark brown on the back, silvery white below. A dark horizontal brown band edged with white runs from the snout to the tail fin along the middle of the body; the horizontal band fades with age.

Small fish found inshore in bays, larger fish further offshore, often associated with floating objects. A good anglers' fish.

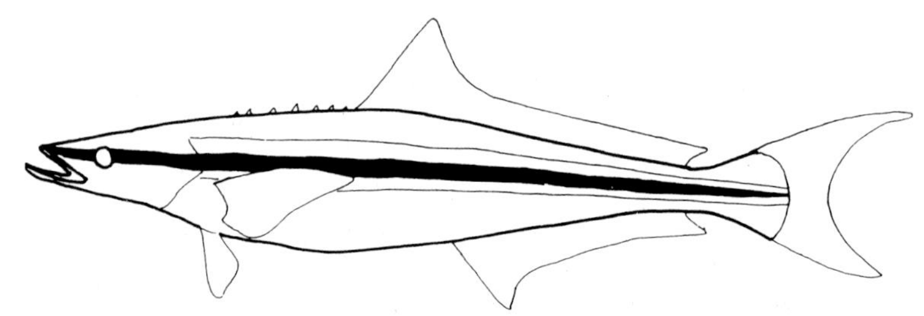

Cobea *Rachycentron canadum*

Body elongate;
low 1st dorsal fin;
dark lateral stripe with white borders;
no sucker on head.
Western Atlantic.
1 m.

Family:

CARANGIDAE

The Horse and Jack Mackerels mostly live near the surface in the open sea. Most are swift swimmers and voracious hunters. Superfically they resemble the Mackerels and Tunas but have no finlets behind the dorsal and anal fins. There are usually two short spines in front of the anal fin and in some species the lateral line is armoured with keeled scales. The tail stalk is narrow.

Horse Mackerel; Scad
Trachurus trachurus (L.)

Mediterranean and Black Sea; eastern Atlantic; English Channel; North Sea; west Baltic.

Body long and slender; *snout* pointed; *mouth* large and slanted downwards; *jaws* prominant extending to beneath the eye; *eyes* large with transparent eyelids at each side; *scales* small and easily rubbed off; *lateral line* has a row of 69–79 wide bony scales each with a point on their rear edge. The lateral line follows the back until the 2nd dorsal fin where it dips downwards. There is a secondary sensory canal which runs from in front of the 1st dorsal fin to near the rear end of the 2nd dorsal fin; *fins* 2 dorsal fins, 1st consists of rays and is higher than the 2nd. 2nd dorsal fin long. Anal fin long and has 2 isolated spines situated in front of it. Pectoral fin long reaching back to the anal fin. Tail fin forked. 1D I-VIII; 2D I, 28–33; A II + I, 25–33.

The back is grey or bluish-green. The flanks are silvery with metallic reflections which are lost after death. Undersurface silver. The gill cover has a small black spot. Young fish are completely silver.

Up to 40 cm.

Horse Mackerel are common fish found swimming in shoals, sometimes very large, from 10–100 m usually in open water. During the summer months they are common near the coasts but during the winter they migrate offshore to deep water sometimes below 500 m. They are particularly common in shallow sandy areas particularly offshore sand banks like those in the North Sea and off the Dutch coast. They are carnivorous. Adult fish feed upon small shoals of fish (herring, sprats, etc.), crustacea and cephalopods. Young fish feed on

Horse Mackerel, *Trachurus sp.* (Leo Collier).

Horse Mackerel, *Trachurus trachurus* (Peter Scoones).

minute larvae and crustacea present in the plankton. Spawning occurs during the summer. Young fish are frequently seen in small groups swimming around the tentacles of jellyfish or around floating debris. They are not considered an important food fish but they are processed for fish meal and are the subject of fairly extensive fisheries off Spain and Portugal.

The False Scad, *Caranx rhonchus*, is remarkably similar to the Horse Mackerel. The differences are that the False Scad has a black blotch on its first dorsal fin and has no enlarged scales on the front curved part of the lateral line.

Horse Mackerel

Trachurus mediterraneus (Steindachner)
= *Caranx trachurus* Steindachner

Very similar to *Trachurus trachurus* except that the body is more flattened laterally, the scales along the lateral line are smaller (between 78–92) and the secondary sensory canal ends under the 3rd or 4th ray of the 2nd dorsal fin. The last rays of both the 2nd dorsal and anal fins end in small distinct membranes. 1D I-VIII; 2D I, 32–33 + I; A II + I, 25 + 1.

The Blue Runner

Caranx crysos (Mitchill)

Western Atlantic, Nova Scotia south to Brazil.

Body elongate, profile of body about equal above and below, snout sharp, eye small; *lateral line* slightly arched on front half of body, straight on rear half, enlarged scales along entire straight portion; *fins* D VIII + I, 23–24; A II + I, 19–20.

A black spot on the gill cover and the tail fin has black lobes.

65 cm.

More abundant further north than other *Caranx*, it is a good food fish.

Crevalle Jack

Caranx hippos (L.)

Eastern Atlantic, Portugal south to Angola and western Mediterranean; western Atlantic, Nova Scotia south to Uruguay.

Body rather deep and flattened with rather square-shaped outline, chest naked; *teeth* canine-like teeth in the upper jaw; *fins* spiny, 1st dorsal fin much lower than soft-rayed 2nd dorsal fin; *lateral line* with 23–39 enlarged scales on rear straight portion.

Greenish to dusky above, silvery below; there is a dark patch on the pectoral fins.

60 cm.

In moderate to large schools on shallow flats, sometimes in brackish water and even into rivers. Larger specimens may occur offshore and down to 350 m.

The **Horse-eye Jack** (*Caranx latus* Agassiz) resembles the Crevalle Jack, and their young often swim together. It differs in

Horse-eye Jack (top), *Caranx latus* (Carl Roessler).

Crevalle Jack (bottom), *Caranx hippos* (Doug Perrine).

having a fully-scaled chest and usually a dark spot on the lobe of the dorsal fin. Western Atlantic.

Round Scad
Decapterus punctatus (Cuvier)

Western Atlantic from Georges bank south to Brazil and Gulf of Mexico; islands of eastern Atlantic and north African coast.

Body long and rounded; *fins* 1st and 2nd dorsal fins both high at front, the last soft ray of 2nd dorsal fin and anal fin are separated from the main body of the fin; *lateral line* front portion gently curved, rear portion straight with 32–42 enlarged bony scales.

Greenish or greyish above, dusky or whitish below, 3–14 blackish spots spaced at regular intervals along curved portion of the lateral line, dark spot on upper angle of gill cover.

25 cm.

Near bottom to mid-water in shallow water, and down to 100 m.

Atlantic Bumper
Chloroscombrus chrysurus (L.)

Western Atlantic, Cape Cod south to Uruguay.

Body curve of bottom profile greater than that of back, tail stalk very narrow, longer than deep; *head* mouth oblique, eye very large, longer than snout; *lateral line* arched in front, straight behind, no enlarged scales; *fins* anal spines are free. 1D VIII + I, 26–28; 2DII + I, 21–23.

Silvery with yellow fins and dark spot on top of the tail stalk.

30 cm.

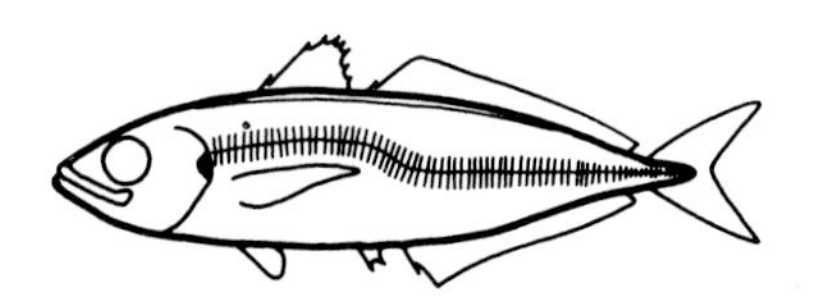

Horse Mackerel; Scad
Trachurus trachurus

69–79 large scales along lateral line;
2nd dorsal canal along back ends near rear of 2nd dorsal fin.
Mediterranean, Eastern Atlantic.
40 cm.

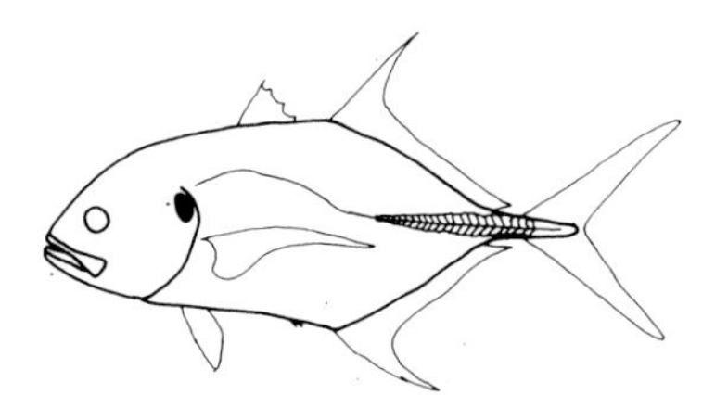

Crevalle Jack *Caranx hippos*

Lateral line, scales 23–39;
dark blotch on pectoral fin.
Warm eastern and western Atlantic.
60 cm.

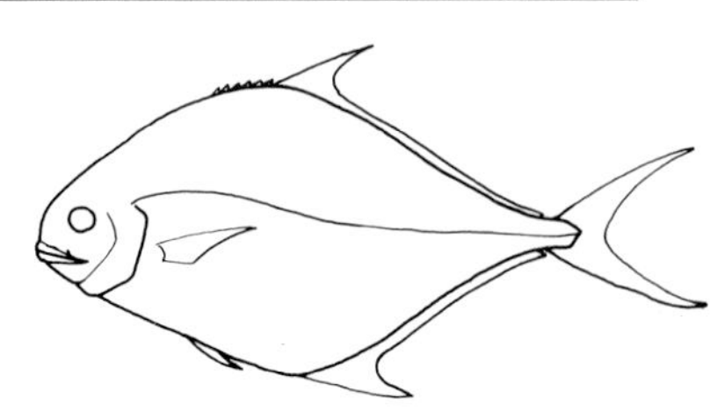

Florida Pompano
Trachinotus carolinus

Snout blunt;
sides silver;
dorsal rays 23–27, anal rays 20–23.
Western Atlantic.
40 cm.

False Scad *Caranx rhonchus*

Straight lateral line, 24–32 enlarged scales, 0–3 on curved lateral line;
black blotch with pale border at front of soft dorsal fin.
Mediterranean.
40 cm.

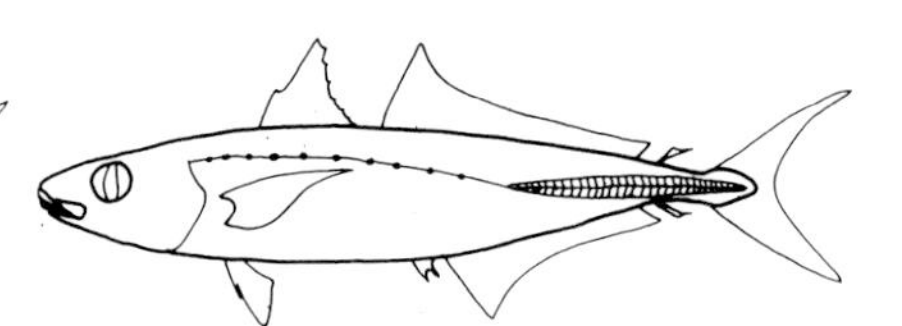

Round Scad *Decapterus punctatus*

Row of spots along body;
dorsal fins both high;
lateral line with enlarged scales at straight rear portion;
last ray of dorsal and anal fin separate.
Western Atlantic;
25 cm.

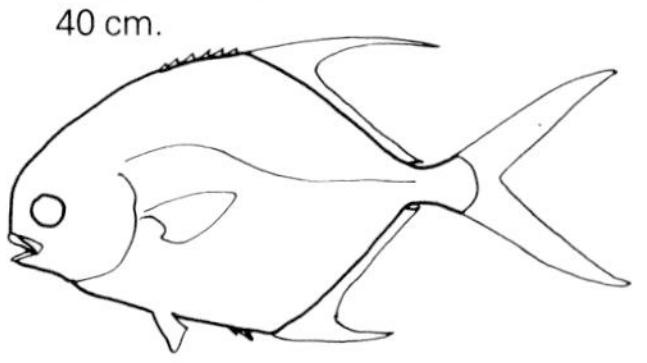

Permit *Trachinotus falcatus*

Deep body, depth more than half length;
young with reddish fins;
no vertical lines on body.
Western Atlantic.
80 cm.

Horse Mackerel
Trachurus mediterraneus

78–92 scales along the lateral line;
2nd sensory canal along back ends under the 3rd or 4th ray of 2nd dorsal fin.
Mediterranean, Eastern Atlantic.
40 cm.

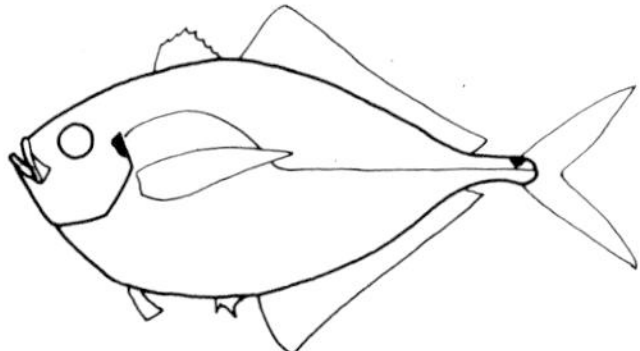

Atlantic Bumper
Chloroscombrus chrysurus

Dark spot on top of tail stalk;
yellow fins;
free anal spines.
Western Atlantic.
30 cm.

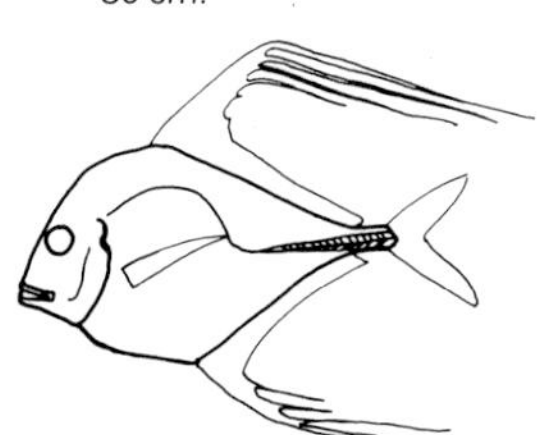

African Pompano *Alectis ciliaris*

Young with filamentous dorsal and anal fins;
adults with spiny dorsal and anal fins overgrown;
enlarged scales on rear lateral line.
Western Atlantic.
60 cm.

Blue Runner *Caranx crysos*

Body rather long;
dark spot on the gill;
lobes of the tail fin are dark;
enlarged scales on lateral line.
Western Atlantic.
65 cm.

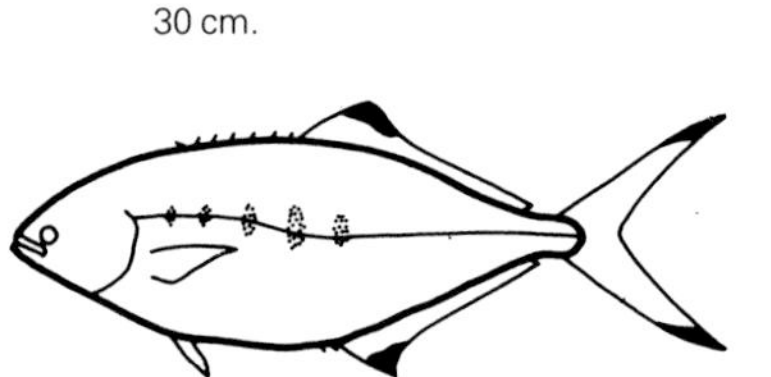

Pompano;
Trachinotus ovatus

4–6 spots on sides.
Eastern Atlantic;
Mediterranean.
30 cm.

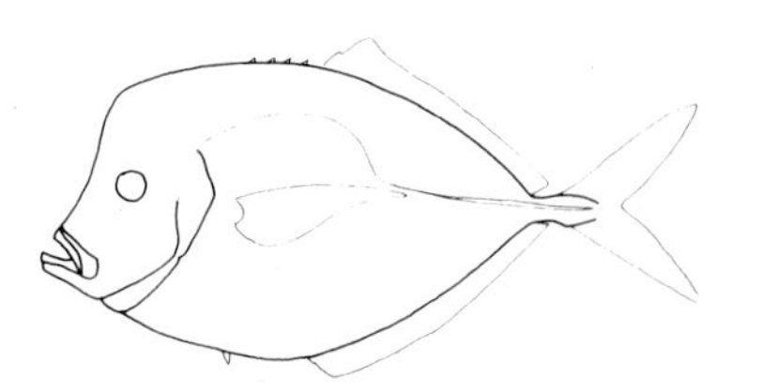

Atlantic Moonfish
Vomer setapinnis

Front profile almost vertical;
no filamentous fins;
young with black spot on side;
Western Atlantic.
35 cm.

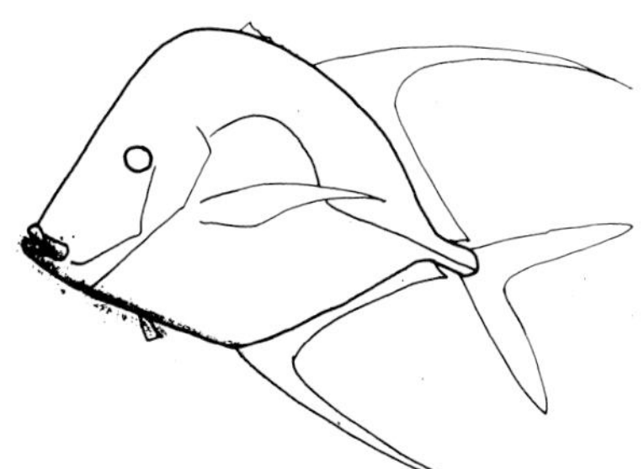

Lookdown *Selene vomer*

Dorsal and anal fins elongate;
back high, forehead oblique;
no enlarged scales on lateral line.
Western Atlantic.
30 cm.

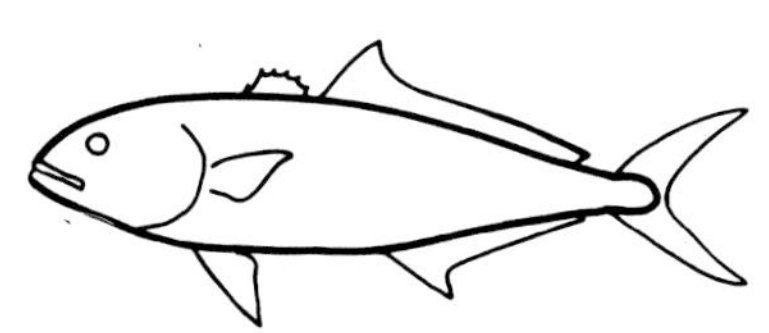

Amberjack; Yellow Tail
Seriola dumerilii

No pre-anal fins or spines;
yellowish sides.
Mediterranean.
2 m.

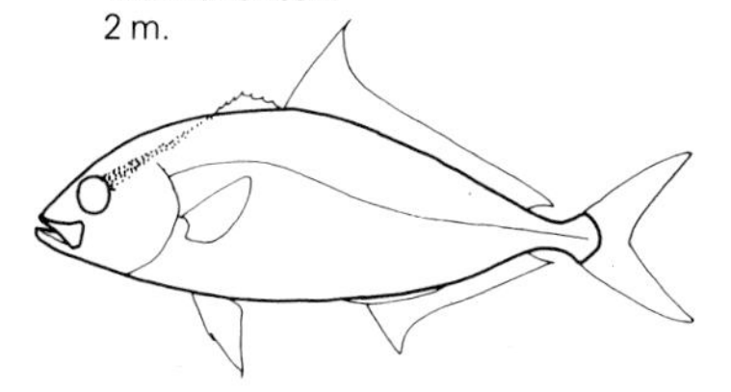

Almaco Jack *Seriola rivoliana*

2nd dorsal and anal fins high;
dark band from eye to 1st dorsal fin;
7 low dorsal spines.
Eastern and western Atlantic.
95 cm.

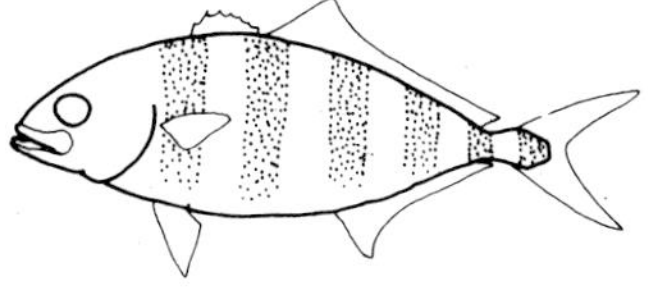

Banded Rudderfish
Seriola zonata

Usually with dark bands;
snout tapering;
at least 7 more dorsal rays than anal;
body depth less than ⅓ standard length.
Western Atlantic.
60 cm.

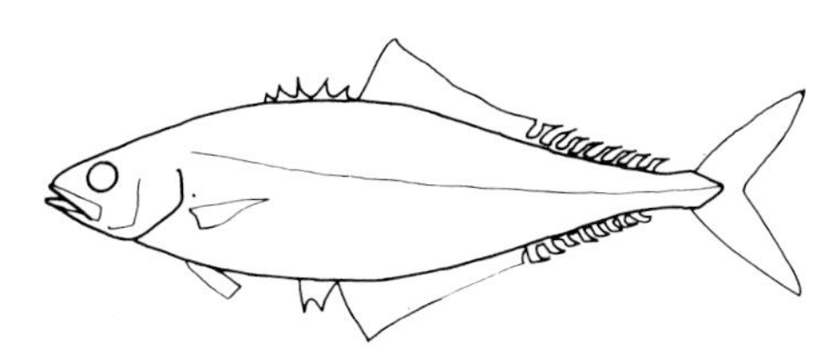

Leatherjacket *Oligoplites saurus*

Soft dorsal and anal rays fan-shaped forming finlets;
yellow dorsal, tail and anal fins.
Western Atlantic.
25 cm.

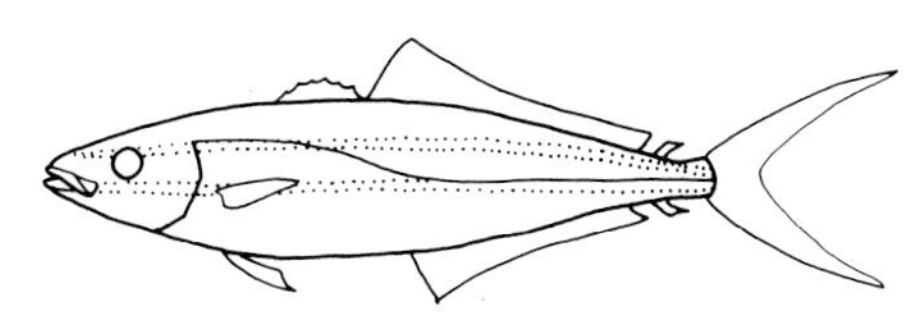

Rainbow Runner *Elagatis bipinnulata*

1 finlet behind dorsal and anal fins;
blue above, yellow below;
2 blue lines run along body.
Western Atlantic.
30 cm.

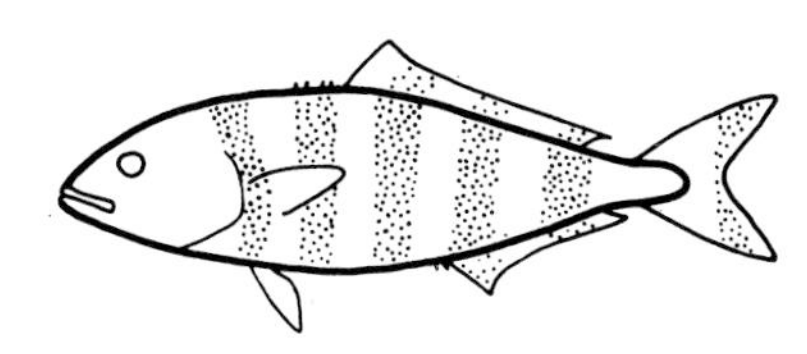

Pilot Fish *Naucrates ductor*

5–7 dark vertical stripes;
Reduced 1st dorsal fin.
Eastern and western Atlantic.
35 cm.

Pompano
Trachinotus ovatus (L.)

Mediterranean; southern Atlantic north to the English Channel, occasionally north to Scandinavia.

Body oval, laterally flattened; *snout* short and rounded; *mouth* small reaching just beyond the front edge of the eye; *scales* small, about 127 along the lateral line; *fins* 2 dorsal fins, 1st fin consists of 5–7 separate spines the 1st being directed forward, 2nd fin long and similar to the anal fin. Anal fin has 2 separate spines in front of it. Tail fin forked. 1D I + V-VI; 2D I, 24–25; A II + I, 23–25.

Back bluish- or greyish-silver, sides and belly pinkish-silver. The sides may have a yellowish tinge. There are 4–6 dark oval spots along the sides. The tips of the 2nd dorsal, anal and tail fins are black. The 2nd dorsal and anal fins are yellowish.

Up to 30 cm, occasionally 50 cm.

Mid-water migratory fish which occasionally approaches shore in small groups. They breed during summer and feed on small fish.

Florida Pompano
Trachinotus carolinus (L.)

Western Atlantic from Massachusetts south to Brazil.

Body moderately deep, its depth more than ⅓ its length, snout blunt; *teeth* none in jaws in adult; *fins* D V-VI + I, 23–27; A II + I, 20–23.

Flanks silvery, pectoral and anal fins perhaps shaded with light orange.

40 cm.

Can be very common on sandy shores in the surf zone, very good flavour and an important game and commercial fish.

Permit
Trachinotus falcatus (L.)

Western Atlantic from Cape Cod south to Brazil.

Body deep, depth more than half its length, the profile from dorsal fin to eye is smoothly convex, from eye to jaw almost vertical, less so in young; *teeth* none in adult; *fins* D VI + I, 18–20; A II + I, 17–18.

Body silvery to dark, young often have reddish fins.

80 cm.

Not common; often in the surf zone like other Pompano.

African pompano
Alectis ciliaris Bloch

Western Atlantic, Massachusetts south to Brazil; tropical Atlantic and Pacific.

Body very compressed square and deep; *lateral line* arched in front with straight portion to rear bearing enlarged scales; *fins* in young, dorsal and anal fins are extended into very long filaments, in adults the spiny fins become embedded and are not easily seen. D VII + I, 18–19; A II + I, 15–16.

Young banded, bluish-silver above, tinted yellow below; the gill cover and fronts of anal and dorsal fins each have a dark blotch.

60 cm.

A mainly tropical species, but individuals are found in our area.

Atlantic Moonfish
Vomer setapinnis (Mitchill)

Western Atlantic, Nova Scotia to Uruguay; eastern Pacific and tropical west Atlantic.

Body deep and square, depth half the length of body, even deeper in young, front profile almost vertical; *lateral line* arched in front, straight behind with weak enlarged scales; *fins* no filamentous fins, dorsal and anal fins very low. D VII + I, 20–23; A II + I, 17–19.

Generally silvery with a black spot in the centre of the body where the lateral line becomes straight.

35 cm.

A schooling fish, may be common in bays in summer in warmer waters.

Lookdown
Selene vomer

Western Atlantic, Nova Scotia south to Argentina; tropical east Atlantic and Pacific.

Body deep, compressed, square shape, front profile very steep and almost straight; *lateral line* arched in front, straight behind but not bearing enlarged scales; *fins* young have filamentous dorsal, anal and tail fins. D VIII + I, 22–23; A II + I, 18–20.

Silvery, sometimes with faint darker bands.

30 cm.

On sandy shores.

Amberjack; Yellow Tail
Seriola dumerilii (Risso)

Mediterranean; western Atlantic, Nova Scotia south to Brazil.

Body long and laterally flattened; *profile* gently curved; *snout* rounded; *jaws* extend to middle of eye; *scales* very small 150–180 along lateral line; *fins* 2 dorsal fins, 1st much smaller than the 2nd. Anal fin is shorter but similar to the 2nd dorsal. Tail fin deeply forked. 1D VI-VIII; 2D 34–39; A II, 18–20.

Back silvery-blue or grey; sides lighter and undersurface silver. The flanks have a golden iridescence and young fish have a distinct yellow eye. There is a diffuse dark stripe running from the shoulder to the eye. Young fish are yellow with dark vertical stripes.

Up to 2 m.

Found in small fast-swimming groups usually around rocks from moderate to deep water. Young fish are sometimes found underneath the bells of jellyfish and occasionally amongst shoals of Saupe (*Boops salpa*). They are carnivorous and spawn in spring and summer.

Lookdown, *Selene vomer* (Ed Brothers).

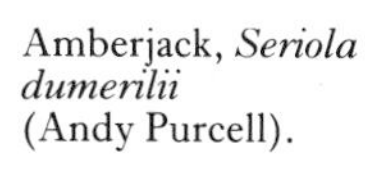

Amberjack, *Seriola dumerilii* (Andy Purcell).

Almaco Jack
Seriola rivoliana Valenciennes

Western Atlantic from New Jersey south to Argentina; Portugal.

Body oval, snout tapering, no enlarged scales on lateral line; *fins* spiny dorsal fin low, always with 7 rays, 2nd dorsal and anal high. D VII + I, 30–32; A II + I 19–20.

Dark bar from eye to beginning of dorsal fin.

95 cm.

General oceanic distribution, rarely found in inshore water.

Banded Rudderfish
Seriola zonata (Mitchill)

Western Atlantic, Nova Scotia south to Brazil.

Body top profile a gentle curve with *head* longer than deep; *mouth* small; *fins* D VII-VIII + I, 33–40; A II+I, 19–21.

About 6 broad black bars on flanks forming 3 large blotches on dorsal fin and 2 on anal fin, dark line runs from spiny dorsal to the eye, bars fading only in older fish.

60 cm.

Mostly offshore near floating objects.

Leatherjacket
Oligoplites saurus (Schneider)

Western Atlantic, Maine south to Uruguay.

Jaw lower jaw not extensible; *fins* many small finlets behind dorsal and anal fins formed from fan-like rays. D V + I, 19–21; A II + I, 18–21.

Dorsal, tail and anal fins yellow in young.

25 cm.

Rainbow Runner
Elagatis bipinnulata (Quoy & Gaimard)

Western Atlantic, New England south to Venezuela.

Body oval, back rather low and snout rather long and pointed; *lateral line* wavy with no enlarged scales; *fins* 1 finlet behind dorsal and anal fins, D VI + I, 25–26; A II + I 16–17.

Body blue above, pale yellow below, with reddish tints, 2 blue bands run along body; fins yellow.

30 cm.

Usually in tropical waters, an esteemed game fish.

Pilot Fish
Naucrates ductor (L.)

Mediterranean; English Channel; eastern Atlantic and all warm and temperate oceans.

Body elongate; *profile* curved; *snout* rounded; *scales* very small and extend onto cheeks; *lateral line* has a keel at each side of the tail stalk; *fins* 2 dorsal fins, the 1st present as 3–5 short separate spines situated in front of the 2nd dorsal which is long. Anal fin has 2 separate spines situated in front of it. Tail fin forked. 1D IV-V; 2D I, 26–28; A II + I, 16–17.

Bluish with 5–7 blue-black, grey or brown-black dark vertical stripes which extend onto the dorsal and anal fins. There is also a dark stripe near the end of the tail fin which also has a white edge.

Up to 35 cm, sometimes up to 70 cm.

An open water fish which is frequently seen in association with sharks, mantas, turtles, boats and driftwood. The young in particular congregate in small groups under and in the neighbourhood of large jellyfish, pieces of wreckage and floating weed. The reasons for this association are not known. Often found in small groups. They are carnivorous and feed mainly on planktonic crustacea, occasionally fish and molluscs.

Family:
CORYPHAENIDAE

The body is long and compressed. The dorsal fin begins on the head and extends to the tail. Teeth very small.

Dolphin Fish
Coryphaena hippurus (L.)

Mediterranean and all warm seas.

Body elongate and very compressed laterally; *profile* in the young fish rounded but with increasing age it becomes progressively steeper and the adult males have a forehead that is almost vertical; *scales* minute, embedded in the skin and not visible to the naked eye; *lateral line* is straight except over the pectoral fin where it is humped; *jaws* lower jaw protrudes beyond the upper, oval patch of teeth on tongue; *fins* there is a single dorsal fin running the length of the body from the eye to the tail fin which is deeply forked. D 55–65; A I, 26–30.

Blue-green above, silvery below with brilliant silver and gold iridescence. Stippled with dark and gold spots.

Up to 190 cm.

A free swimming species of open water which undertakes long and regular breeding migrations. They are often associated with other species such as Pilot Fish and seem to like the shade cast by flotsam and small boats. Feed on fishes and may be

Dolphin Fish, *Coryphaena hippurus* (Norbert Wu).

caught by trolling. At night they are one of those fishes attracted to a bright light. Very good to eat.

A less common related fish, *Coryphaena equisalis* L., differs in having a deeper body and teeth covering most of the tongue.

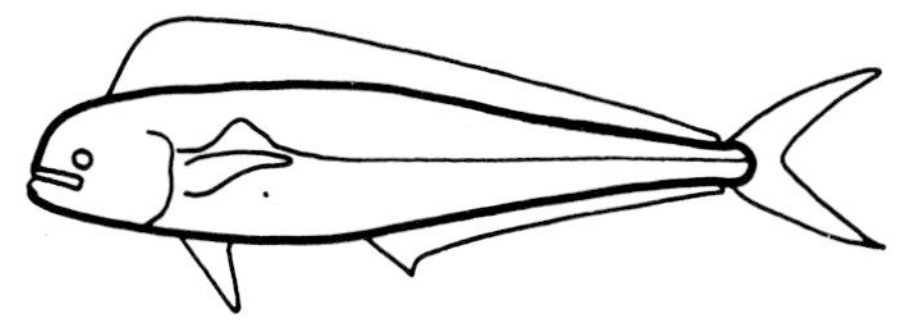

Dolphin Fish *Coryphaena hippurus*

Long flattened body;
long dorsal fin;
steep forehead.
190 cm.

Family:

LOBOTIDAE

Tripletail

Lobotes surinamensis (Bloch)

Warm Atlantic; in the east, Atlantic islands and Mediterranean; in the west, from Cape Cod south to Argentina.

Body deep oval, compressed about half as deep as long, profile of head concave, upper jaw very extensible; *front gill cover* toothed with no spine; *fins* soft rays of dorsal and anal fins longer than spiny rays and reaching back to the tail rays, tail fin rounded. D XI-XII, 14–16; A III, 11.

In adults blackish above shading to silvery below, tail with a pale margin; juveniles, yellow with dark blotches.

1 m.

Often found in the coastal zone. Small fish associate with floating objects and also live round pier pilings, etc. Observed to float flat on the surface like the Sunfish.

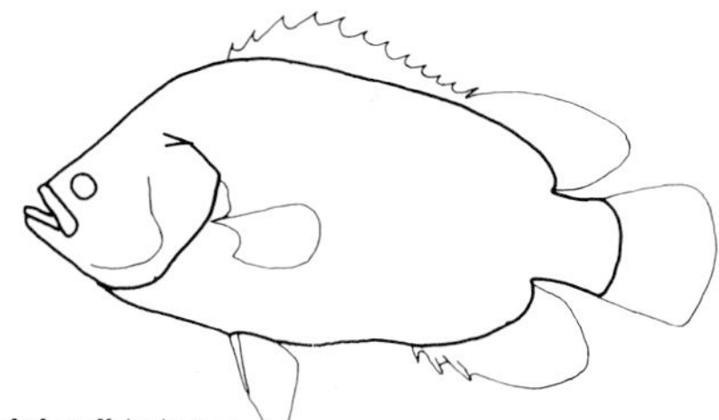

Tripletail *Lobotes surinamensis*

Body rounded, mouth oblique;
1st and 2nd dorsal fin continuous;
adults: blackish grey;
juveniles: yellow with dark blotches.
Western and eastern Atlantic; Mediterranean.
1 m.

Family:
SCIAENIDAE

Perciform fishes with a wide tail stalk, 2 dorsal fins and a lateral line that extends onto the tail fin. The otoliths are very large. The swimbladder is well-developed and can be caused to oscillate by the vibration of special muscles. This makes the loud grunting noise for which many of these fishes are famous.

Meagre
Argyrosomus regius (Asso)

Mediterranean; eastern Atlantic north to Britain, south to Guinea; English Channel; occasionally North Sea.

Body long and slender; *snout* rounded; *mouth* long; *jaws* of equal length or lower jaw slightly longer than the upper; *scales* large on body, smaller on head, between 50–55 along the lateral line. The scales appear to run in oblique rows; *lateral line* continues onto tail; *barbels* none; *fins* 2 dorsal fins distinct from each other but not separated by a space. Anal fin short. 1D IX-X; 2 D I, 27–29; A II, 7–8 *Swimbladder* very large, it occupies nearly the whole of the abdominal cavity.

Back silvery-grey or brown, sides lighter with golden and silver reflections. Undersurface silver. The fins are darker, grey or brown. The inside of the mouth is golden. Sometimes there is a darker spot on the gill cover.

Up to 200 cm.

Found in shallow water and is able to withstand brackish water. Young fish particularly are found in estuaries. Usually seen amongst rocks. They are carnivorous and hunt shoals of small fish. The large swimbladder is used to produce deep sounds which may be heard over several metres. Day (1880) reports that the Dutch were said to perceive an image or representation of the Virgin on each scale.

Brown Meagre; Corb
Sciaena umbra (L.)
= *Corvina nigra* Valenciennes

Mediterranean and Black Sea; eastern Atlantic from Gulf of Gascony to Senegal.

Body deep and large; *profile* curved; *snout* rounded; *jaws* large extending to the rear edge of the eye; *scales* fairly large; *fins* 2 dorsal fins distinct but joined by a fine membrane. Anal fin short, the 2nd spine large. Tail fin square-cut in the adult, slightly indented in the young. 1D X; 2D I, 23–25; A II, 7–8.

Back and sides brown-bronze with golden reflections. Undersurface lighter and silvery. The fins are dark but the spines of the pelvic and anal fins are white and appear very conspicuous underwater.

Up to 40 cm, sometimes up to 70 cm.

These fish are found in small shoals from 5–20 m. They are seen amongst rocks often in caves and cracks and amongst sea-grass (*Posidonia*) where they sit quietly with only their tails slowly moving. Spawning in the Mediterranean occurs in late spring and summer. Carnivorous, they feed on fish, crustacea and molluscs.

Corb
Umbrina cirrosa (L.)

Mediterranean and Black Sea; eastern Atlantic north to Biscay, south to Senegal; Red Sea.

Body elongate and laterally flattened; *profile* gently curved; *snout* rounded; *jaws* upper jaw longer than lower; *barbels* there is a small fleshly barbel at the extreme tip of the lower jaw; *scales* 50–60 along the lateral line; *lateral line* follows the curve of the back and extends onto the tail; *fins* 2 dorsal fins distinct but not separated from each other. Tail fin slightly concave in upper half. 1D X; 2D I, 21–25; A II, 7.

Back and sides silvery with diagonal golden stripes, edged with fine brown-violet stripes. Undersurface silvery. The rear edge of the gill cover is black.

Up to 1 m.

These fish are solitary and found over mud and sand and in rocky areas. Young fish may be seen in estuaries. They are carnivorous and feed on molluscs, worms and crustacea. Spawning occurs in June in the Mediterranean.

Brown Meagre; Corb, *Sciaena umbra* (Andrea Ghisotti).

Brown Meagre; Corb, *Sciaena umbra* juvenile (Andy Purcell).

Meagre *Argyrosomus regius*

Long slender body;
silvery in colour.
Eastern Atlantic,
Mediterranean.
2 m.

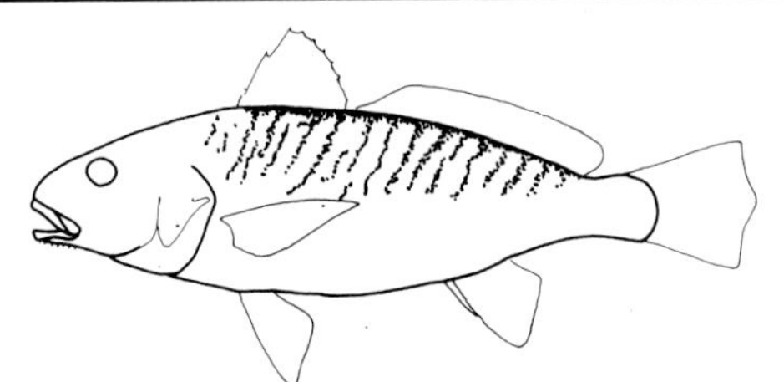

Atlantic Croaker
Micropogonias undulatus

Thin bars on body;
many small barbels on jaw.
Western Atlantic.
50 cm.

Northern Kingfish
Menticirrhus saxatilis

2nd spine of dorsal fin very long;
V-shaped blotch on side;
barbels on chin.
Western Atlantic.
30 cm.

Brown Meagre; Corb
Sciaena umbra

Heavy body;
dark bronze colour;
conspicuous white spines on anal and pelvic fins.
40 cm.

Black Drum *Pogonias cromis*

Body deep;
4–5 black bars;
barbels long.
Western Atlantic.
170 cm.

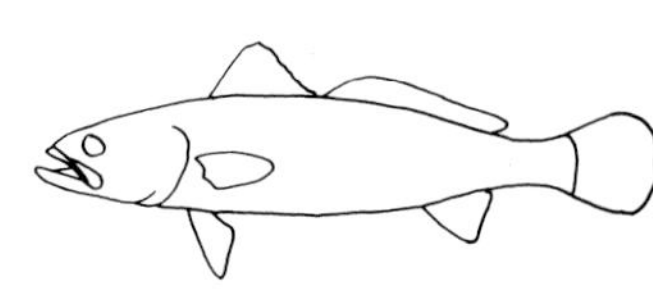

Silver Seatrout
Cynoscion nothus

Body long;
lower jaw projecting;
pale, no conspicuous markings.
Western Atlantic.
30 cm.

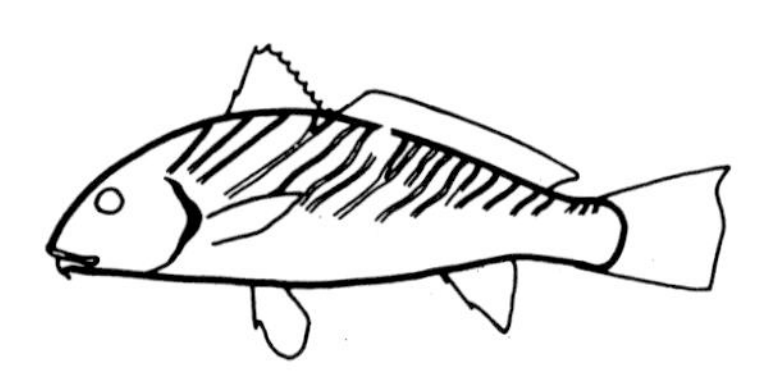

Corb *Umbrina cirrosa*

Diagonal golden striping edged with brown-violet;
Eastern Atlantic,
Mediterranean.
1 m.

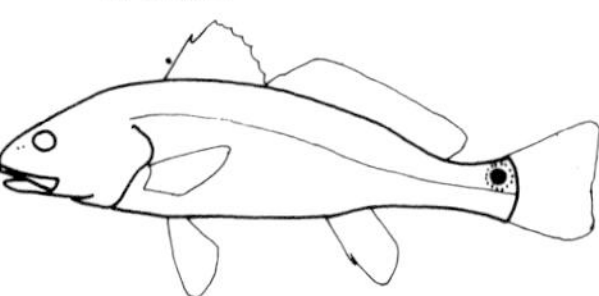

Red Drum
Sciaenops ocellatus

'Eye' spot high on tail stalk;
no barbels on chin.
Western Atlantic.
130 cm.

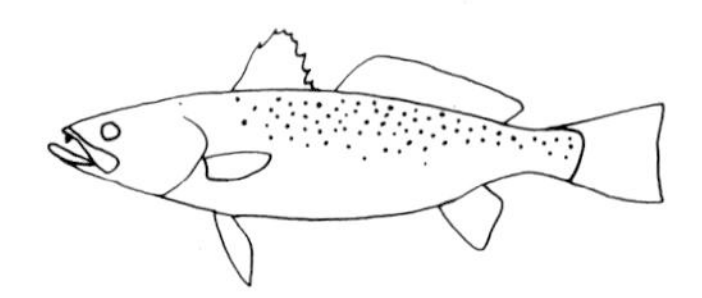

Spotted Seatrout
Cynoscion nebulosus

Body long;
lower jaw projecting;
spots on upper sides and median fins.
Western Atlantic.
90 cm.

Atlantic Croaker
Micropogonias undulatus (L.)

Western Atlantic from Rhode Island south to Cape Kennedy and Gulf of Mexico.

Head many small barbels in a line along the inner edge of the lower jaw, sometimes worn away in larger fish, strong spine on the angle of the front gill cover; *scales* large, there are 7 between dorsal origin and the lateral line; *fins* 2nd spine on anal fin, less than ⅔ length of longest ray. D X+I, 28–29; A II, 5–6.

Body silvery-grey, becoming bronze or golden in the breeding season; spots on the scales which form thin dark diagonal bars which are sometimes interrupted.

50 cm.

Move into bays in spring and leave for deeper water in autumn.

Black Drum
Pogonias cromis (L.)

Western Atlantic, Nova Scotia south to Argentina, most often from New York southwards.

Body deep and back strongly curved at the shoulder, bottom profile rather flat; *head* mouth small, set low with many long

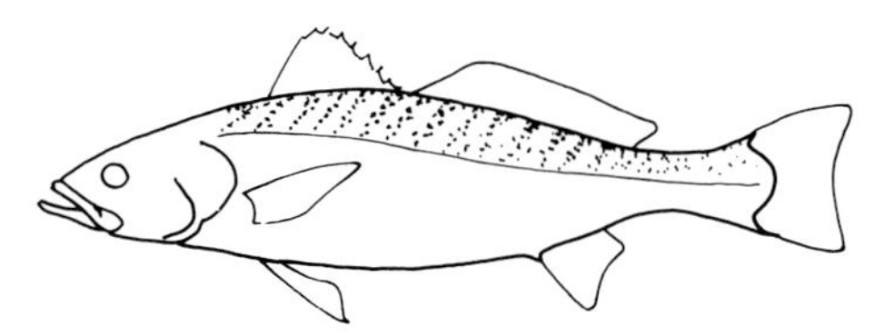

Weakfish *Cynoscion regalis*

Long body;
many small brown spots forming wavy lines.
Western Atlantic.
90 cm.

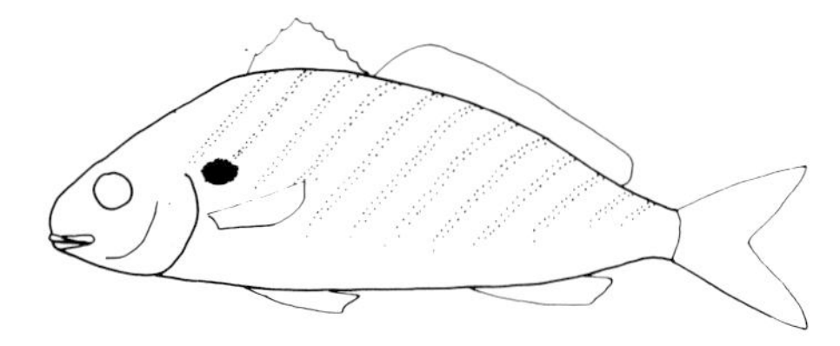

Spot *Leiostomus xanthurus*

Tail fin forked;
dark spot on shoulder;
12–15 diagonal lines on body.
Western Atlantic.
35 cm.

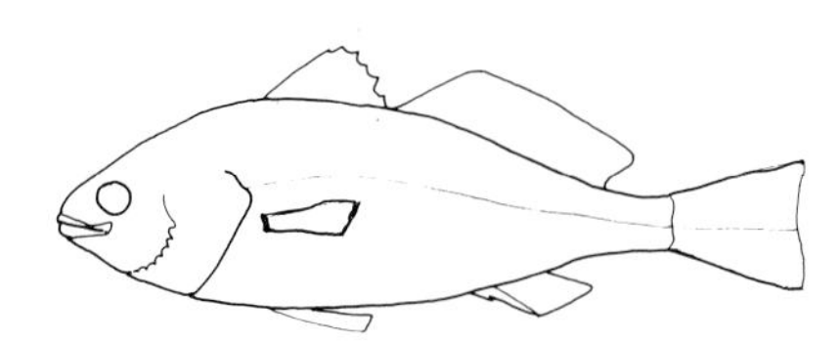

Silver Perch *Bairdiella chrysoura*

Small and silvery;
mouth oblique;
fins yellow.
Western Atlantic.
23 cm.

barbels on the chin arranged in more than 1 row, some short barbels on the snout, upper jaw projects beyond lower; *fins* anal spine slightly longer than the 1st soft ray. D X+I, 21; A II, 5–6.

Body silvery-grey with brassy sheen, 4–5 dark bars in young fish, becoming more uniformly dark in older fish, fins blackish. No black spot on the base of tail fin.

170 cm.

A shoreline fish, often found near piers and in the surf. Lives near the bottom, where it feeds on crustacea and molluscs. This is perhaps the largest fish of the family.

Red Drum

Sciaenops ocellata (L.)

Western Atlantic from Massachusetts south to Texas.

Body long, robust with rather steep profile; *head* the upper jaw is longest, no barbels on chin, front gill cover strongly toothed in young; *teeth* outer series in upper jaws much enlarged; *scales* 40–45 scales along the lateral line; *fins* D X+1, 24; A II, 8.

Silver when young becoming copper-bronze. All ages have a black spot with a paler ring set high on the tail stalk; there may be smaller spots to the front of the main eye spot.

130 cm.

An important fish for anglers and esteemed as food. Young found inshore in large aggregations; adults tend to be solitary, but some remain in shallow water.

Northern Kingfish

Menticirrhus saxatilis (Bloch & Schneider)

Western Atlantic from Cape Cod south to Florida.

Head slightly concave above the eyes; *teeth* outer series in upper jaw not greatly enlarged; *fins* 2nd dorsal spine greatly elevated, reaching back past the front of the soft dorsal fin. D X+1, 26–27; A I, 8.

Dark grey to blackish, strongly marked oblique cross-bands make V-shape on the nape; dark streak divides the grey upper half from the paler belly.

30 cm.

Found mostly on sandy bottoms.

Silver Seatrout

Cynoscion nothus (Holbrook)

Western Atlantic from New York south to Florida, Gulf of Mexico.

Body long and streamlined; *head* snout shorter than the least depth of the tail stalk, lower jaw projecting; *teeth* one pair of canine teeth in upper jaw; *eye* large; *scales* easily detached; *fins* 2nd dorsal fin is rather long, having 27–29 soft rays.

Upper part light or dark straw colour and silvery below. Markings are faint or absent.

Mouth orange inside.

30 cm.

More common in the ocean than close inshore and only enters bays in the cooler months. The related Sand Seatrout (*Scynoscion arenarius*) has 10–11 rays in its anal fin and the scales are firmly attached, the range of the Sand Seatrout is also more southerly.

Spotted Seatrout
Cynoscion nebulosus (Cuvier)

Western Atlantic from New York south to Florida.

Body long and streamlined; *head* long, lower jaw is the longest; *teeth* 2 large canines in upper jaw; *scales* firmly attached; *fins* the dorsal fins may be joined or separate. D X+I, 24–26; A II, 10–11.

Dark grey above with blue reflexions, silvery below. Many round dark spots on the upper part of body, tail fin and dorsal fin; inside of mouth orange. Very young fish have a broad band running along the centre of the body.

90 cm.

A very popular sport fish. Spawns in March to November within bays and lagoons. Juveniles develop in areas sheltered by aquatic vegetation within 50 m of the shore. As winter approaches the young move into deeper water. In very cold weather many fish that stay in coastal water are killed.

Weakfish
Cynoscion regalis (Bloch & Schneider)

Western Atlantic south to Florida.

Resembles the Spotted Seatrout (*Cynoscion nebulosus*) in general form. *Fins* soft rays scaled at base. D X+I, 26–29; A II, 11.

Dark olive or greenish blue above, upper part of body with many small spots sometimes joined together into diagonal undulating lines; dorsal fin dusky tinged with yellow; ventral, pectoral and tail fins tinged yellow.

90 cm.

A schooling fish that prefers shallow sandy areas. Frequent in surf, sounds, bays, channels and saltwater creeks. Enters brackish water, but not fresh water. May feed right at the surface. Food includes a variety of animals, such as worms, sand lances and other small fish, molluscs, etc. Said to get the name 'weakfish' because anglers' hooks tear easily from the jaw.

Spot
Leiostomus xanthurus Lacépède

Western Atlantic from Cape Cod south to Gulf of Mexico, most common south of New Jersey.

Body short and deep; *head* blunt front profile, no barbels on lower jaw, mouth small, lower jaw short; *teeth* in adults minute in upper jaw, absent in lower jaw; *fins* the tail fin is as long as the head and strongly forked. D X+1, 31; A II, 12.

Bluish grey with 12–15 diagonal brownish-yellow bars, fins olive except for tail fin. There is a dark spot on the lateral line below the shoulder, but no spot at the base of the tail fin.

35 cm.

Generally found over mud, sand and shell bottoms over a very wide range of depths from the shallows when young down to 300 m. Feeds on small plankton and bottom-dwelling crustacea, and is sometimes abundant. Able to tolerate a wide range of salinity from twice normal seawater strength to pure fresh water.

Silver Perch
Bairdiella chrysoura (Lacepede)

Western Atlantic from New York south to Mexico.

Body oblong, compressed, back rather high; *head* blunt snout, profile slightly concave over the eye, front gill cover serrated; *jaw* oblique; *teeth* on upper jaw with an outer series of small curved canines; *fins* 2nd anal spine long, but not as long as soft rays; soft dorsal and anal fins scaled to at least half way up. D XI+I, 22; A II, 10.

Greenish above, silvery below, back and sides often speckled with many small dark spots; fins yellow.

23 cm.

Often found in sea-grass beds; does not enter brackish water.

Family:
MULLIDAE

There are 2 sensory barbels on the chin; the forehead is steep; the body is covered with large scales and there are 2 well-separated dorsal fins.

Red Mullet
Mullus surmuletus L.

Mediterranean; eastern Atlantic from Scotland south to the Canaries.

Body rather long and flattened; *profile* steep; *barbel* there is a pair of sensory barbels beneath the jaw; *scales* large, there are two large scales on the cheek beneath the eye; *fins* 1D VII-VIII; 2D 8–9; A II, 6–7.

Colour varies with depth, emotion and time of day. In the Mediterranean, fishes living shallower than about 15 m are basically yellow-brown; those living deeper are red. In both cases the scales have dark edges. During the daytime, especially when the fishes are shoaling, they have a pronounced dark red or brown longitudinal stripe from the eye to the tail and 4 or 5 longitudinal yellow stripes. At night the pattern breaks up into indistinct marbling.

Up to 20 cm, rarely 40 cm.

Swim either singly or in groups of up to about 50. The young browse on algal-covered rocks as shallow as 1 m, but the adults live over sand and mud from about 3 m down to about 90 m. They feed on small animals buried in the sand which they locate with their sensory barbels and may excavate holes as deep as themselves in their search for food.

The Red Mullet has been a highly prized food fish since ancient times. The Romans

Red Mullet, *Mullus surmuletus* night camouflage coloration (Christian Petron).

Red Mullet, *Mullus surmuletus* daytime coloration (Christian Petron).

indeed would pay more for a good specimen than they would for the fisherman who caught it. A mullet in a bowl was also used as a pre-banquet entertainment so that the guests might marvel at the rapid changes in colour of the dying fish.

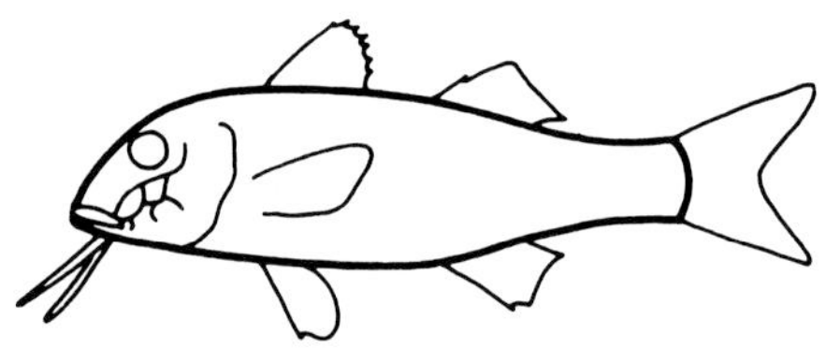

Red Mullet *Mullus surmuletus*

Sloping profile;
barbels longer than pectoral fin.
Eastern Atlantic;
Mediterranean.
40 cm
A similar species, the Red Goatfish in the western Atlantic.

Red Mullet
Mullus barbatus L.

Mullus barbatus has a steeper profile, 3 scales beneath the eye and the longitudinal lines are not present.

Generally lives deeper than *Mullus surmuletus*, and may be caught from 300 m. It is an important bottom fish of the Mediterranean coast of Israel, but in the Eastern Atlantic it probably does not occur further north than Biscay.

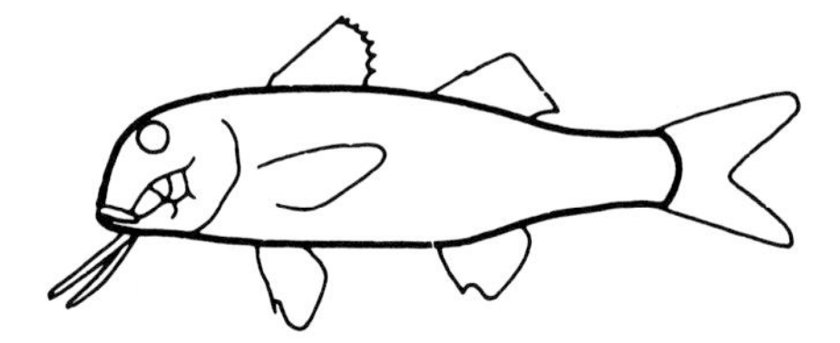

Red Mullet *Mullus barbatus*

Steep profile;
barbels shorter or equal to pectoral fin.
Eastern Atlantic;
Mediterranean.
40 cm.

Red Goatfish
Mullus auratus Jordan & Gilbert

Western Atlantic from Cape Cod south to Florida.

Resembles the European Red Mullet (*Mullus surmuletus*), and differs in appearance only in the lower dorsal and anal fins and in having a yellow rather than a black stripe on the dorsal fin.

Family:
SPARIDAE

The **Sea Bream** have oval, flattened bodies with a moderate or small mouth set low in the head. Both the body and head have distinct, large scales. There is a single dorsal fin, the front part having strong spiny rays, the rear half having soft branched rays. The pelvic fin always has 1 spiny ray and 5 soft rays; the anal fin always has 3 spiny rays but a variable number of soft rays. There are no spines on the gill cover and the pre-operculum is never toothed. The teeth in the jaws are well developed and specialized to serve the particular feeding habits of the fish. Thus there may be chisel-shaped incisors for scraping algae and small animals off rock and weeds, fang-like canines for catching fish or grinding molar for crushing shellfish. There are no teeth on the roof of the mouth.

Their scientific classification relies strongly on their teeth but most bream can be identified from their markings, especially by their dark spots and longitudinal bands. Some have a characteristic head profile, either very steep, or concave with a protruding mouth. Many appear pink or reddish-brown on the surface, but underwater the red tints are not visible and the fish appear a silver grey.

Dentex
Dentex dentex (L.)

Mediterranean and eastern Atlantic from Senegal to Biscay; Rare on south coast of Great Britain.

Body oval; *profile* convex and rather steep. Mature males have a fleshy lump over the eyes; *head* massive; *eye* small and set high in the head; *teeth* there are 4–6 long and well-developed canine-like teeth in the front of both jaws followed by many smaller teeth of the same shape; *fins* D XI-XIII, 11–12; A III, 7–9.

Colour variable. In specimens up to 1 m in length the back is bluish-silver above and silvery below with 4–5 indistinct darker cross-bands. The flanks are dotted with small blue spots that lose their colour after death. The pectoral fins have a rosy tint. The very large specimens, longer than 1 m, may be a uniform dull red.

Up to at least 1 m, usually 50 cm.

Lives over rocky ground from about 10 m to 200 m. In the spring it may come close in to the coasts but in winter retires to deeper water. An active hunter, it feeds on fishes and cephalopods. Dentex is a much-prized food fish.

Gilthead
Sparus aurata L.

Mediterranean and eastern Atlantic from north Biscay to Ghana.

Body deep and oval; *profile* rather steep and smoothly convex; *jaws* upper jaw slightly longer than the lower; *lips* are thick; *teeth* 4–6 strong conical teeth in the front of each jaw followed by 4–5 rows of molar-like teeth in the upper jaw and 3–4 rows in the lower jaw; *fins* D XI, 13; A III, 11–12.

Colour variable, usually grey or gunmetal above, silvery below. The belly is white. There is a dark blotch at the origin of the lateral line.

The edge of the gill cover has a scarlet patch. In life the most characteristic feature is the golden band running between the eyes but the colour fades after death.

Up to 70 cm, usually 35 cm.

A shallow-water species that does not generally go deeper than 30 m and tolerates, or even prefers, brackish water. Usually found in small groups over mud or sand in the shadow of large rocks, but in the spring they congregate in large numbers where the water is brackish. They remain inshore all summer and in the winter return to deeper water to breed. The Gilthead is very sensitive to cold and may die in cold weather.

Uses its strong jaws to crush crustacea and molluscs. The Romans (who dedicated the fish to Venus) fed it upon oysters, which were deemed to improved both its colour and taste.

Common Sea Bream; Red Porgy
Pagrus pagrus (L.)

Mediterranean; eastern Atlantic south to Senegal and north to Biscay; western Atlantic, New York south to Argentina.

Body oval, *profile* steep with a sharp change in slope above eye; *mouth* terminal and set low in the head; *teeth* 4–6 strong fang-like teeth in the front of both jaws followed by sharp curved teeth and then by 2 rows of grinding teeth in each jaw; *fins* D XII, 9–11; A III, 8–9.

Silvery colour with rosy-buff tinted fins, tail fin with a darker rim in the centre often with white tips. In life there is a small red-brown spot at the base of the last ray of the dorsal fin. Face dark.

Up to 50 cm, rarely 75 cm.

Not very common in our area. In the summer it can be seen at depths greater than 20 m over sand and sea-grass, also near algal-covered rocks. In winter it migrates into the deeper water of the Continental Shelf.

Spawns in the summer, the eggs and larvae are pelagic. At 5–9 mm the larvae have a characteristic bony crest above the eye and a spiny gill cover. At 1.5 cm the young adopt the adult form and migrate nearer coasts.

It is carnivorous, feeding chiefly on crustacea but may also eat algae.

Red Sea Beam
Pagellus bogaraveo (Brünnich)

Eastern Atlantic from Senegal to Ireland, rare north to Norway and North Sea.

Body oval; *profile* convex with a slight concavity above the eye; *eye* very large, its diameter at least as long as the space in front of the eye; *teeth* in the front of the jaw are short, recurved and sharp, those behind are small and rounded; *fins* D XII-XIII, 11–13; A III, 11–13.

Greyish or rose-tinted. There is a large black spot on the front end of the lateral line beneath the first dorsal ray. In fishes shorter than 20 cm this spot may be absent.

Up to 35 cm, sometimes up to 50 cm.

This is a schooling species often found in very large shoals. The younger specimens may enter water as shallow as 40 m where they favour rocky coasts and wrecks. The larger specimens are most common from about 150 to 500 m.

Although this fish is sufficiently abundant to have some commercial importance it is rarely seen by divers.

Pandora, *Pagellus erythrinus* (Guido Picchetti).

Pandora
Pagellus erythrinus (L.)

Mediterranean; eastern Atlantic from northern Biscay to southern Angola.

Body oval; *profile* gently convex, sloping down from the front of the dorsal fin to the snout which is pointed; *nostrils* the two pairs are set close together, the hind ones are enclosed by a flap of skin which is extended forward to protect the front nostrils; *eye* not large; in the adult the eye diameter is less than the length of the space in front of the eye; *teeth* the teeth in the front of the jaw are small and pointed. Those in the side of the jaw are small and rounded; *fins* D XII, 10: A III, 9–10.

Uniform rosy-red above, paler flanks and silvery below. The gill flaps have a red rim and the mouth and gill cavities are black.

Up to 25 cm, sometimes up to 60 cm.

Lives near the bottom over sand or mud particularly near rocks at 15–120 m deep.

Spanish Bream; Axillary Bream
Pagellus acarne Risso

Mediterranean; eastern Atlantic from Senegal to Biscay.

Body long, *profile* convex with a rather blunt snout; *eye* diameter about equal to or slightly less than the length of the space in front of the eye; *teeth* 5 comb-like teeth in the front of the jaw, irregular rows of molar-like teeth in the sides of the jaw; *fins* the spiny dorsal rays are of progressively decreasing height after the 2nd spine. D XII, 10–11, A II, 9–10.

Rose-coloured above, silvery below with a black spot at the base of the pectoral fin. The mouth cavity is golden or reddish-orange.

Up to 25 cm, occasionally up to 35 cm.

Found in small shoals from 20–100 m, although the young fish may be found near the coasts in shallower water. They are mostly caught in springtime over soft bottoms but are rarely seen by divers.

Marmora
Lithognathus mormyrus (L.)

Mediterranean and all the warm waters of the eastern Atlantic.

Body long and laterally flattened; *profile* rather rounded; *jaws* set low in the head and strengthened at the front; *lips* rather thick and wrasse-like; *teeth* there are several rows of fine teeth in the front of the jaw of which the outer ones are the larger. These are followed by 3–4 rows of molars in the upper jaw, 2–3 rows in the lower; *fins* D XI-XII, 12; A III, 10–11.

Silver with a characteristic pattern of usually 6 strongly marked, dark cross-bands alternating with an equal number of narrow bands.

Up to 30 cm.

A gregarious species swimming tightly disciplined schools at depths down to about 20 m. Favour soft bottoms where they feed mostly on small bottom-living animals. Particularly common where there are accumulations of decomposing sea-grass (*Posidonia*). They fill their mouths with the organic débris, rise half-way to the surface, spit out the débris but retain the small crustacea it contains. They are very tolerant of changes in salinity and are able to enter the mouths of rivers as well as to survive in salt lagoons made very saline by evaporation.

Annular Bream, *Diplodus annularis* (Andy Purcell).

Sheepshead Bream
Puntazzo puntazzo (Cetti)

Mediterranean; eastern Atlantic north to Biscay and south to the Congo.

Body oval with an elongate snout; *profile* strongly concave; *mouth* small; *teeth* there is a single row of forward-sloping chisel-like teeth in each jaw followed by a single row of minute grinding teeth; *fins* D XI, 13–14, A III, 12.

Body silvery-grey with from 7 to 11 dark vertical stripes and a black spot on the tail stalk. There is no yellow on the fins.

Up to 30 cm, rarely 45 cm.

This bream does not form large shoals like most other bream but tends to be more solitary. It lives to a depth of 50 m amongst algal-covered rocks and reefs or near meadows of sea-grass.

The eggs float and are laid in September and October. By the following spring the young have grown to about 5 cm and are found in swarms near the coast. It feeds upon algae and also upon the felt of small animals and diatoms that grow on rocks and on algal fronds.

Annular Bream
Diplodus annularis (L.)

Mediterranean; eastern Atlantic south to Senegal.

Body oval and flattened; *profile* slightly convex; *jaws* equal in length; *teeth* prominent and chisel-shaped in the front of each jaw with robust rounded grinding teeth behind; *fins* D XI, 12–13; A III, 10–11.

Greyish- or brownish-silver with a dark spot on the tail stalk and another at the base of the pectoral fin. There may be 4 or 5 darker cross-bands on the flanks. The pelvic fins are yellow.

Up to 10 cm, rarely 20 cm.

One of the commonest of the Mediterranean bream, it normally shoals around rocky coasts especially near deep water.

White Bream
Diplodus sargus (L.)

Mediterranean; eastern Atlantic south to Angola and north to Biscay.

Body oval; *profile* gently convex; *mouth* rather small; *teeth* 8 chisel-like teeth in the front of each jaw with rounded grinding teeth behind; *fins* D XI-XII, 12–15; A III, 13–14.

Silver-grey with 7–8 dark vertical bands and a dark rim on the tail. The pelvic fin is grey. There is a dark spot on the tail stalk.

Up to 45 cm, usually 15 cm.

This bream is common near rocky coasts especially where fallen rocks form a slope containing many holes suitable for refuge. Found all the year round but is especially common in spring and summer. It is usually found between about 2 and 20 m. The White Bream often forms shoals although the larger individuals are usually solitary. In early summer it may enter brackish water, but returns to the sea in the autumn.

The eggs which float are laid in April or June and the young fish live a pelagic life for the remainder of the summer.

It feeds on crustacea, molluscs and echinoderms, which it crunches up with its strong grinding teeth.

Two-banded Bream
Diplodus vulgaris (Geoffroy)

Mediterranean; eastern Atlantic south to Senegal.

Body oval; *profile* slightly concave above eye; *mouth* very small; *teeth* 8 chisel-shaped forward-sloping teeth in the front of each jaw, followed by 2 rows of grinding teeth; *fins* D XI-XII, 14–15; A III, 14.

Two-banded Bream, *Diplodus vulgaris* (B. Picton).

There are 2 conspicuous black saddle marks, one between the dorsal fin and the eye, the other on the tail stalk. Generally golden-brown or grey above and silvery below. There are 15 to 16 longitudinal golden lines on each flank.

Up to 15 cm, rarely 40 cm.

A very common fish especially near algal covered rocks where there is sand close by. It is found from about 2 to 20 m.

The floating eggs are laid in October and the young live a pelagic life until they are 1 or 2 cm long when they move close in-shore. By late autumn they have grown to about 3 cm, but do not aquire the black saddles so characteristic of the adult until they are 4 cm long.

They are chiefly carnivorous feeding on small worms, crustacea, etc.

Zebra Sea Bream
Diplodus cervinus (Lowe)

Eastern Atlantic north to Spain, south to South Africa; Rare in Western Mediterranean.

Body oval; *profile* slightly concave; *mouth* small with fleshy lips; *teeth* in the upper jaw there are 10–12 chisel-like teeth followed by 2 rows of grinding teeth, in the lower jaw there are 9 chisel-like teeth followed by 3 rows of grinding teeth on each side; *scales* large, there being only 56–59 along the lateral line; *fins* D XI-XII, 12–14; A III, 11–12.

There are 4 or 5 very strongly marked,

dark brown cross-bars extending across the whole flank. The middle cross-bars may be divided into 2 on the lower part of the flank.

Up to 50 cm, perhaps even more.

It is found near the coast over rocky ground from shallow water to at least 300 m.

Bogue

Boops boops L.

Mediterranean; eastern Atlantic south to the Canary Islands, north to Biscay; occasionally in the Irish Sea and North Sea.

Body long; *profile* smoothly convex; *mouth* small, terminal and oblique; *eye* very large; *teeth* small, a single row in each jaw. The upper teeth are notched with 4 sharp points, the lower teeth have 5 points of which the centre one is the largest; *fins* D XIV, 15–16; A III, 13–14.

Blue-green or blue-grey with a yellow tinge. There is a small dark spot at the base of the pectoral fin. The curved lateral line is dark, and when seen in life, there are 2–3 straight dark lines beneath.

Up to 25 cm, sometimes up to 30 cm.

A common species which lives in shoals often in mid-water. They frequently approach rocky coasts or sea-grass (*Posidonia*) meadows where they may be as shallow as 2 m. In foul weather, however, they may be found as deep as 150 m.

Spawning in summer, the eggs float and the young feed chiefly on planktonic crustacea.

The young have a characteristic salmon tint. At about 10 cm they come close inshore. An important food fish in the Mediterranean, the shoals are often attracted to very bright lights carried in small boats by fishermen who then surround them in nets.

Saupe

Sarpa salpa (L.)
= *Boops salpa* (L.)

Mediterranean; eastern Atlantic from Biscay to South Africa.

Body oval and symmetrical; *profile* smoothly convex; *mouth* terminal and small; *teeth* a single row of cutting teeth in each jaw. The teeth in the upper jaw are notched, those in the lower jaw are triangular and serrated; *fins* D XI-XII, 14–16; A III, 13–15.

Greyish-silver with 10–12 longitudinal golden stripes and a yellow eye. There is a black spot just above the base of the

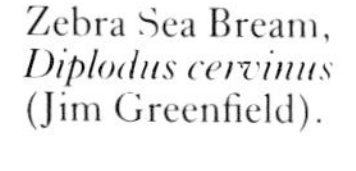

Zebra Sea Bream, *Diplodus cervinus* (Jim Greenfield).

Saupe, *Sarpa salpa* (Andy Purcell).

Saddled Bream, *Oblada melanura* (Andy Purcell).

pectoral fin.

Up to 30 cm, sometimes up to 45 cm.

The Saupe is very common in the Mediterranean. Shoals of this species are instantly recognizable by their tight packing and disciplined behaviour. They feed off the algae and associated diatoms that cover rocks and weeds. A shallow-living bream, it is never found deeper than 15 m and the shoals of young may feed in water less than 1 m deep, where they are dragged back and forth by wave action. Schools of Saupe are often accompanied by other species such as the Amberjack (*Seriola dumerili*) and the Bogue (*Boops boops*).

Spawning is in early spring and late autumn. Fishes of 10–15 cm are about 1 year old.

Saddled Bream
Oblada melanura (L.)

Mediterranean; eastern Atlantic from Portugal to Angola.

Body oval; *profile* gently convex; *jaws* equal in length; *teeth* there are several rows of teeth in each jaw. In the outermost row there are incisor-like teeth at the front and sharp conical teeth at the sides. At the front of each jaw are 4 rows of small granular teeth; *fins* D XI, 14; A III, 13–14.

A silvery fish with faint longitudinal dark stippled stripes. On the tail stalk is a characteristic black spot outlined in white along its front and rear margin.

Up to 20 cm, sometimes up to 30 cm.

Although this is a strictly coastal species living near rocks, it is usually seen swimming in small shoals 2–3 m below the surface in much more open water than most of the other bream. Breeding in late spring, the larvae remain in the plankton until late summer. Diet is varied, consisting of algae and all manner of small bottom-living animals.

Black Bream: Old Wife
Spondyliosoma cantharus (L.)
= *Cantharus lineatus* Günther

Mediterranean; eastern Atlantic north to Scotland and south to Senegal.

Body elliptical and rather deep. The mature males have a humped shoulder and concave forehead. The young, 10–15 cm in length have a convex forehead and a sharp snout; *mouth* terminal and small set rather low, not reaching back to the level of the eye; *teeth* there is a single row of slightly curved small pointed teeth in each jaw. The teeth in the front of the jaws are larger than the rest; *fins* D XI, 12–13; A III, 9–11.

The coloration of this fish is very variable, the young up to about 20 cm have a grey or yellowish back with silvery flanks bearing numerous longitudinal light stippled stripes. The fins are sometimes spotted with white and there is a dark rim on the tail fin. The adults are dark grey or blue-grey sometimes very dark. There may be 6–9 lighter vertical cross bars on the sides. The spawning males have an iridescent blue-grey band between the eyes. On the nest the male is almost black.

Up to 30 cm, sometimes up to 50 cm.

In the Mediterranean it is generally found over sand and sea-grass (*Posidonia*) meadows at depths greater than 15 m. In northern Europe it is commonly fished from around wrecks and rock areas and it is frequently seen by divers. This is one of the few bream that lays its eggs on the bottom, generally in early summer. The male follows the female and digs an oval pit in the sand with his tail and the female then deposits her slightly sticky eggs into it in a single layer. The young keep to the vicinity of the nest until they are about 7–8 cm long when they leave the coast for deeper water.

It is also seen searching for food amongst sea-grass. It is found throughout the year and in spring may enter brackish water.

The eggs float and are laid in April and June. The young, from 9–15 cm in length, may be found some kilometres from the shore amongst floating seaweed, etc., but when they reach a length of 3 cm they come close inshore.

Feeds on small worms, crustacea, etc.

Black Bream (top), *Spondyliosoma cantharus* (Christian Petron).

Jolthead Porgy (bottom), *Calamus bajonado* (Carl Roessler).

Jolthead Porgy
Calamus bajonado (Schneider)

Western Atlantic from Rhode Island south to Brazil, but mainly Caribbean; eastern Pacific.

Body with steep profile, rear nostril slit-like; *teeth* conical in front, molar-like behind; *scales* 50–57 along lateral line. D XII, 12; A III, 10.

Silvery or brassy with violet and blue tints, numerous chestnut blotches on body, blue line along lower rim of eye, and the corner of the mouth is orange.

45 cm.

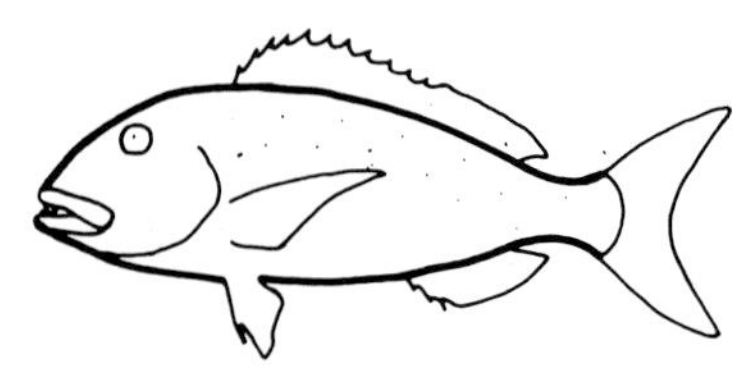

Dentex *Dentex dentex*

Powerful jaw;
large teeth;
blue spots on flanks.
Eastern Atlantic;
Mediterranean.
50 cm.

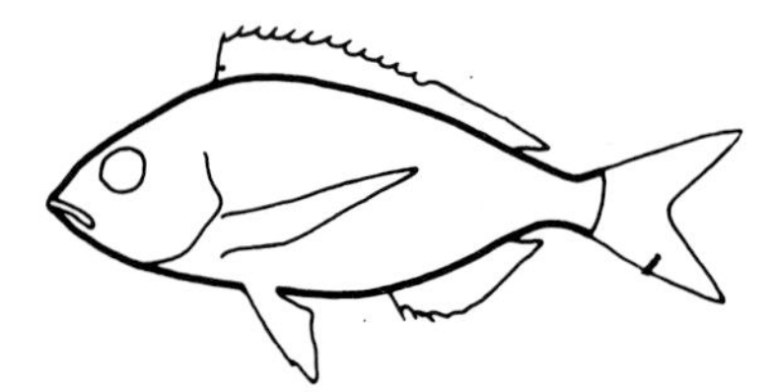

Pandora *Pagellus erythrinus*

Pointed snout;
Red edge to gill cover.
Eastern Atlantic;
Mediterranean.
25 cm.

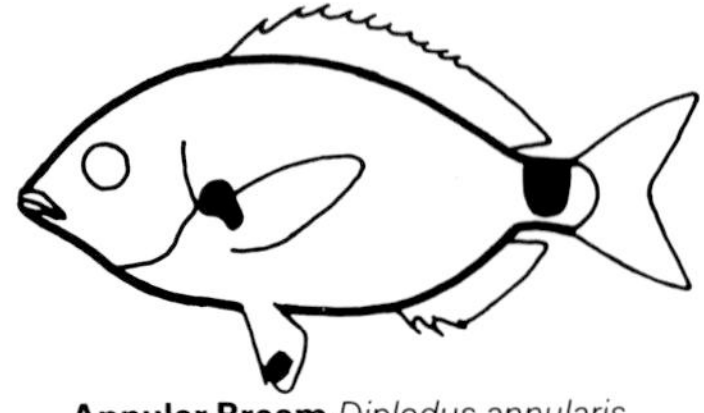

Annular Bream *Diplodus annularis*

Black spot on tail stalk;
pelvic fin yellow
Eastern Atlantic;
Mediterranean.
10 cm.

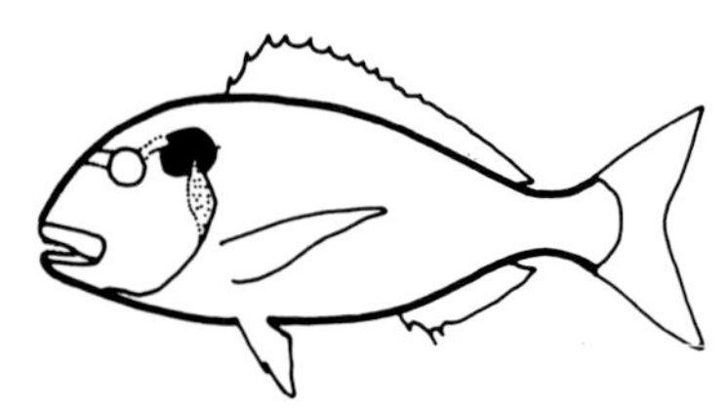

Gilthead *Sparus aurata*

Golden band between eyes;
red spot on gill cover.
Eastern Atlantic; Mediterranean.
35 cm.

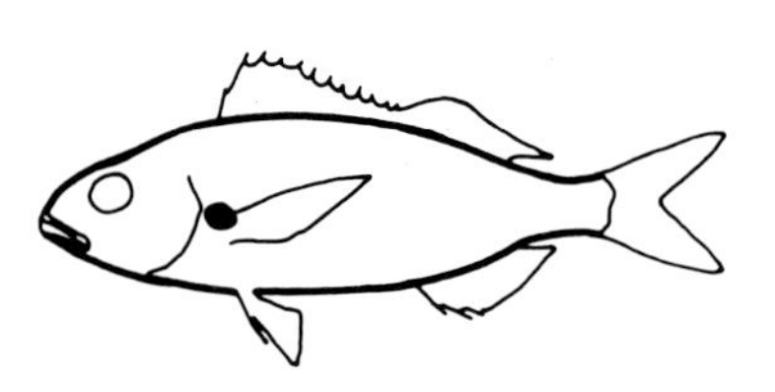

Spanish Bream; Axilary Bream
Pagellus acarne

Black spot at base of pectoral fin;
depression in dorsal fin.
Eastern Atlantic;
Mediterranean.
25 cm.

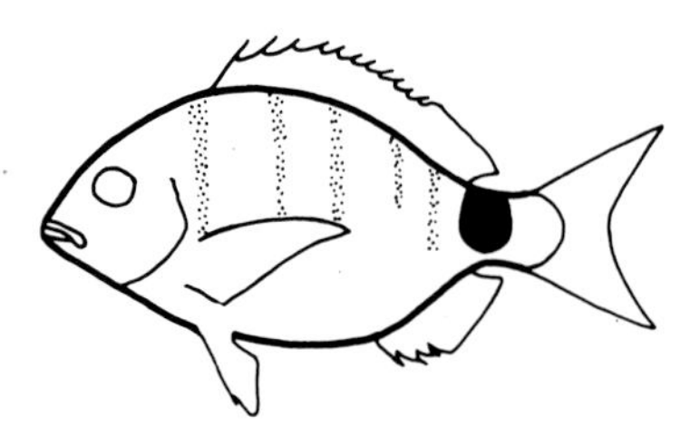

White Bream *Diplodus sargus*

Black spot on tail stalk;
vertical bands on body;
pelvic fin not yellow.
Eastern Atlantic.
15 cm.

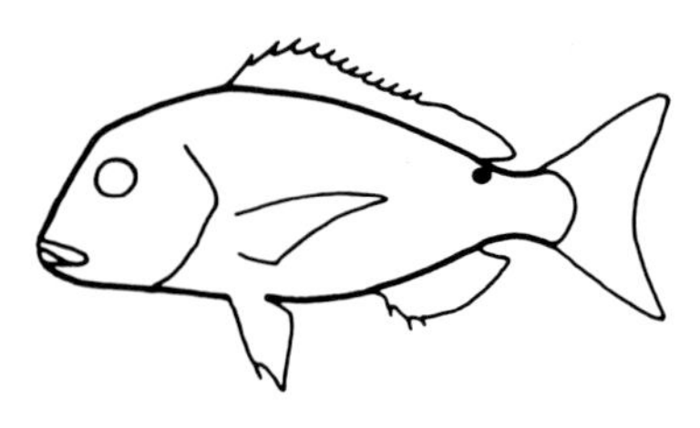

Common Sea Bream; Red Porgy
Pagrus pagrus

Steep forehead; face dark;
brown spot at base of dorsal fin.
Eastern Atlantic; western
Atlantic; Mediterranean.
50 cm.

Marmora *Lithognathus mormyrus*

About 12 dark cross bands on body;
long body.
Eastern Atlantic;
Mediterranean.
25 cm.

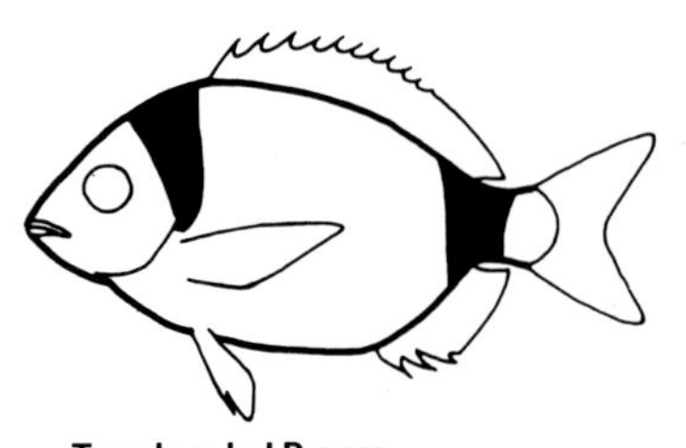

Two-banded Bream
Diplodus vulgaris

Two black saddle-like
blotches on body.
Eastern Atlantic;
Mediterranean.
25 cm.

Red Sea Bream *Pagellus bogaraveo*

Large eye;
often with black spot at
beginning of lateral line.
Eastern Atlantic; Mediterranean.
35 cm.

Sheepshead Bream
Puntazzo puntazzo

Elongate snout;
dark cross bands on body;
black spot on tail stalk.
Mediterranean.
30 cm.

Zebra Sea Bream
Diplodus cervinus

4–5 conspicuous black
cross bands on body;
lips fleshy.
Eastern Atlantic;
Mediterranean.
35 cm.

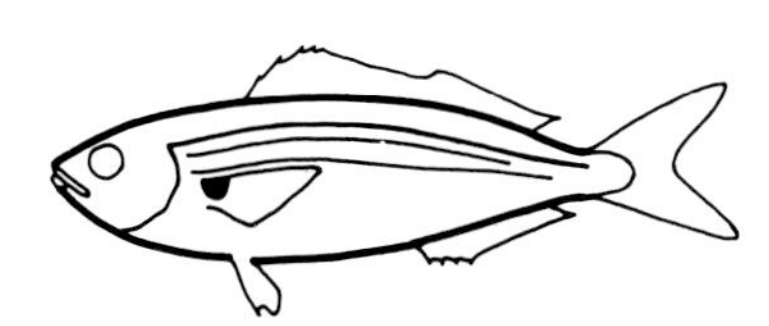

Bogue *Boops boops*

Long body;
2 dark parallel stripes along body;
dark lateral line.
Eastern Atlantic;
Mediterranean.
25 cm.

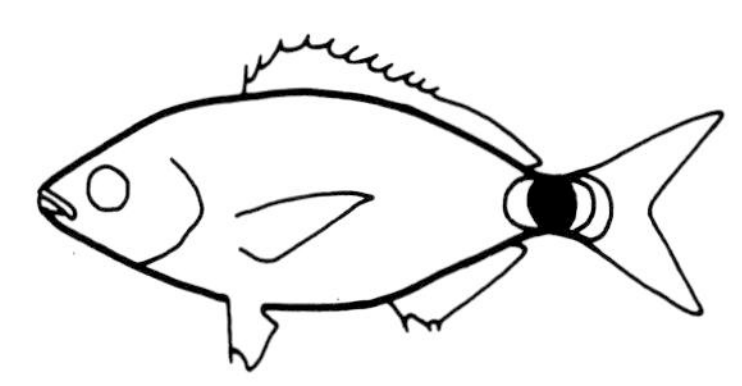

Saupe *Sarpa salpa*

Always in schools;
10–12 golden stripes along body.
Eastern Atlantic;
Mediterranean.
30 cm.

Saddled Bream *Oblada melanura*

Black spot on tail stalk ringed with white.
Eastern Atlantic;
Mediterranean.
20 cm.

Black Bream; Old Wife
Spondyliosoma cantharus

Blue-black rim on tail;
adult males with humped shoulder.
Eastern Atlantic;
Mediterranean.
30 cm.

Jolthead Porgy *Calamus bajonado*

Steep profile;
chestnut brown blotches.
Western Atlantic.
45 cm.

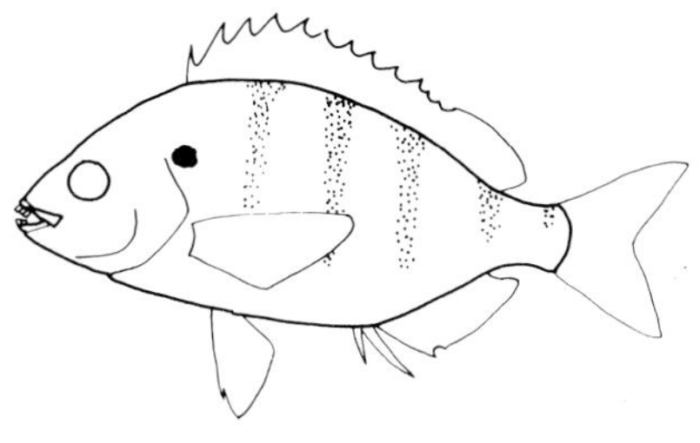

Pinfish *Lagodon rhomboides*

Spot on beginning of lateral line;
bluish silver with 4–6 darker bars.
Western Atlantic.
20 cm.

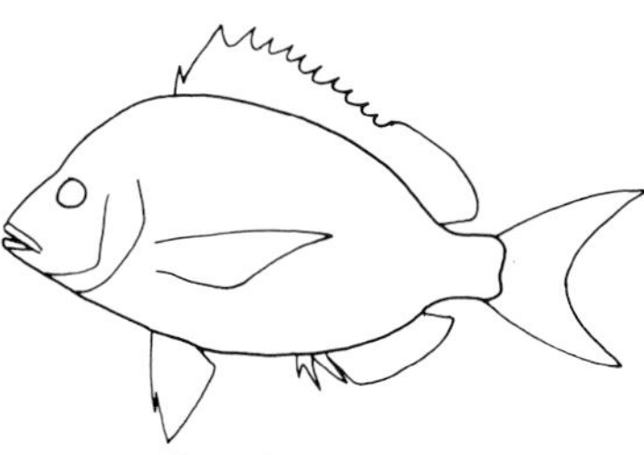

Scup *Stenotomus chrysops*

Silvery, not strongly marked;
body has oval outline;
tail forked with pointed tips.
Western Atlantic.
45 cm.

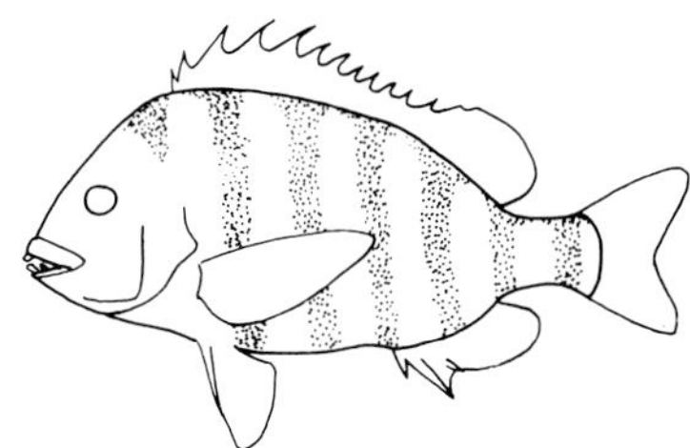

Sheepshead
Archosargus probatocephalus

Silver with 5–6 vertical bars;
dark bar across nape;
tail slightly forked with round tips.
Western Atlantic.
90 cm.

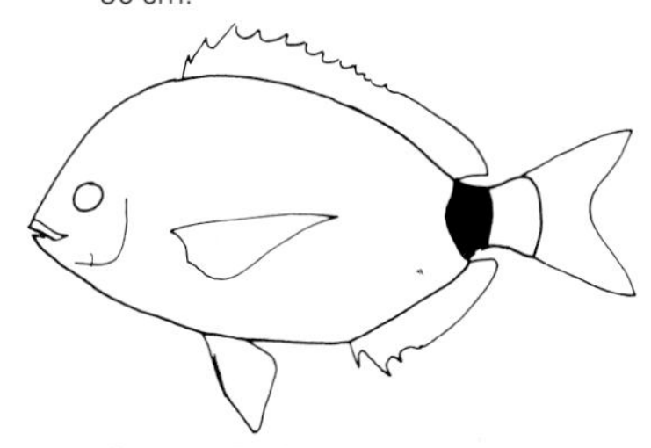

Spottail Pinfish *Diplodus holbrooki*

Steep profile;
crossbar on tail stalk.
Western Atlantic.
45 cm.

Pinfish
Lagodon rhomboides (L.)

Western Atlantic from Cape Cod south to Yucatan Peninsula.

Body front profile not steep, line between pectoral and anal fins smoothly curved; *teeth* notched incisor-like teeth in front of jaw; *fins* D XII, 11; A III, 11.

Bluish silver with 4–6 darker cross-bars and many longitudinal yellow stripes, dark spot on lateral line just behind the gill cover.

20 cm.

Frequents shallow, marine, grassy flats, but may enter brackish and even fresh water, where it grazes small invertebrate animals. Larger adults are found in water that is too deep to permit plant growth. Abundant and an important source of food for larger fish.

Scup
Stenotomus chrysops

Western Atlantic, Nova Scotia south to Florida.

Body oval in outline; *head* profile moderately steep and slightly concave, no spines on front gill cover; *jaws* equal in length; *teeth* small, conical, molars in 2 rows above; *scales* large; *fins* continuous dorsal fin, spiny portion highest, tail fin forked with pointed tips. D XII, 12; A III, 11–12.

Silvery body with 12–15 indistinct stripes, the central one being the most marked.

45 cm.

Common, especially in the northern part of its range, and is often taken by anglers. Usually occurs in schools inshore in summer, but moves into deeper water in winter.

Sheepshead
Archosargus probatocephalus

Western Atlantic, Nova Scotia south to Brazil; absent from West Indies and Bahamas.

Body deep; *head* front profile steep, not concave; *mouth* small, slit rear nostril; *teeth* well developed, those in front conical or flat incisor-like, teeth at sides of jaw rounded; *fins* tail fin moderately forked but with rounded tips. D XI-XII, 11–13; A III, 10–11.

Sheepshead, *Archosargus probatocephalus* (Phil Lobel).

Variable reddish, yellowish or bluish-silver with 5–6 dark vertical bars, which may be interrupted into blotches; there is a dark bar on the nape of the head.

90 cm.

In bays and estuaries, it is often seen around pilings. Enters brackish water.

Spottail Pinfish
Diplodus holbrooki (Bean)

Western Atlantic, Chesapeake Bay south to Florida; north-west Gulf of Mexico.

Body regularly elliptical, with steep front profile, rear nostril a slit; *scales* moderately large, about 56 in the lateral line; *teeth* incisor-like front teeth, molars in 3 series above and 2 below; *fins* tail fin deeply forked. D XII, 14–15; A III, 13.

Body bluish-brown above, silvery below; 8 faint bars on the body alternating long and short, edge of gill cover blackish; black spot on the upper part of pectoral base; pelvic and anal fins dusky brown.

45 cm.

Inshore on sea-grass beds.

Family:
CENTRACANTHIDAE

Closely resemble the Sea Bream (Sparidae) but the teeth are minute and the mouth is protrusible. There is usually a rectangular black spot on the centre of the flank.

Picarel
Spicara smaris (L.)
= *Maena smaris* (L.)

Mediterranean; eastern Atlantic from the Canary Islands to Biscay.

Body rather long; the length from the snout to the rear edge of the gill cover more or less equals the maximum height of the body; *profile* straight or slightly concave above eye, the snout is pointed; *scales* 80–94 along the lateral line; *jaws* the upper jaw slightly longer than the lower; *fins* the dorsal fin is of equal height throughout its length but varies in proportion with sex and age. D XI, 11–12; A III, 8–10.

Colour varies with age and season. Generally bluish or brownish-grey above, silvery below. Longitudinal lines of blue spots run along the flanks. There is a rectangular spot in the middle of the flanks. The fins are yellowish or grey. The male has a single row of blue spots along centre of anal fin and in a band across tail.

Females up to 15 cm, males up to 20 cm.

A gregarious fish living in 15–100 m, usually over sea-grass beds, but not close to the bottom. In late winter they gather in huge shoals to breed.

Picarel
Spicara flexuosa Rafinesque
= *Maena chryselis* Valenciennes

Mediterranean.

Body oval, length between the snout and the rear of the gill cover equals the maximum height of the body; *profile* almost straight above the eyes; *scales* there are 70–80 scales along the lateral line; *fins* maximum height of spiny dorsal rays greater than maximum height of the soft dorsal rays. There is a distinct dip in the middle of the dorsal fin. D XI, 11–12; A III, 9–10.

Colour varies with age, sex and season. Generally bluish-grey above and silver below with a pronounced rectangular spot in the centre of the flank. There is a brownish spot at the base of the pectoral fin. The fins are generally greyish or orange-yellow.

Females up to 18 cm, males up to 21 cm.

A non-migratory gregarious fish that is usually found from 70–130 m. They appear to prefer muddy bottoms. Spawn in spring.

Picarel, *Spicara smaris* (Christian Petron).

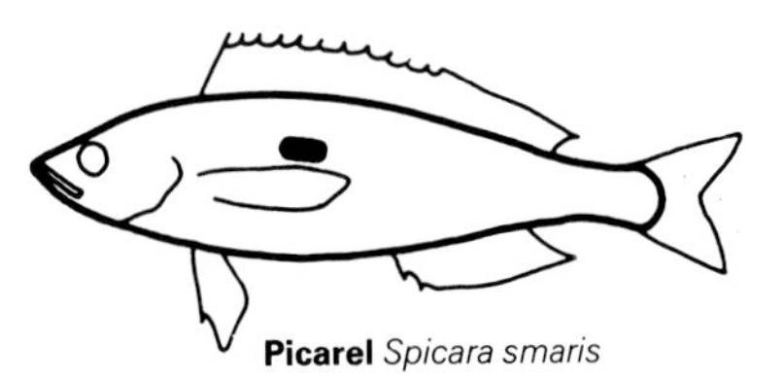

Picarel *Spicara smaris*

Rectangular spot on body;
long body;
sharp snout;
dorsal fin of uniform height.
Mediterranean, eastern Atlantic.
20 cm.

Picarel *Spicara flexuosa*

Rectangular spot on body;
dorsal fin in 2 lobes.
Mediterranean
21 cm.

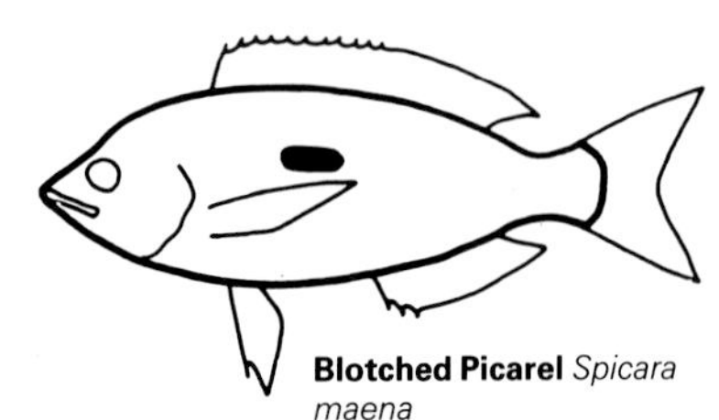

Blotched Picarel *Spicara maena*

Rectangular spot on body;
distinct shoulder;
deep body;
dorsal fin of uniform height.
Mediterranean, eastern Atlantic.
24 cm.

Blotched Picarel

Spicara maena (L.)
= *Maena maena* (L.)

Mediterranean except Black Sea; eastern Atlantic south to Canary Isles and north to Portugal.

Body rather deep, the length from snout to the rear edge of the gill cover less than greatest height of body. The proportions of the body varies with age and sex; *profile* concave above the head; *scales* 70–75 rows of scales along the lateral line; *fins* dorsal fin of uniform height throughout its length. D XI, 12; A III, 10.

Colour varies with age, sex and season. Always with a rectangular black patch on the flank, 2 scales high and 7–8 scales long in the young, 3 scales high and 6–7 scales long in the adults. Often blue-grey above, silvery below. The dorsal, anal and tail fins are brownish-blue with irregular blue spots. The pectoral fin is brownish-yellow. Adult males may have brilliant blue bands on the head.

Female up to 21 cm, male up to 24 cm.

A non-migratory fish which tends to swim in mid-water. Spawns in late summer at about 10–20 m depth. The male excavates a circular nest in the sand where the female deposits her sticky eggs. These are immediately fertilized by the male.

Family:
CHAETODONTIDAE

The **Angelfish** and **Butterflyfish** are an important tropical family especially around coral reefs. One species regularly comes far enough north to be included.

Gray Angelfish

Pomacanthus arcuatus (L.)

Western Atlantic from New England south to Brazil.

Body rounded, disc-shaped, compressed; *mouth* small; strong spine at base of front gill cover; *fins* soft dorsal fin very high in front. D VIII-IX, 30–32, A III, 24.

The adults are grey, but each scale has a black spot and the mouth and snout is white; juveniles banded black and yellow, median yellow stripe on forehead extends onto the chin.

35 cm.

Gray Angelfish, *Pomacanthus arcuatus* (Carl Roessler).

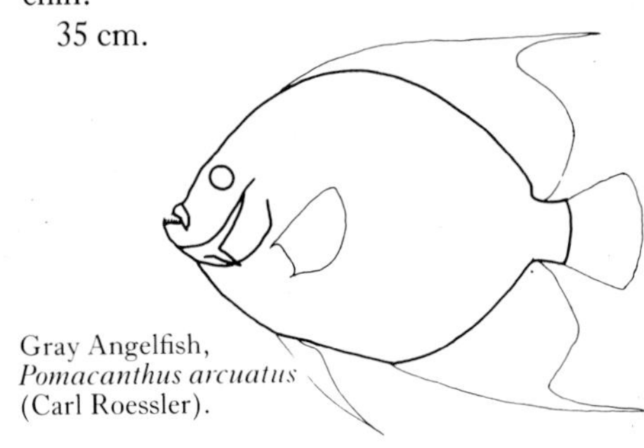

Gray Angelfish
Pomacanthus arcuatus

Disc-like shape; spine on front gill cover.
Western Atlantic.

Family:
EPHIPPIDAE

Resemble the Angelfish in body outline, but there is no spine on the front gill cover and the strongly spiny 1st dorsal fin is clearly differentiated from the 2nd dorsal fin.

Atlantic Spadefish
Chaetodipterus faber (Broussonet)

Western Atlantic from Cape Cod south to Brazil.

Body deep and round in profile, compressed, *mouth* small and front profile steep; *scales* small, covering the body and the bases of the soft dorsal and anal fins; *fins* 2 dorsal fins, the front with strong spiny rays ending next to the origin of the soft-rayed 2nd dorsal fin; the anal has 3 spiny rays, the middle one longest; soft dorsal and anal fins extended to acute points (falcate) in adults, the median fins are low in young.

Greyish, with at least 4 darker, broad vertical bands which fade in adults; lower fins black. The juveniles are black.

70 cm.

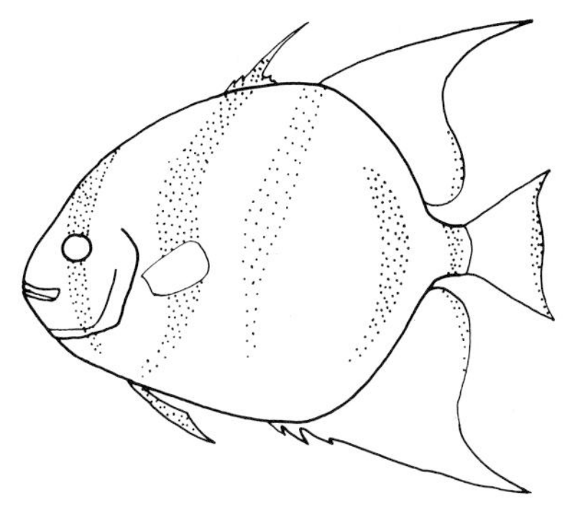

Atlantic Spadefish
Chaetodipterus faber

Disc-like shape; no spine on gill cover. Strong spiny rays in first dorsal fin.

A schooling fish around jetties and in open water off beaches, tending to move to offshore reefs as they grow older. The young, which are round and completely black, tend to float on the surface, imitating blackened floating leaves. More abundant on the Atlantic coasts of the southern United States.

Atlantic Spadefish, *Chaetodipterus faber* (Doug Perrine).

Family:
POMACENTRIDAE

There are many species of Pomacentrid fishes in tropical waters but there is only one representative in our area.

The body is small, rather deep and laterally flattened. The mouth is small. There is a single dorsal fin having a long section of spiny rays and a short section of soft rays. The soft rays, however, are taller than the spiny ones.

Damsel Fish
Chromis chromis (L.)

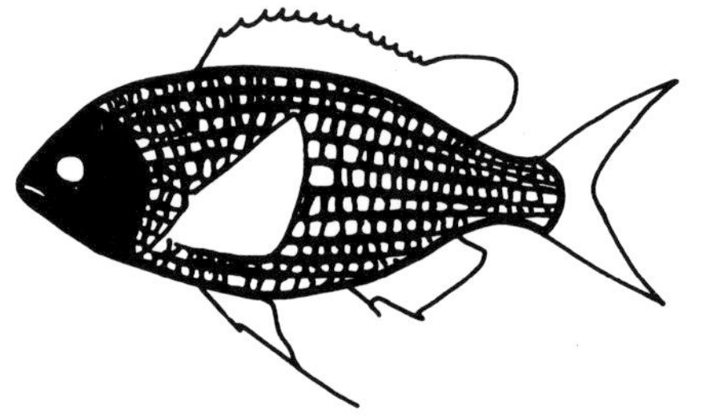

Damsel Fish *Chromis chromis*

Adults very dark colour; forked tail; young, brilliant iridescent blue. 15 cm.

Mediterranean; eastern Atlantic from Portugal to Guinea.

Body oval and laterally flattened; *mouth* small, terminal and oblique; *scales* very large, 24–30 along the lateral line; *fins* a single dorsal fin with two distinct lobes, the soft rays being the longer. D XIV, 10–11; A II, 10–12.

The adults are dark brown, the scales being outlined in darker brown. The young up to about 10 mm have brilliant iridescent blue stripes running along the head and flanks. The blue colour gradually disappears as the fish gets bigger and the adults have no blue colour.

Up to 15 cm.

One of the most common fishes in the Mediterranean. They are generally seen in huge almost stationary shoals near rocky coasts in mid-water. Do not usually go deeper than 25 m.

Spawning is in summer. Mixed shoals choose a spawning site on a rocky slope where there are flat depressions or large rocks; preferable in depths of 2–15 m. Occasionally they will choose a sandy area in deeper water. The male appears to attract the female by fanning his tail and making jumping movements. The females hover in a swarm nearby occasionally descending to the spawning places. Fertilization takes place after the eggs are laid. After three days the females leave the area but the males continue to guard the spawn.

The brilliant-blue young hover in depressions and sheltered crannies in the rocks and are a common sight in late summer.

Damsel Fish, *Chromis chromis*, juvenile (Christian Petron).

Damsel Fish, *Chromis chromis*, adult (Andy Purcell).

Family:

LABRIDAE

The **Wrasse** have a terminal mouth with fleshy lips. The teeth are well developed. The main gill cover is not spined although the front gill cover may be finely toothed. The scales are large and easily visible. The lateral line is complete. The single dorsal fin is distinctly divided into a hard and soft-rayed portion.

Cuckoo Wrasse
Labrus bimaculatus (L.)
= *Labrus mixtus*.

Eastern Atlantic from Senegal to northern Scotland; Mediterranean.

Body rather elongate; *mouth* reaches back almost to the eye and the lips are thick; *teeth* large and conical arranged in a single row, the largest being in front; *front gill cover* smooth; *lateral line* parallel to the back; *fins* dorsal of almost equal height throughout its length. D XVI-XIX, 11–14; A III, 9–12.

In the mature fish the sexes differ greatly in colour and pattern. The females and immature males are yellow or reddish-orange with 3 dark blotches on the back, 2 below the soft dorsal rays, the 3rd on the tail stalk. The dark markings are often spaced with lighter markings making the pattern very conspicuous. The males have brilliant blue heads with darker horizontal bands extending onto the flanks which are yellow or orange. There are no blotches on the back of the male. During courtship the males have a white patch on the forehead.

Up to 35 cm.

Common during the summer around rocks at depths greater than about 10 m. The 'nest' is a depression in the gravel bottom and the male drives off other males that come too close. The eggs are laid during the summer, the younger fish tend-

Cuckoo Wrasse, *Labrus bimaculatus*, male (B. Picton).

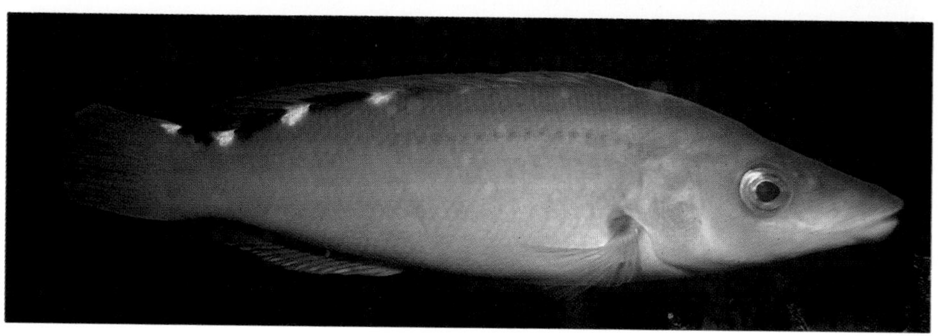

Cuckoo Wrasse, *Labrus bimaculatus*, female (Jim Greenfield).

ing to live in the shallower water. Their food appears to be chiefly bottom-living animals such as crabs and whelks. A long-lived fish; a specimen of 28 cm may be 17 years old. The older females may take on the patterns and colour of the adult male and it is possible that they achieve the full sexual function of the male as well.

Green Wrasse
Labrus viridis L.

Mediterranean.

Body long, the greatest depth of the body is less than the length between the snout and the hind corner of the gill cover; *lips* fleshy; *teeth* conical in a single row, 3 plates of pharyngeal teeth; *front gill cover* not toothed; *scales* 41–49 along lateral line; *fins* D XVII-XIX, 10–14; A III, 9–12.

The colour is very variable and has led to much confusion. Sometime completely green with emerald tints or yellow-green with a white longitudinal band running from eye to tail, or mottled wine-red or orange without any green but with white stripes and spots.

Up to 45 cm.

This is a coastal species living near rocks and sea-grass (*Posidonia*). The larger specimens live deeper than 15 m. Has the habit of lying on its side on the bottom and in this condition it is easy to approach. Most specimens shorter than 27 cm are female; above 38 cm they are chiefly male. Spawns in winter and spring. The food is made up of bottom-dwelling animals of all kinds.

Brown Wrasse
Labrus merula (L.)

Mediterranean; eastern Atlantic from Portugal to the Azores.

Body oval, the maximum height of body is greater than the distance from the snout to rear corner of the gill cover; *mouth* very small; *lips* fleshy; *teeth* are conical, arranged in a single series in each jaw, no pharyngeal teeth; *front gill cover* smooth; *scales* distinct but rather small, there being 40–48 along lateral line; *lateral line* runs parallel to the back; *fins* tail fin often has a distinct notch in the centre. D XVII-XIX, 11–14; A III, 8–12.

Less variable in colour than other members of the genus, usually olive or bluish-grey sometimes with a blue spot in the centre of each scale. Dorsal, anal and tail fins have a blue rim. The pectoral fin often has orange-brown rays. In the breeding season all the blue tints are accentuated.

Up to 45 cm.

A coastal species living in water of moderate depth near rocky reefs and sea-grass (*Posidonia*) meadows. They are not swift swimmers and do not stray far from some refuge amongst rocks or in clumps of sea-grass. The breeding season is in late

Brown Wrasse, *Labrus merula* (Andy Purcell).

winter or early spring, the male behaving very pugnaciously towards possible rivals. The eggs are generally laid on sea-grass fronds. They feed on echinoids, molluscs, crustacea and polychaetes.

Ballan Wrasse
Labrus bergylta Ascanius

North-east Atlantic from Canary Islands to North Scotland; rare in the western basin of the Mediterranean.

Body rather long but massive with a large head; *lips* thick and fleshy; *teeth* strong and conical of moderate size, a single row in each jaw; *front gill cover* smooth; *scales* rather large, 41–47 along lateral line; *lateral line* runs parallel to the back; *fins* D XVIII-XXI, 9–13; A III, 8–12.

The colour depends on age, maturity, 'emotional state', reproductive condition and habitat, and may be greenish-brown, sometimes reddish and occasionally conspicuously mottled with white spots. Each scale has a lighter centre with a darker hind rim which makes them appear very prominent. The fins tend to be the same as the body colour but with white spots; some specimens have a light lateral stripe. The young fish may be emerald green.

Up to 40 cm, rarely up to 60 cm.

Most often found near steep underwater rocky cliffs and reefs down to some 20 m. They rarely venture far from crevices in the rocks. The young are occasionally found in rock pools.

Judging by its stomach contents and teeth this wrasse feeds on encrusting molluscs etc., which it bites and sucks from the rocks. Spawning is in early summer. The young fishes are common inshore in early autumn. Sudden periods of cold may kill large numbers of Ballan Wrasse.

Axillary Wrasse
Symphodus mediterraneus (L.)

Mediterranean.

Mouth small; *jaws* upper slightly longer than lower; *lips* moderate; *teeth* two large conical teeth jutting forward in the front of each jaw, those of the lower jaw being somewhat smaller than the upper. On each side of the upper jaw there is a small conical tooth (sometimes absent), and in the sides of the lower jaw there are 3–4 teeth; *front gill cover* toothed; *scales* 30–35 along the lateral line; *lateral line* follows parallel to the back; *fins* D XV-XVIII, 9–11; A III, 8–12.

It has 2 conspicuous dark spots, one at the base of the pectoral, the other on the upper part of the tail stalk. In the male the pectoral spot is blue with yellow margin, in the female it is dark brown. The normal background colour of the male is orange-brown, in the female it is slightly less red. The urinogenital papilla in the female is prominent and black; in the male it is white. In the breeding season the male is a

Ballan Wrasse (top), *Labrus bergylta* (Jim Greenfield).

Ballan Wrasse (bottom), *Labrus bergylta* (B. Picton).

Axillary Wrasse, *Symphodus mediterraneus* (Andy Purcell).

violet-red with iridescent blue spots and stripes.

Up to 15 cm.

Lives around the deeper meadows of sea-grasses (*Posidonia* and *Zostera*). Little else is known about it save that it probably breeds in late spring and summer.

The young live near the coast but move off into deeper water in winter and as they get older.

Black-tailed Wrasse
Symphodus melanocercus (Risso)

Mediterranean.

Body oval; *eyes* large equalling in diameter the length of the snout; *mouth* small; *teeth* 6–14 in the upper jaw, 12–18 in the lower; *front gill cover* deeply toothed on rear edge; *scales* rather large; about 30–38 along the lateral line, three rows of scales on the cheek below the eye; *lateral line* follows the contour of the back; *fins* D XV-XVII, 6–10; A III, 8–11.

The young and females are pinkish-brown above, pinkish-yellow beneath. The males are dull blue above shading to yellowish beneath. The characteristic marking of this fish is the broad dark vertical band covering the rear ⅔ of the tail fin which is edged with white.

Up to 14 cm.

Over sea-grass (*Posidonia*) not deeper than 10 m. They have been observed to clean other fishes by removing parasites, fungal growths, etc. Breed in spring and summer.

Corkwing Wrasse
Symphodus melops (L.)

Mediterranean; eastern Atlantic south to Azores, north to Faroes; English Channel.

Body rather deep; *front gill cover* toothed; *scales* large (31–37 along lateral line); 5–6 rows of scales on the cheeks below the eye; *teeth* in each jaw pointed, slightly curved and arranged in a single row; *fins* D XIV-XVII, 8–11; A III, 8–11.

The coloration varies with sex, maturity and season. There is always a dark spot on

Corkwing Wrasse, *Symphodus melops* (Andy Purcell).

Ocellated Wrasse, *Symphodus ocellatus* (Evan Jones).

the tail stalk either on, or slightly below the lateral line, but this spot may be partly obscured in the darker males. The sexes differ in colour. The females are a dull mottled brown. The males are often a dark red-brown with iridescent blue bars on the gill covers.

Up to 6 cm, occasionally up to 25 cm.

The Corkwing is the most commonly seen Wrasse of northern European waters. It is frequent in shore pools but is more often found amongst weed below the low-tide mark. In the winter it moves into slightly deeper water. The food is chiefly small crustacea and molluscs. The nest is built out of algae and spawning is in the spring and summer. Fishes of 5–8 cm are about one year old. The three-year-olds are about 15–17 cm and are mature.

Ocellated Wrasse
Symphodus ocellatus (Forsskål)

Mediterranean.

Body oval; *mouth* small with prominent lips and the upper jaw slightly longer than the lower; *lateral line* follows the contour of the back to the level of the last dorsal ray where there is a sharp downward step and from thence the lateral line is straight to the tail fin; *scales* rather large there being 30–34 along the lateral line; there are 3 rows of scales on the cheek below the eye; *front gill cover* finely toothed; *fins* D XIII-XV, 8–11; A III, 8–11.

There is always a spot on the gill cover and another dark spot at the base of the tail stalk. The colours are exceedingly variable; the ground colour is usually greenish or brownish. In the adults the spot on the gill cover is rimmed with red and (probably in the male) may be an iridescent blue.

Up to 13 cm.

Lives at moderate depths near rocks and sand. Probably breeds in late spring and early summer, building a nest of seaweed. The eggs are adhesive. These fish have been observed to clean parasites from the bodies of other fishes.

Painted Wrasse
Symphodus tinca (L.)

Mediterranean.

Body oval and flattened; *eye* small; snout long, 3–4 times diameter of eye; *profile* long, concave above the eye with a slight convex hump on the snout; *lips* very fleshy and protruberant; *teeth* small and almost hidden by soft tissue, there are 8–24 in the upper jaw and 10 or more in the lower; *front gill cover* finely toothed; *scales* large with 33–38 scales along the lateral line and 5 series of scales on cheek below the eye;

fins soft rays of the dorsal and anal fins are much longer than the spiny rays. D XIV-XVI, 10–12; A III, 8–12.

Very variable in colour. Usually the ground colour is yellow-olive with red and blue stippling arranged in longitudinal bands on the flanks. There is a dark spot at the base of the tail fin and another at the base of the pectoral. A dark band runs between the eyes across the snout just above the level of the upper lip.

Males up to 33 cm, females up to 25 cm.

This species is known to be active only during the day, but most wrasse behave in this way. Lives near rocks and sea-grass (*Posidonia*) meadows. Spawn in summer; the male makes a nest out of algae, the female attaches her eggs to the fronds and the male fertilizes them. They feed on small bottom-dwelling animals.

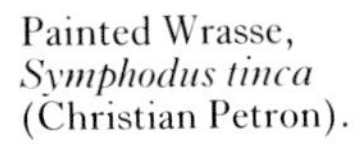

Painted Wrasse, *Symphodus tinca* (Christian Petron).

Five-spotted Wrasse
Symphodus roissali (Risso)

Mediterranean.

Body deep, its greatest depth 3–3½ times total length; *teeth* conical, 10–12 in the upper jaw, 10–16 in lower; *front gill cover* finely toothed; *scales* very large, 30–33 along lateral line, 4 series of scales beneath the eye; *lateral line* follows contour of back until the last few rays of the dorsal fin when there is an abrupt downward step and thereafter is straight to the tail; *fins* D XIV-XV, 9; A III, 8–9.

Colour very variable; generally whitish-brown with large darker mottling, or it is distinctly dark and light spotted. The male has red lips and the body is flecked with greenish-blue. There is often a black spot at the base of the tail stalk on the lateral line. During spawning the female genital papilla is bright blue.

Up to 15 cm.

Lives on rocky bottoms sometimes in water less than 1 m deep even where there

Five-spotted Wrasse, *Symphodus roissali* (Andy Purcell).

is considerable surge. Breeds in early summer. The nest is built by the male, who will not tolerate the presence of the female at this time. The nest is a hemispherical mound with a cavity near the bottom. When the nest is built the female enters the cavity, lays her eggs and the male fertilizes them. He then chases off the female. The exterior of the nest is covered with the flat fronds of the sea-lettuce (*Ulva*) and after the eggs are laid the whole structure is covered with gravel. The male guards the eggs.

Symphodus doderleini Jordan

Mediterranean.

Body long, its length being about 4 times its maximum height; *mouth* considerably longer than eye diameter; *teeth* small and conical, about 12 in the upper jaw, 14 in the lower; *front gill cover* finely and shallowly toothed; *scales* large, 30–36 along the lateral line and 3 rows of smaller scales beneath the eye; *lateral line* follows contour of back along it entire length; *fins* D XIII-XVI, 9–11; A III, 8–10.

Reddish or chestnut brown, sometimes with white spots. There is a longitudinal dark band running from the snout to the tail stalk with a lighter band beneath. There is a dark spot on the tail stalk above the lateral line.

Up to 10 cm.

Rare, lives at about 3–12 m near rocks and sea-grass (*Posidonia*). Breeds in spring.

Grey Wrasse
Symphodus cinereus (Bonnaterre)

Mediterranean.

Body oval, head as long as the maximum depth of body; *mouth* small; *teeth* 20–28 in the upper jaw of which the front ones slant forwards, there are 28 teeth in the lower jaw; *front gill cover* finely toothed; *scales* rather large, 31–35 along the lateral line; 2 rows of large scales beneath the eyes; *lateral line* smoothly curved but the hind end lies higher on the flank than the front end; *fins* D XII-XV, 9–11; A III, 8–10.

Usually grey with white spots arranged in a longitudinal band; often mottled brown or with longitudinal brown bands. There is a dark spot set low on the tail stalk below the lateral line. The mature males have golden tints on the flanks and throat and the cheeks have gold and blue stripes. The urinogenital papilla is colourless in the male but bright blue in the female.

Up to 8 cm, sometimes up to 16 cm.

Common on sandy and muddy bottoms and seems to prefer estuaries and enclosed bays where there is plenty of vegetation and detritus. Spawns in early summer, the nest being similar to that of *Crenilabrus quinquemaculatus* but is more firmly glued together. Feeds on crustacea and molluscs.

Symphodus doderleini (Andrea Ghisotti).

Long-snouted Wrasse
Symphodus rostratus (Günther)

Mediterranean.

Body rather elongate; *snout* very long giving a concave profile; *mouth* small; *teeth* a single file of small sharp teeth in each jaw; *front gill cover* finely toothed on its hind edge; *scales* rather large, 30–35 along the lateral line; 3–4 rows of scales beneath the eye; *lateral line* follows the contour of the back to the last dorsal ray when there is an abrupt downward step and thereafter it is straight to the tail; *fins* D XIV-XVI, 9–12; A III, 9–11.

Colour variable; reddish-brown, mottled brown and black or green. The sexes are not greatly different in colour.

Up to 12 cm.

Lives in water deeper than 15 m near

sea-grass (*Posidonia*). The eggs are laid on the bottom in early summer.

Rock Cock
Centrolabrus exoletus (L.)

North-east Atlantic from Biscay to north Scotland and Norway. It is not present in the southern North Sea.

Mouth small reaching half-way back to eye; *teeth* conical in a single row in each jaw; *front gill cover* toothed on rear and lower edge; *lateral line* 33–37 scales along lateral lines; *fins* usually 5 anal spines. D XVIII-XX, 5–7; A IV-VI, 6–8.

Usually a warm red-brown above, yellowish-silver on the flanks and silver below. Males are flecked with iridescent blue and have two iridescent blue bands passing from the eye to the angle of the mouth. On the tail there is a broad dark band; the rim of the tail is white.

Rarely up to 15 cm.

Usually found amongst kelp-covered rocks beneath low-tide mark. Probably spawns in summer.

Goldsinny
Ctenolabrus rupestris Cuv. et V.
= *Ctenolabrus suillus* (L.)

Mediterranean; eastern Atlantic north to Norway; English Channel.

Body rather long; *teeth* several rows in each jaw, the front ones are curved and inclined slightly forward; *front gill cover* very finely

Long-snouted Wrasse (top), *Symphodus rostratus* (Andy Purcell).

Rock Cock (middle), *Centrolabrus exoletus* (B. Picton).

Goldsinny, *Ctenolabrus rupestris* (B. Picton).

toothed but smooth on the lower edge; *scales* 34–40 along lateral line; *lateral line* parallel to back; *fins* the dorsal fin is of uniform height throughout its length. D XVI-XVIII, 8–10; A III, 7–10.

The adults are a red-brown colour; the young may be a fawn to dull green. There is always a dark spot on the top edge of the tail stalk.

Up to 18 cm.

The Goldsinny does not come into very shallow water, being found chiefly amongst rocks. They are said to move into deeper water in winter but it is more likely that they are just more difficult to find.

They spawn in summer and feed on bottom-living crustacea, molluscs, etc.

Rainbow Wrasse
Coris julis (L.)

Mediterranean and eastern Atlantic from Guinea to Biscay.

Body long and sinuous; *snout* long and pointed; *teeth* sharp and pointed, inclined foward in the front of the jaws; *eye* small; *scales* very small (73–80 along the lateral line); no scales beneath the eye; *fins* first 2–3 spiny dorsal rays are elongate in the male. Dorsals and anals are long and of uniform height throughout their length. D VIII-IX, 12; A III, 11–12.

There are two distinct colour forms; the smaller female phase has a blue spot on the lower edge of the gill cover. Specimens caught from water less than about 20 m have a dark olive-brown ground colour, those from deeper water are red or red-brown. There is a longitudinal yellow stripe running from nose to tail fin. To the diver the coloration appears the same irrespective of depth as the increased red-dening of the fish compensates for the loss of red light penetrating the water. The male phase has a black and orange spot on the elongated dorsal rays. There is a zig-zag orange mark along the flanks and a lozenge-shaped black mark behind the pectoral fin.

Intermediate colour forms are common and the female form somtimes turns out to be the male and vice versa.

Males up to 25 cm, females up to 18 cm.

Very abundant in the Mediterranean,

Rainbow Wrasse, *Coris julis*, male (Andy Purcell).

Rainbow Wrasse, *Coris julis*, female or young (B. Picton).

they live around weed-covered rocks and over sea-grass (*Posidonia*) from near the surface to at least 120 m. In winter they move off into deeper water.

Like most (perhaps all) wrasse they are active only during the daytime. The young of this species has been observed to clean parasites from other fishes.

This is an hermaphrodite species and females often become males. They breed in early summer and the eggs float.

They feed on small animals.

Turkish Wrasse; Rainbow Wrasse; Peacock Wrasse
Thalassoma pavo L.

Mediterranean; North-east Atlantic from Biscay to Guinea.

Body somewhat elongate; *mouth* small; *teeth* 1 row of small sharp teeth in each jaw; *scales* large, 26–31 along the lateral line; *lateral line* follows the contour of the back until the 10th–11th soft dorsal ray when there is an abrupt downward step and from thence to the tail fin it is straight; *fins* D VIII, 12–13; A III, 10–12.

Two main colour forms. The most common has a black spot on the back beneath the centre of the dorsal fin; it is generally moss-green in colour with lighter vertical bands. A less common form (probably the male) has a diagonal red and blue band running from the front of the dorsal fin to the pelvic. The flanks are green and the head is dark and red with blue veins. An uncommon intermediate form has neither a dark dorsal spot nor diagonal blue and red lines.

Up to 20 cm.

Often very common living around weed-covered rocks to about 20 m.

Spawns in summer and the eggs float. The sex of an individual fish may change in this species; the change from female to male may be induced by warmer weather.

Cleaver Wrasse
Xyrichthys novacula (L.)

Mediterranean; eastern Atlantic from Biscay to Guinea.

Body deep and laterally flattened; *profile* almost vertical forehead; *eye* is small, set high in the head; *mouth* set very low; *teeth* in the front of the jaw are very sharp and are slanted forward; there are grinding teeth in the back of the jaw; *scales* large,

Turkish Wrasse, *Thalassoma pavo*, adult male (Andy Purcell).

Turkish Wrasse, *Thalassoma pavo*, female or young (Jim Greenfield).

24–29 along the lateral line; *lateral line* follows the contour of the back to the last dorsal rays where there is a pronounced downward step and from thence continues straight to the tail; *fins* D IX-X, 11–12; A III, 11–13.

The females are predominantly red and the males chiefly grey-green. They may be very pale in colour over sand.

Up to 20 cm, sometimes up to 30 cm.

Lives on sand and mud bottoms at around 5–10 m. They may take refuge in holes and are also able to bury themselves very rapidly in the sand. Spawn in summer and move into deeper water during the winter.

Vielle
Symphodus bailloni (Valenciennes)
= *Crenilabrus bailloni* (Valenciennes)

Eastern Atlantic from North Sea south to Mauritania; occasionally Mediterranean.

Body oval; *head* length equal to, or shorter than, greatest body depth; *mouth* small; *teeth* small and numerous (50–130); *scales* along lateral line 33–38, 2–3 rows of scales above lateral line, 2–3 rows behind eye; *fins* D XIV-XV, 9–11; A III, 9–11.

Both sexes have a dark spot just below the lateral line on the tail stalk, and another dark brownish-black or blue-black spot in the middle of the dorsal fin over the first soft rays; there is a blue arc on the base of the pectoral fins; most have 5 irregular bars on sides. Juveniles and females are brown with 3 dark stripes on upper part of flanks and a dark bar on the snout; males are greenish, reddish or brown, the head is brownish-green with orange stripes.

18 cm.

Found from the surface down to about 50 m over rocks and sea-grass beds.

Tautog
Tautoga onitis (L.)

Western Atlantic from Nova Scotia to South Carolina.

Body stout, deeply oval with a steep profile; *head* rounded, mouth with thick lips; *teeth* 2 rows of stout conical teeth in each jaw and 2 groups of rounded crushing teeth at the rear; *scales* membrane of dorsal and anal fins partly scaled, lower and rear part of gill cover without scales; *fins* D VII, 11; A III, 8.

Grey with irregular dark mottled markings, adults have whitish chins.

90 cm.

Adults are usually found in rocky coastal areas where the water is relatively warm, and sometimes enter brackish water. The young are often found amongst sea-grass off sandy beaches. They feed on molluscs and crustacea. In the northern part of their range they feed extensively on mussels. Like many wrasse Tautog are only active in the daytime.

Tautog, *Tautoga onitis* (Ed Brothers).

Cuckoo Wrasse *Labrus bimaculatus*

Adult female and young; 3 dark blotches on rear back.
Eastern Atlantic; Mediterranean.
35 cm.

Ballan Wrasse *Labrus bergylta*

Scales usually with a light centre and darker rim;
young fish usually green;
front gill cover not toothed.
Eastern Atlantic.
40 cm.

Ocellated Wrasse *Symphodus ocellatus*

Spot outlined in red on gill cover;
black spot on base of tail stalk.
Mediterranean.
13 cm.

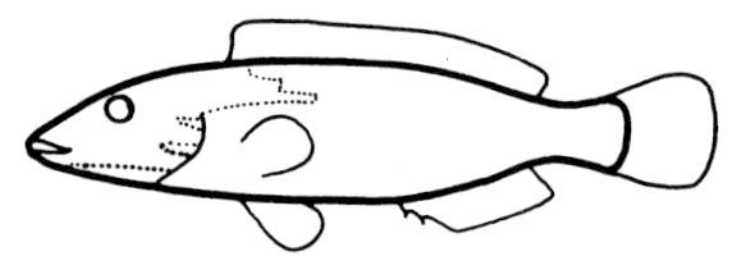

Cuckoo Wrasse *Labrus bimaculatus*

Adult male: blue head.
Eastern Atlantic;
Mediterranean.
35 cm.

Axillary Wrasse *Symphodus mediterraneus*

Black spot at base of tail stalk above lateral line;
dark spot at base of pectorals.
Mediterranean.
15 cm.

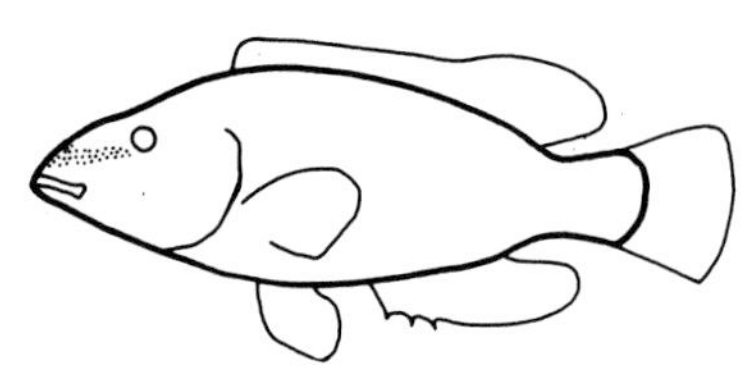

Painted Wrasse *Symphodus tinca*

Slight hump on snout;
dark band between eyes.
Mediterranean.
30 cm.

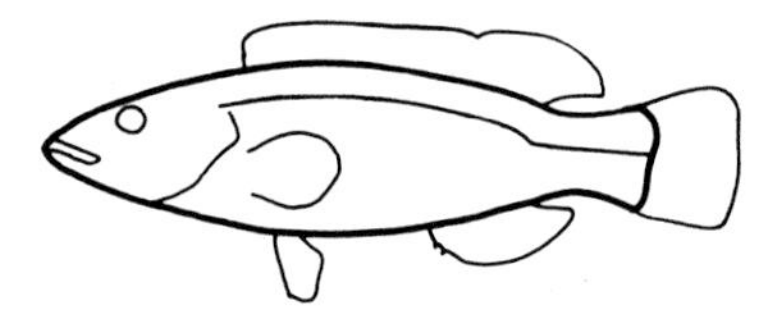

Green Wrasse *Labrus viridis*

Long body;
step in lateral line.
Mediterranean.
45 cm.

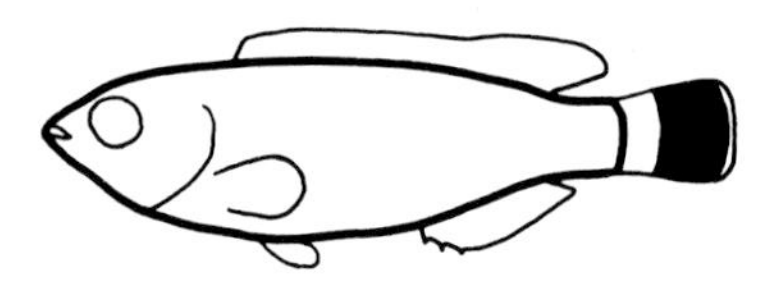

Black-tailed Wrasse *Symphodus melanocercus*

Dark band on tail;
3 anal fin spines.
Mediterranean.
14 cm.

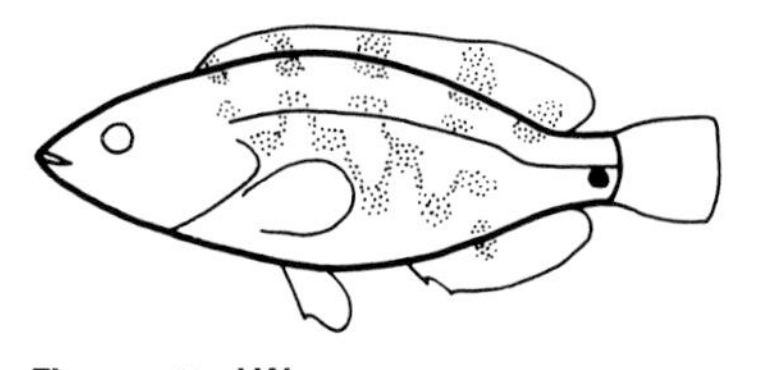

Five-spotted Wrasse *Symphodus roissali*

Deep body;
large scales;
often a black spot on tail stalk below lateral line.
Mediterranean.
15 cm.

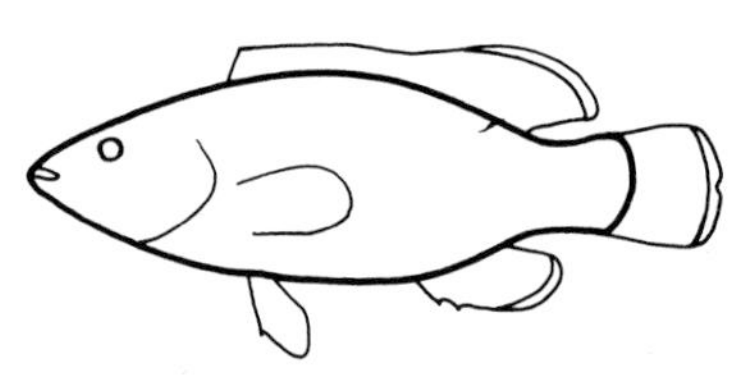

Brown Wrasse *Labrus merula*

Mouth very small;
dorsal, anal and tail fins with blue rims.
Eastern Atlantic;
Mediterranean.
45 cm.

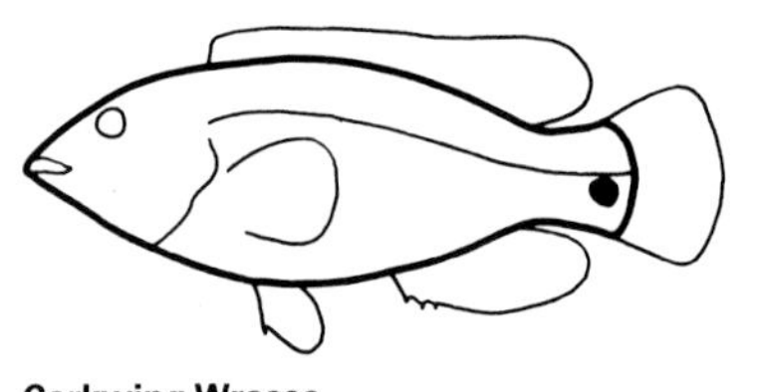

Corkwing Wrasse *Symphodus melops*

Black spot on tail stalk on or below lateral line,
sometimes obscured in adult male.
Eastern Atlantic;
Mediterranean.
6 cm.

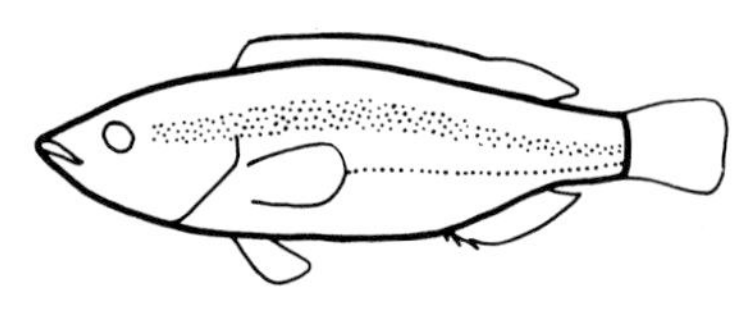

Symphodus doderleini

Long body;
dark and light longitudinal band along body;
Mediterranean.
10 cm.

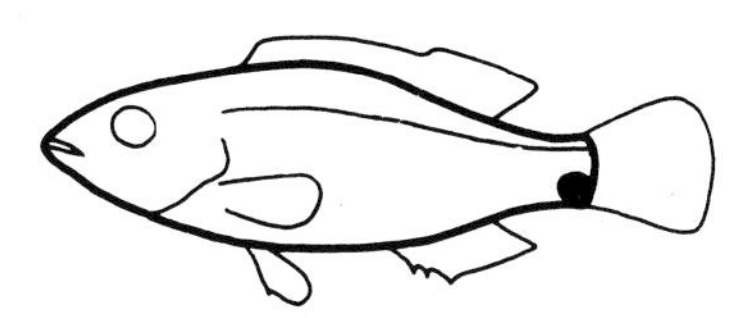

Grey Wrasse *Symphodus cinereus*

Lateral line higher at rear than at front;
dark spot set low on tail stalk.
Mediterranean.
8 cm.

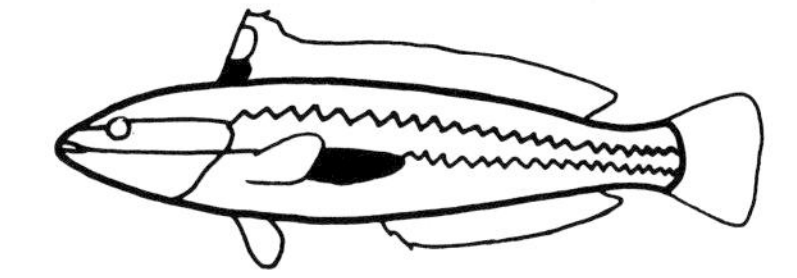

Rainbow Wrasse *Coris julis*

Adult male: long body;
black and orange spot on elongate dorsal rays;
zig-zag orange stripe on flanks.
Eastern Atlantic;
Mediterranean.
25 cm.

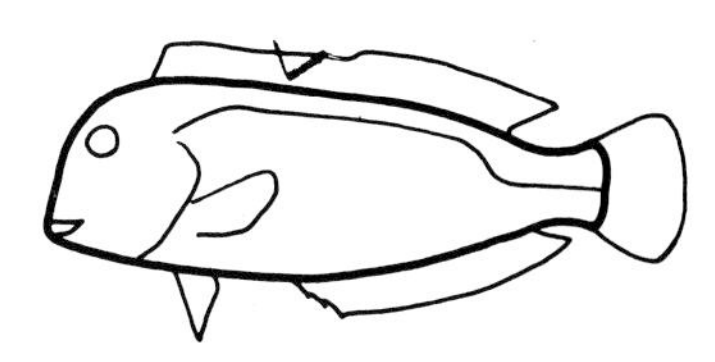

Cleaver Wrasse *Xyrichthys novacula*

Steep forehead;
body laterally flattened.
Eastern Atlantic;
Mediterranean.
20 cm.

Long-snouted Wrasse *Symphodus rostratus*

Very long snout;
concave profile.
Mediterranean.
12 cm.

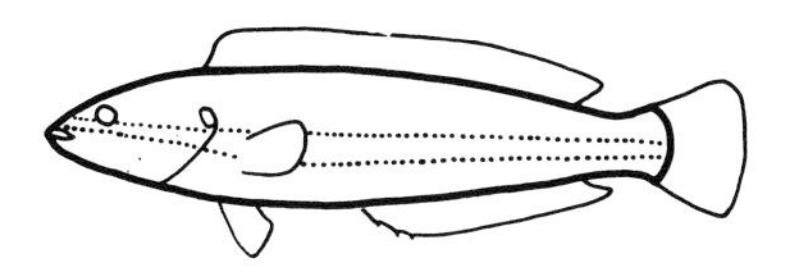

Rainbow Wrasse *Coris julis*

Female and young: long body;
blue spot on gill cover;
light stripe along flanks.
Eastern Atlantic;
Mediterranean.
18 cm.

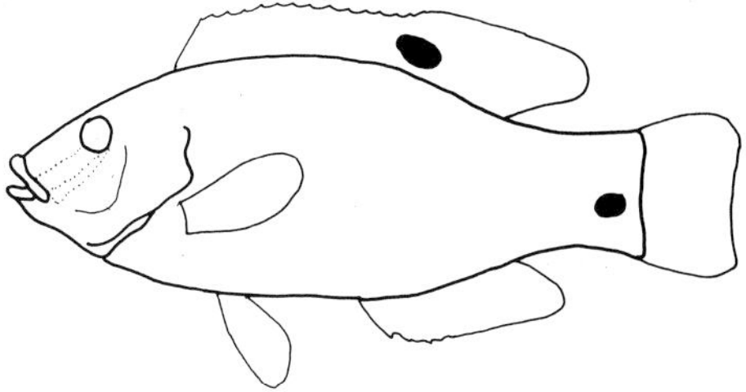

Vielle *Symphodus bailloni*

Dark spot below lateral line on tail stalk;
black spot on dorsal fin.
Western Atlantic.
20 cm.

Rock Cock *Centrolabrus exoletus*

Black bar on tail;
IV-VI anal fin spines.
Eastern Atlantic.
15 cm.

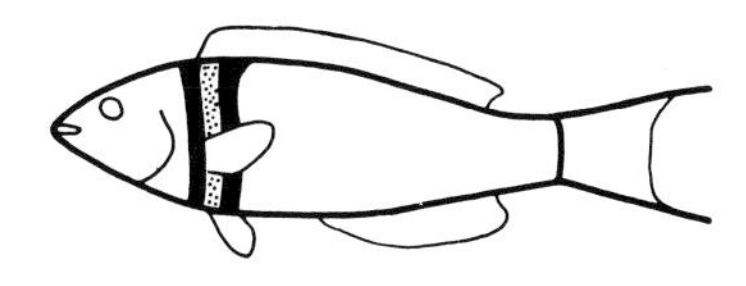

Turkish Wrasse; *Thalassoma pavo*

Adult male: diagonal red and blue stripe across front of body.
Eastern Atlantic;
Mediterranean.
20 cm.

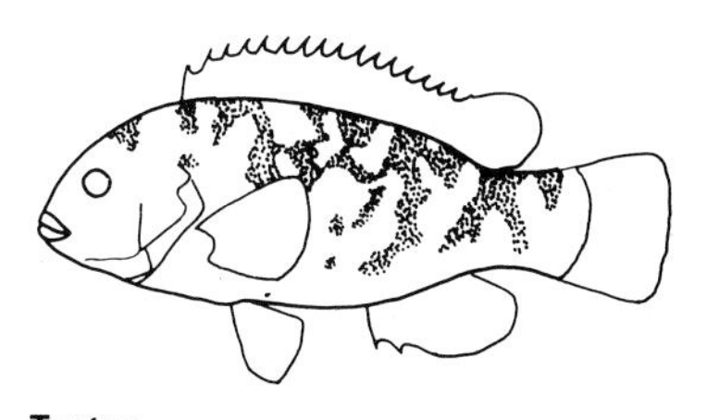

Tautog *Tautoga onitis*

Adults with white chin;
steep profile;
crushing teeth.
Western Atlantic.
90 cm.

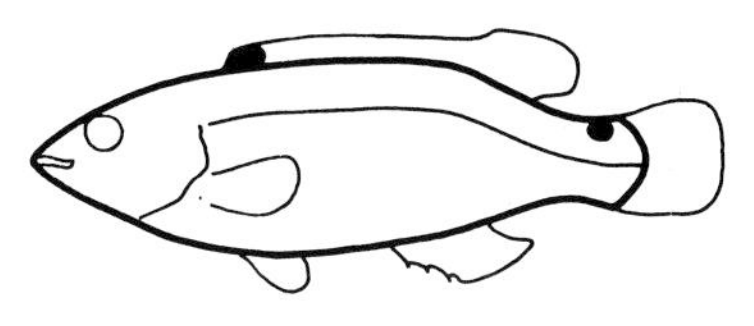

Goldsinny *Ctenolabrus rupestris*

Red-brown colour;
black spot high on tail stalk.
Eastern Atlantic;
Mediterranean.
18 cm.

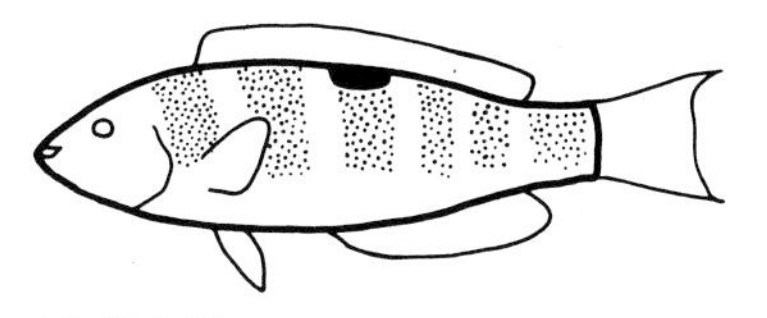

Turkish Wrasse *Thalassoma pavo*

Adult female and young: black spot below mid-point of dorsal fin;
light vertical bands.
Eastern Atlantic;
Mediterranean.
20 cm.

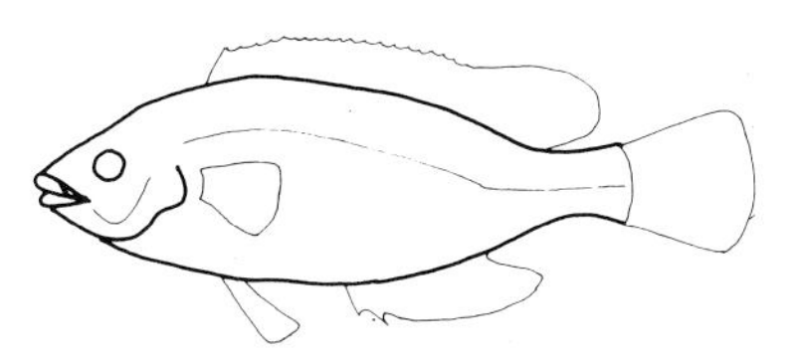

Cunner *Tautogolabrus adspersus*

Deep tail stalk;
head pointed;
up to 10 cm, with spot at front of soft dorsal fin.
Western Atlantic.
40 cm.

Cunner

Tautogolabrus adspersus (Walbaum)

Western Atlantic from Newfoundland south to Chesapeake Bay.

Body rather long with deep tail stalk; *head* pointed, both the upper and lower profiles being rather straight; *teeth* in several rows, conical, the outer rows being the strongest; *scales* large, covering the body and small scales on the gill covers; *fins* D XVII, 9–10; A III, 9.

Varies according to the background and mottled with reddish, blue, olive-green. In young specimens of less than about 10 cm there is a conspicuous black spot in the centre of the dorsal fin on the first soft rays.

40 cm.

Inshore water close to the bottom, where it often occurs in large groups around wrecks, wharves, etc.; do not enter brackish water. Appear to stay in small home territories. In winter they remain inshore sheltering under rocks, etc. In severely cold weather many die. Cunner are omnivorous feeders consuming barnacles, sea urchins, molluscs, worms and eelgrass.

Cunner, *Tautogolabrus adspersus* (Jon D. Witman).

Family:
SCARIDAE

In our area *Sparisoma cretense* is the only representative of this family, but in tropical waters there are numerous species, many of which are common. The teeth in each jaw are fused into a parrot-like beak and are used to scrape the hard surface of rocks and coral. On tropical reefs the Parrot Fishes are a significant force in grinding the coral to sand.

Parrot Fish *Sparisoma cretense*

Teeth fused into parrot-like beak;
very large scales.
Two colour forms:
1, grey with black spot behind gill cover;
2, red with yellow spot at top of tail stalk.
Grey patch on shoulder.
50 cm.

Parrot Fish

Sparisoma cretense Jordan & Gunn

Mediterranean; eastern Atlantic from central Portugal to the Canary Islands.

Body oval; *snout* blunt; *teeth* are fused together in each jaw giving a parrot-like beak; *scales* are very large only 23–25 along the lateral line. A single row of scales on the cheek below the eye. D X, 10; A II, 10.

There are two colour forms. One (probably the male) is grey with a large black blotch behind the gill cover. The other (probably the female) has a red ground-colour but with a large grey patch on the shoulder with yellow blotches on the gill cover in front of the pectorals and another on the top of the tail stalk. The eye is also yellow.

Up to 50 cm, commonly 30 cm.

Usually seen in groups of 3 or 4 which may contain individuals of both colour forms. They frequent rugged rocky areas, especially underwater cliffs. They are not strong swimmers and scull themselves along with their pectoral fins. Parrot Fish scrape the encrusting algae from the rocks with their strong beak and are also said to scrape the algal growth from sea-grass (*Posidonia*) fronds.

Parrot Fish, *Sparisoma cretense*, male (John Lythgoe).

Sub-order:
AMMODYTOIDEA

Perciform fishes with elongated bodies and no spiny rays in the fins.

Family:
AMMODYTIDAE

The **Sand Eels** can be extremely abundant especially in more northern waters over sandy bottoms. They are important as they provide one of the main sources of food for the carnivorous fishes that are eaten by man. The Sand Eels have elongate eel-like bodies with a single soft-rayed dorsal fin set in a groove along the back. The upper jaw, which is extensible, is shorter than the lower. The tail fin is forked and free from the dorsal and anal fins.

The schools are tight and well-disciplined and give a shimmering appearance underwater as each fish is swimming with an eel-like motion.

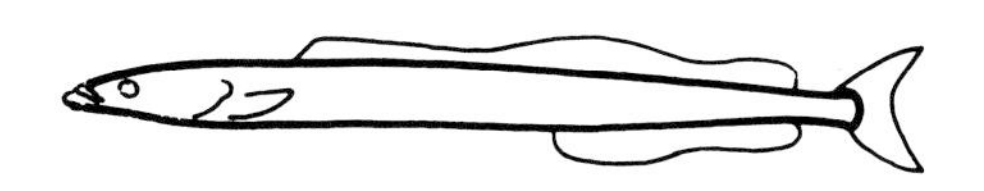

Sand Eels *Ammodytidae*
Silvery eel-like fishes; swim in shoals. About 20 cm.

Smooth Sand Eel
Gymnammodytes semisquamatus (Jourdain)

Eastern Atlantic from Southern Norway south to Biscay; English Channel; North Sea.

Body long and slender; *snout* pointed; *jaws* lower jaw longer than upper. Upper jaw protrusible; *lateral line* has short cross cannals running at right angles from the main canal. For each canal running upwards there are 2 directed downwards; *scales* small, only present on the rear of the body; *grooves* there is a groove running below the pectoral fin which is only slightly longer than the fin; *fins* 1 long dorsal fin which starts behind the pectoral fin and has 2 shallow dips along its length. Anal fin less than ½ the length of the dorsal fin and has 1 shallow dip. Tail fin forked. D 56–59; A 28–32; vertebrae 65–70.

Back golden or greenish-brown. Yellowish sides and silver underneath. These fish

Sand Eel, *Ammodytes tobianus* (Jim Greenfield).

appear silver and are extremely well camouflaged with only the eye being conspicuous.

Up to 23.5 cm.

Usually found from 20–200 m, often on shell or gravel bottoms in which they often bury themselves. They are often encountered swimming in large shoals. Spawning occurs in the winter in the English Channel and progressively later northwards so that in the North Sea spawning occurs in late spring and summer. The eggs are laid on the sea bed. The adults feed on very small crustacea and worms.

Sand Lance

Gymnammodytes cicerelus (Rafinesque)

Northern Mediterranean.

This species was for many years confused with the Atlantic species the Smooth Sand Eel (*Gymnammodytes semisquamatus*). The Sand Lance is, however, the only common Sand Eel in the Mediterranean and may be distinguished by the lower cross canals of the lateral line which are very short. D 53–59; A 27–32.

Back greenish, silvery flanks and belly. There is a blue-black spot on the head.

Up to 16 cm.

Found in large shoals over or in sand. Spawns in winter.

Sand Eel

Ammodytes tobianus L.
= *Ammodytes lancea* Cuvier

Eastern Atlantic from 69°N, south to Gibraltar; English Channel; North Sea; Iceland; west Baltic; occasionally Mediterranean.

Body long and slender; *snout* pointed; *jaw* upper jaw protrusible, lower jaw longer than upper; *lateral line* without any cross canals; *scales* over whole body and are situated in diagonal skin folds and appear in regular rows. There are patches of scales on the base of each tail fin lobe; *fins* dorsal fin starts in front of the tip of the pectoral fin. There are no dips in the dorsal and anal fins; *grooves* the groove below the pectoral fin reaches past the beginning of the anal fin. D 50–56; A 25–31; vertebrae 60–66.

Yellowish-green or bluish on back, yellowish sides, silver belly.

Up to 20 cm.

Sand Eel, *Ammodytes marinus* (Alan Stanley).

Very common from shallow water to 30 m. Usually found over sand in large shoals. Spawning occurs during the winter in the English Channel and the eggs are laid amonst the sand. In the Baltic there are thought to be 2 races, one spring spawning, the other autumn spawning. They feed on small fish and worms.

Sand Eel
Ammodytes marinus Raitt

Eastern Atlantic south to English Channel north beyond Iceland; North Sea; west Baltic.

This species is very similar to *Ammodytes tobianus* but may be distinguished by the scales on the belly, which are not in regular rows; no scales at the base of the tail fin lobes and a greater number of rays in the dorsal and anal fins. D 55–67; A 26–35; vertebrae 66–72.

Back greenish-blue, flanks bluish-silver, belly silver.

Up to 25 cm.

These fish are not found as shallow as *Ammodytes tobianus*, usually not above 30 m, but are very common. Spawning occurs during winter in the southern part of their distribution and early spring in their northern range off Iceland. They feed on small worms and very small crustacea, larvae and eggs. Commercially important as they are a food source for Herring, and are also eaten by sea birds.

Greater Sand Eel
Hyperoplus lanceolatus (Le Sauvage)

Eastern Atlantic; English Channel; North Sea; west Baltic.

This fish may be most easily identified by the black spot on the sides of the snout. Other characters are the scales which are not situated in skin folds; 2 teeth on the roof of the mouth and the upper jaw which can swing forward but is not truely protrusible. D 52–61; A 28–33; vertebrae 65–69.

Back bluish-green or brownish-green, sides bluish-green, belly silver. There is a dark spot on the sides of the snout.

Up to 32 cm, it is the largest of the Sand Eels.

These fish are found from shallow water to 150 m, usually over or in sand. Breeding occurs in water between 20–100 m deep during spring and summer. They feed on small crustacea, larvae, eggs and small fishes.

Hyperoplus immaculatus Corbin

West coast of the British Isles; German North Sea coast.

This fish is easily confused with the Greater Sand Eel *Hyperoplus lanceolatus* and it may be more extensive than records suggest.

It may be identified by the uniformly dark snout and the greater number of vertebrae.

D 59–61; A 31–34; vertebrae 70–74.

Back bluish-green, sides lighter, belly silvery. The snout is dark.

Up to 30 cm.

Very little is known about this species. They are caught on sand, gravel and shell bottoms both near and offshore. Spawning occurs in winter and spring.

American Sand Lance
Ammodytes americanus De Kay

Western Atlantic from Hudson Bay south to Virginia.

Closely resembles the other members of the genus, but the North American species require further study. D 54–60; A 24–33; vertebrae 64–69.

Inhabits shallow inshore waters, sometimes intertidal. It forms large schools but spends long periods buried in the sand, sometimes with head protruding. An important source of food for other fishes, sea birds and mammals.

A closely related species, the Northern Sand Lance (*Ammodytes dubius*) Reinhardt, differs in the number of fin rays and vertebrae: D 61–69; A 31–36; vertebrae 71–78. The Northern Sand Lance has a range from Greenland south to the Scotian Shelf. It is found over offshore banks of sand and fine gravel at depths not greater than 91 m. Due to the difficulty of distinguishing the two species, the exact range of the two North American species is not known.

Family:
TRACHINIDAE

Elongate bottom-living fish which have venom glands at the base of the spiny rays in their 1st dorsal fin and at the base of a strong spine on their gill cover. The scales are arranged in diagonal rows.

There are 4 species in our area. They all spend most of the day buried in the sand with only their head and 1st dorsal fin exposed. When alarmed the 1st dorsal fin is raised and can cause a wound so painful that it can be incapacitating and is thus dangerous. Most cases of permanent injury, however, are the result of the wound becoming septic. The best treatment known at present is to hold the affected part in as hot water as can possibly be borne for at least half an hour. Afterwards the wound should be cleaned, disinfected and covered in the normal way. This hot water treatment is effective because the venom from the Weever is destroyed by heat.

The Weevers feed on small fishes of all kinds and it is thought that they may come out more into the open at night.

Spotted Weever
Trachinus araneus Cuvier

Mediterranean.

Body long and laterally flattened; *mouth* almost vertical in the head; *spines* there is a pair of small spines on the head just in front of the eyes; *fins* there are 7 poisonous spine rays on the 1st dorsal. 1D VII; 2D 26–29; A II, 29–31.

Brown or yellow-brown above, paler below with 7–11 darker spots in a longitudinal series along the flanks. Front half of 1st dorsal fin black.

Up to 25 cm, exceptionally up to 40 cm.

A shallow-living species. They are generally found half buried in sandy bottoms especially where there are rocks or sea-grass nearby.

Greater Weever
Trachinus draco L.

Mediterranean and eastern Atlantic from Morocco to Norway.

Body elongate, shallow and laterally flatened; *mouth* terminal and oblique; *spines* a small spine on head in front of each eye. A solitary venomous spine on each gill cover; *fins* the 1st dorsal fin usually has 5 or 6 strong venomous spines. 1D V-VII; 2D 29–32; A II, 28–34.

Greyish-yellow above, paler below with characteristic narrow diagonal streaks on the flanks. The dorsal fin has a large black spot.

Up to 40 cm.

Lives on sandy bottoms down to 100 m or more, but may occasionally come into water more shallow than 10 m.

Breeds in late spring and summer. The eggs are free-floating.

Streaked Weever
Trachinus radiatus Cuvier

Mediterranean; eastern Atlantic from Morocco to Senegal.

Body long and laterally flattened; *head* on the head behind the eyes are 3 large bony plaques with radiating ridges; *mouth* terminal and nearly horizontal; *spines* there are 2 or 3 small spines on the head just in front of the eyes and a large venomous spine on each gill cover; *fins* the 1st dorsal

Greater Weever, *Trachinus draco* (Jim Greenfield).

fin has 6 venomous spines. 1D VI; 2D 24–27; A I, 26–29.

Brown above and yellow-brown below. The area beneath the head is whitish. There is a characteristic pattern of dark rings and spots. The membranes between the first 4 dorsal spines are black.

Frequents sandy bottoms. Probably lives down to about 200 m and is generally a deeper-living fish than the other Weevers.

Streaked Weever, *Trachinus radiatus* (Andrea Ghisotti).

Lesser Weever

Echiichthys vipera (Cuvier)

Mediterranean; eastern Atlantic from Senegal to Scotland.

Body long and laterally flattened but the body is rather deeper than the other weevers; *mouth* terminal and oblique; *spines* there are no spines on the head. There is a long venomous spine on the gill cover; *fins* the 1st dorsal fin normally has 6 spines. 1D V-VII; 2D 21–24; A I, 24–26.

Yellowish-brown above, lighter below. No conspicuous pattern but there are small dark spots on the head and back. The dorsal fin is entirely black.

Up to 14 cm.

Found on sandy bottoms from about 1 m down to 50 m. During the daytime they lie with their eyes and dorsal fins exposed and can cause excruciating pain to a bather if he treads on one. A more serious menace is to inshore trawlermen, especially shrimpers, who get stung when sorting their catch. There is a report that this fish has actually attacked a swimmer.

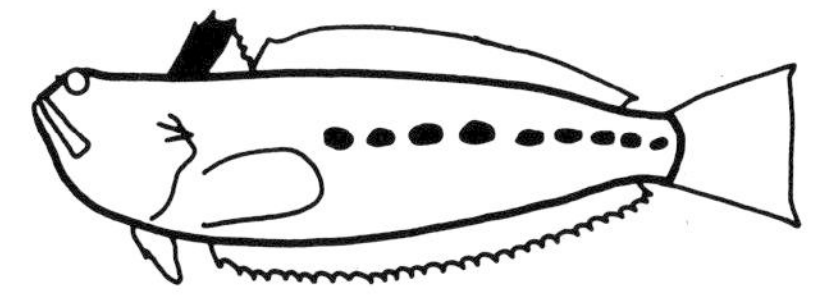

Spotted Weever
Trachinus araneus

7–11 dark spots along flanks; 7 rays in 1st dorsal fin. Mediterranean. 25 cm.

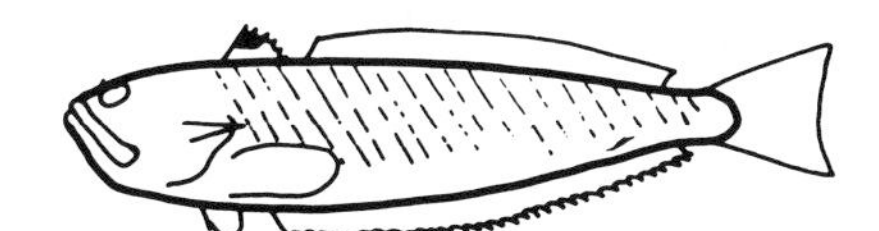

Greater Weever
Trachinus draco

Oblique pattern on flanks; 5 rays in 1st dorsal fin. Eastern Atlantic; Mediterranean. 40 cm.

Streaked Weever
Trachinus radiatus

Irregular dark spots and circles; plaques with radiating grooves on head. Eastern Atlantic; Mediterranean. 25 cm.

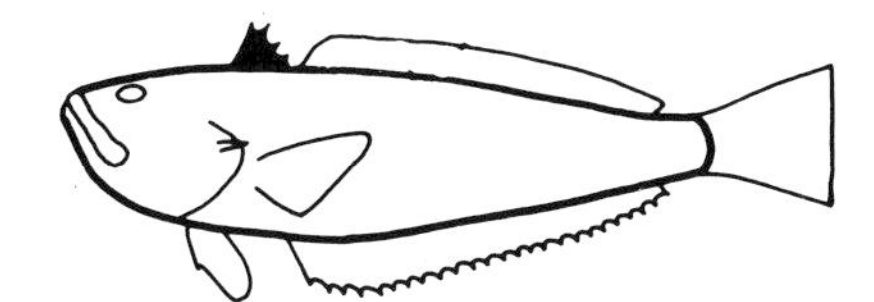

Lesser Weever
Echiichthys vipera

Markings faint or absent; dorsal fin black; no small spines on head. Eastern Atlantic; Mediterranean. 14 cm.

Lesser Weever, *Echiichthys vipera* (David Maitland).

Family:
URANOSCOPIDAE

Body has a solid conical form with a flat top on the head. The eyes are set on the upper surface of the head. There is a strong spine behind the gill cover and an electric organ behind each eye.

Star Gazer
Uranoscopus scaber

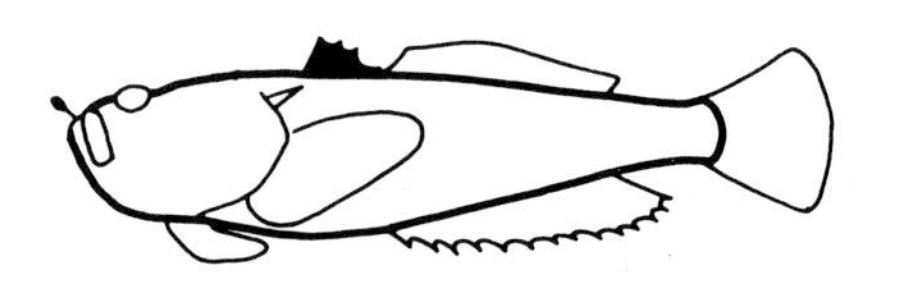

Star Gazer *Uranoscopus scaber*

Stiff conical body; flat top to head; mouth nearly vertical. 25 cm.

Mediterranean; eastern Atlantic from Spain to Senegal.

Body elongate and rounded; *head* laterally flattened with rough bony plates; *eye* very small and set on the top of the head; *mouth* large and almost vertical. On the tip of the lower jaw is a small appendage; *spines* there is a sharp venomous spine behind the gill cover; *fins* two dorsal fins, the 1st has 4 spiny rays. 1D IV; 2D 13–14; A I, 13–14.

Dull brown in colour, darker above, lighter beneath.

Up to 25 cm, rarely 30 cm.

The Star Gazer lies almost buried in the sand with only its mouth and eye visible. The small appendage on the lower jaw can be made to vibrate, luring small fishes within reach of its jaw.

Behind the eye on each side is an oval electric organ. Out of water the Star Gazer is able to give an unpleasantly powerful electric shock. Underwater this shock would not be so strong. Breeds in spring and summer. The eggs are free-floating.

Star Gazer, *Uranoscopus scaber* (Jon Kenfield).

Family:
SCOMBRIDAE

Strong and active swimmers that may undertake long migrations. The **Mackerels** are chiefly known as free-swimming fishes living in large shoals near the surface, where they feed on small fishes. They are abundant and good to eat, but like their close relatives the Bonitos and Tunnies, they should always be cooked soon after capture for their flesh seems particularly prone to a type of bacteria that makes it poisonous. The scales are very small but cover the body. Between the head and about the centre of the first dorsal fin there is an irregular band where the scales are larger (the corselet) but this is not well developed in the Mackerel (*Scomber scombrus*). The 2 dorsal fins are well separated and behind both the anal and 2nd dorsal fins there is a series of finlets. The tail fin is divided into 2 distinct lobes.

Atlantic Mackerel
Scomber scombrus L.

Eastern Atlantic from English Channel south to North Africa; western Atlantic from Newfoundland south to the Carolinas; Mediterranean.

Body rather long and streamlined; *eye* large with vertical transparent eyelids; *keels* none along the mid-line of the tail stalk but there are 2 small keels on the base of each lobe of the tail fin. No keel between the 2 dorsal fins; *teeth* small; *lateral line* gently curved; *scales* no corselet; *fins* the 2 dorsal fins are well spaced, there is a series of 5 pinules behind the 2nd dorsal and the anal fins. Tail fin in 2 lobes. D X-XIII; D 11–13 + 5 finlets; A I, 11–13 + 5 finlets.

Iridescent green and blue above, whitish-silver below. There are irregular, zebra-like dark stripes on the back and upper flanks.

Up to 50 cm, sometimes much more.

In the late spring through the summer to early autumn great shoals of Mackerel often come close inshore. In the winter they move into deeper water offshore, usually to the edge of the continental slope, where they keep close to the bottom and probably eat very little.

In the warmer months their food is very variable, consisting of swimming crustacea, fish fry, young herrings, etc. They will snap at almost anything and a wide variety of objects such as feathers and silver paper can be used to hook them.

Chub Mackerel; Spanish Mackerel
Scomber japonicus Houttuyn
Scomber colias Gmelin

Eastern Atlantic from English Channel south to North Africa; western Atlantic from Novia Scotia south to Venezuela; Mediterranean.

Resembles the Atlantic Mackerel (*Scomber scombrus*) in general body form but differs in having no swimbladder, 7–9 spiny dorsal rays and a corselet around the head and pectoral base.

The flanks and back are marked with spots rather than stripes, and there is a golden band running from the gill cover to the tail.

Up to 30 cm, occasionally 60 cm.

Lives in somewhat deeper water than the mackerel. At night they can be attracted by surface lights perhaps because the lights also attract swarms of pelagic crustacea on which they feed.

A shoaling pelagic species, they can undertake long migrations. They chiefly feed on small fishes and swimming crustacea. They come into breeding condition in summer but the exact time depends on the water temperature. In the Mediterranean the breeding grounds are in the Sea of Marmora. They probably do not breed in the northern part of their range.

Tunny; Blue-fin Tuna
Thunnus thynnus (L.)

Mediterranean; both sides of the Atlantic south from Newfoundland and Norway. A similar species inhabits the Pacific and Indian Oceans.

Body rather deeper than the mackerels; *jaws* rather small and extend back to the level of the eye; *teeth* small, sharp and conical arranged in a single file in each jaw; *keels* there is a well-developed keel on each side of the tail stalk; *scales* small on most of the body but large in the region of the corselet; *fins* the 2 dorsal fins set closer together than the eyes's diameter. Pectoral fin short and does not reach further back than the 12th dorsal spine. 1D XIII-XV; 2D I, 13–15 + 8–10 finlets; A 11–15 + 8–9 finlets.

Blue-black above, silvery-white below. The fins are dark except for the anal fins and finlets, which are dull yellow. Sometimes the pelvic fins are yellow as well.

Up to 3 m.

The Tunny together with its near relative in the Pacific is of great commercial importance. They embark on long migrations and the presence of large shoals in particular places are predictable. They are known to cross the Atlantic and in the summer some migrate from the Mediterranean to Norway and the North Sea. They do not inhabit water colder than about 10–14°C and are not found in northern European waters during the winter. In the summer they school near the surface and may come close inshore but in winter they live deeper, sometimes to 180 m.

Tunny, *Thunnus thynnus* (Andrea Ghisotti).

Albacore; Long-finned Tunny
Thunnus alalunga (Bonnaterre)

Mediterranean and all warm oceans; rare further north than Biscay.

Resembles the Tunny (*Thunnus thynnus*) except that the pectoral fin is very long, extending back at least as far as the rear edge of the 2nd dorsal fin.

Bluish-black or bronze-brown above, whitish below with an iridescent blue band running along the flanks. The margin of the tail fin is light and may show up well underwater. The finlets are yellow.

Up to 100 cm.

Like the Tuna these are a social and migratory species but keep to water warmer than 14–18°C. During their summer migrations they do not normally go further north than Biscay but there they are common. They do not come close inshore and can be caught from as deep as 200 m off the Florida coast. Tuna have a body temperature warmer than the surrounding water.

An extremely important commercial species in southern European waters. They are both caught on lines and trapped in nets.

Yellowfin Tuna
Thunnus albacares (Bonnaterre)

Eastern Atlantic from Portugal to West Africa; not in Mediterranean; western Atlantic from Massachusetts south to Brazil.

Dark blue-black above, silvery below; dorsal and anal fins and all finlets yellow with black edges; often with about 20 incomplete narrow vertical lines on belly.

170 cm.

In open water, sometimes schooling at the surface, sometimes beneath the thermocline.

The Bigeye Tuna (*Thunnus obesus*) (Lowe) is a similar species with a similar distribution but the lower surface of the liver has no striations as in the other true tunas.

Blackfin Tuna
Thunnus atlanticus (Lesson)

Western Atlantic from Massachusetts south to Brazil.

Scales over entire body; *fins* dorsal fins set close together, no fins greatly extended; dorsal finlets 7–9, anal finlets 7–8. D XIII-XIV, 12–15; A II, 11–15.

Body blue-black above, silvery below, pectoral dorsal and anal fins black, finlets yellow with black edges.

90 cm.

Atlantic Bonito
Sarda sarda (Bloch)

Mediterranean and tropical Atlantic; rarely north to Sweden.

Body elongate; *jaws* long, reaching to the hind edge of eye. The upper jaw being slightly the more prominent; *teeth* robust;

Tunny, *Thunnus thynnus* (Gilbert van Rijckevorsel).

keels there is a keel on each side of the tail stalk flanked on each side by small longitudinal ridges converging towards the tail; *scales* well-developed corselet present; *lateral line* wavy; *fins* the 2 dorsal fins are adjacent. 1D XXI-XXIV; 2D 12–16 + 7–10 finlets; A 11–15 + 6–8 finlets.

Azure or ultramarine on the back, silvery-white below. Upper part of flanks are marked with longitudinal or oblique stripes.

Up to 60 cm in northern waters, 80 cm in the south.

A strongly migratory fish swimming in shoals, usually near the surface and some miles offshore. They are only caught from waters between 15° and 24°C. Do not spawn off northern Europe, but in the Mediterranean they spawn in the winter between November and June.

Little Tunny; Bonito
Euthynnus alletteratus (Rafinesque)

Mediterranean and the warm waters of the Atlantic.

Body resembles the other Tunnies in general shape; *scales* no scales except in the region of the corselet and along the lateral line; *keels* 1 distinct keel on each side of the tail stalk; *fins* the 2 dorsal fins are separated by less than the eye's diameter. The pectoral fin is short and the pelvic fins are joined at the base by the so-called interpelvic processes. 1D XIII-XVI; 2D 11–14 + 8 finlets; A 12–14 + 7 finlets.

Ground colour is blue-back above, whitish below. There is a panel of mackerel-like lines and spots on the back behind the tip of the pectoral fin. There is often a pattern of about 6 dark spots behind and below the pectoral fin.

Up to 80 cm.

A gregarious warm-water species chiefly feeding on pelagic fishes.

Skipjack Tuna
Katsuwonus pelamis Kishinouye
= *Euthynnus pelamis* (L).

Mediterranean and all warm oceans.

Body oval and rather deep, of the typical Tunny shape; *scales* there are no scales except in the region of the corselet; *fins* the two dorsal fins are set close together. The pectoral is short, not extending beyond the 9th dorsal ray. 1D XV-XI; 2D 11–16 + 8 finlets; A 11–16 + 7 finlets.

Blue-black above, whitish below with 4 to 6 dark longitudinal stripes below the lateral line and behind the pectoral fin. Occasionally there is a series of diffuse vertical bands along the flanks.

Up to 65 cm, occasionally 90 cm.

A free swimming and gregarious species, normally prefers deep ocean water from 18–20°C but occasionally come close inshore, generally in the summer.

Frigate Mackerel; Bullet Mackerel

Auxis rochei (Lacépède)

Mediterranean and warm parts of all oceans.

Body slightly longer than the other Tunnies; *eye* rather small and set forward in the head, the head being between 5 and 6 times the diameter of the eye; *keel* there is a well developed keel along the mid-line of the tail stalk; *scales* there are no scales on the body except for a corselet of large scales that extend in two lobes, one along the back, the other along the lateral line to the level of the second dorsal fin; *fins* the 2 dorsal fins are widely spaced and rather small. Between the two pelvic fins is a large triangular flap of skin characteristic of the genus. 1D X-XI; 2D 11–12 + 8–9 finlets; A 12–15 + 7–8 finlets.

Blue or blue-green above, silvery beneath. Above the lateral line and behind the corselet is a conspicuous mackerel-like panel of darker spots and lines.

Up to 60 cm, occasionally 100 cm.

A schooling pelagic species that sometimes comes close inshore during the summer. Not sufficiently common to have any commercial importance.

Spanish Mackerel

Scomberomorus maculatus (Mitchill)

Western Atlantic from Cape Cod south to Florida and Gulf of Mexico.

Body long and streamlined, profile equal above and below; *lateral line* dips gently below the 2nd dorsal fin; *fins* soft dorsal fin inserted in front of anal fin, no scales on pectoral fin. D XVII+I, 18; A II 17. There are 8–9 dorsal finlets and 9 anal finlets.

Silvery with 3 rows of yellow-orange

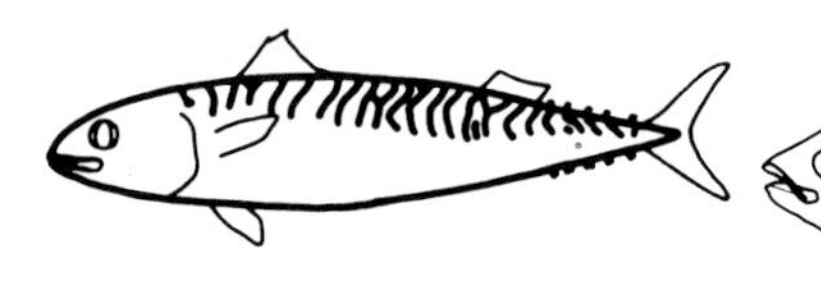

Atlantic Mackerel
Scomber scombrus

'Mackerel-stripes' along entire back;
widely spaced dorsal fins;
no corselet.
Eastern and western Atlantic; Mediterranean.
50 cm.

Yellowfin Tuna *Thunnus albacares*

2nd dorsal and anal fins very long, yellow;
belly often with thin vertical lines.
Eastern and western Atlantic.
1.7 m.

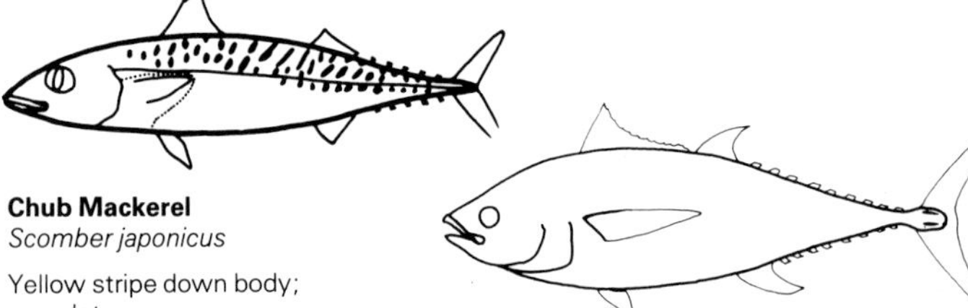

Chub Mackerel
Scomber japonicus

Yellow stripe down body;
corselet;
widely spaced dorsal fins.
Eastern and western Atlantic; Mediterranean.
30 cm.

Blackfin Tuna *Thunnus atlanticus*

Pectoral, dorsal and anal fins black;
finlets yellow.
Western Atlantic.
90 cm.

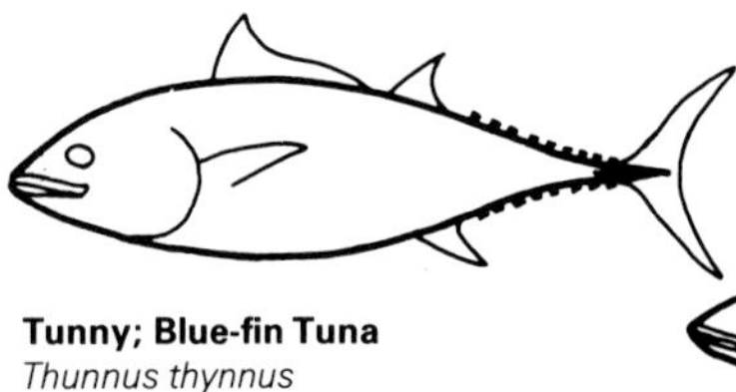

Tunny; Blue-fin Tuna
Thunnus thynnus

Dorsal fins set close together;
no markings on body;
pectoral fin short.
Atlantic; Mediterranean.
3 m.

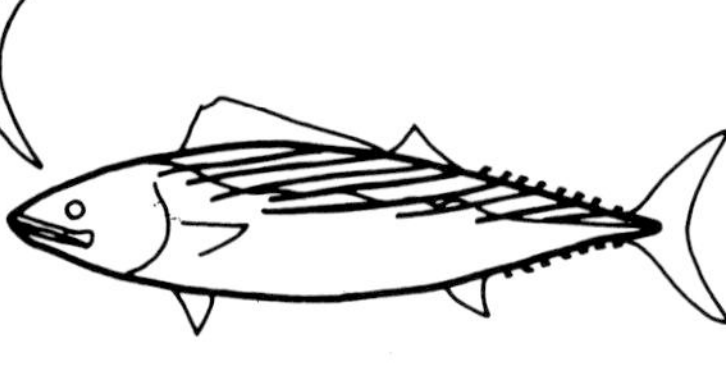

Atlantic Bonito
Sarda Sarda

Longitudinal stripes on upper flanks;
wavy lateral line;
adjacent dorsal fins.
Eastern Atlantic; Mediterranean.
80 cm.

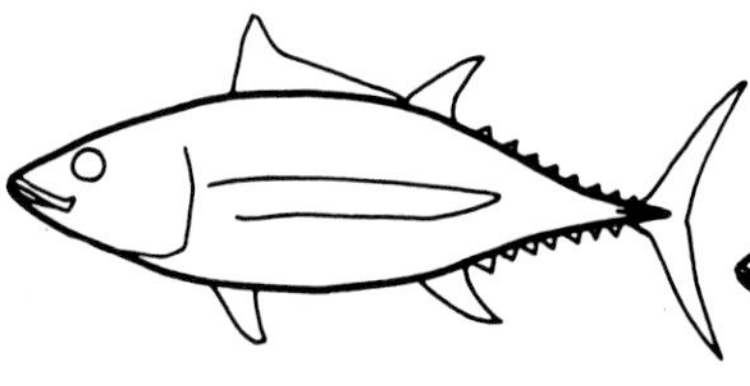

Albacore; Long-finned Tuna *Thunnus alalunga*

Close-set dorsal fins;
long pectoral fin.
Eastern and western Atlantic.
1 m.

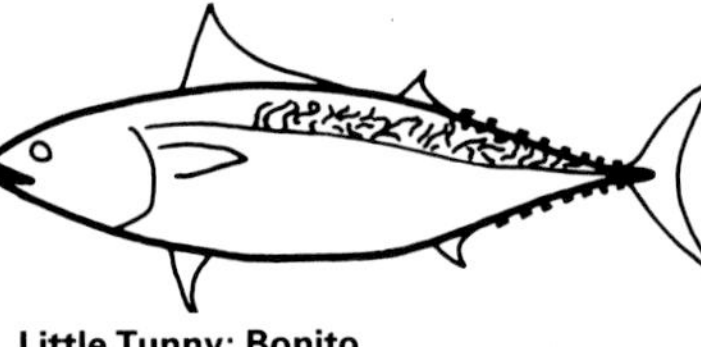

Little Tunny; Bonito
Euthynnus allettaratus

Dorsal fins close together;
mackerel-like marking above lateral line and behind pectoral fin.
Eastern and western Atlantic; Mediterranean.
80 cm.

Skipjack Tuna; Ocean Bonito *Katsuwonus pelamis*

Dorsal fins close together;
longitudinal stripes behind pectoral fin and below lateral line.
Eastern and western Atlantic; Mediterranean.
65 cm.

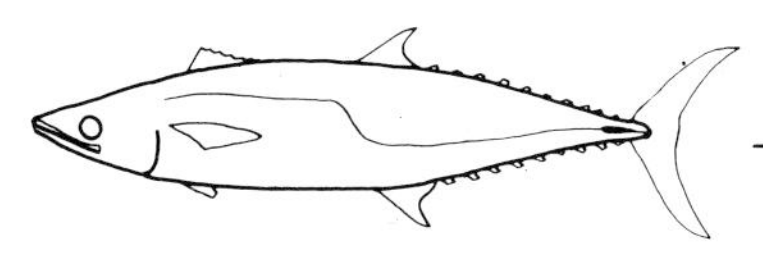

King Mackerel; Kingfish *Scomberomorus cavalla*

Uniform grey in adults;
lateral line dips sharply below 2nd dorsal fin;
no blotch on 1st dorsal fin.
Western Atlantic.
1.5 m.

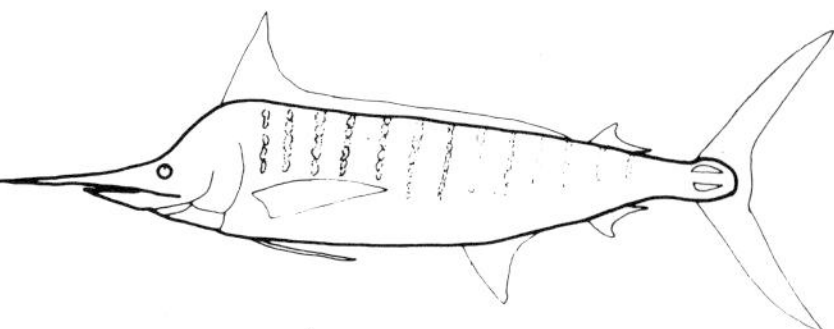

Blue Marlin *Makaira nigricans*

Bill rounded in cross section;
2 keels on tail stalk;
front lobe of 1st dorsal fin pointed;
dorsal always lower than body depth.
Warm eastern and western Atlantic.
4 m.

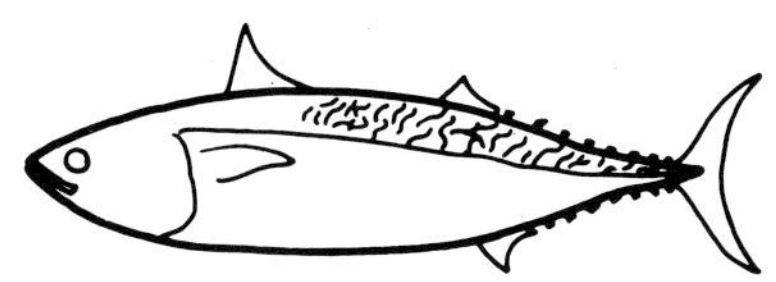

Frigate Mackerel; Bullet Mackerel *Auxis rochei*

Widely spaced dorsal fins;
mackerel-like markings behind 1st dorsal fin and above lateral line.
Eastern and Western Atlantic; Mediterranean.
60 cm.

Wahoo *Acanthocybium solandri*

Body long, rounded, streamlined;
rear of jaw hidden by bone below eye;
dark vertical bands.
Western Atlantic.
2 m.

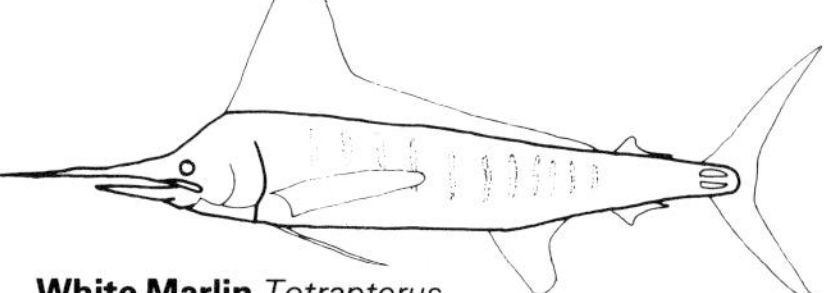

White Marlin *Tetrapterus albidus*

Bill rounded in cross section;
2 keels on tail stalk;
front lobe of 1st dorsal fin rounded.
Warm eastern and western Atlantic; Mediterranean.
3 m.

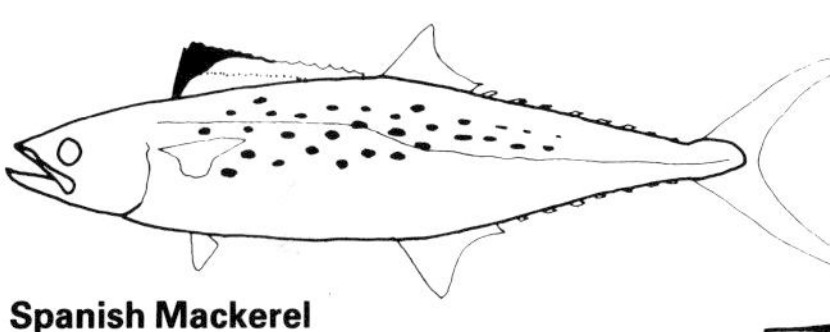

Spanish Mackerel *Scomberomorus maculatus*

Orange spots at all ages;
1st dorsal white below, black blotch at front;
lateral line dips gently below 2nd dorsal fin.
Western Atlantic.
60 cm.

Swordfish *Xiphias gladius*

Sword flattened;
2 dorsal fins, 2nd small;
single keel on tail stalk.
Eastern and western Atlantic; Mediterranean.
4.5 m.

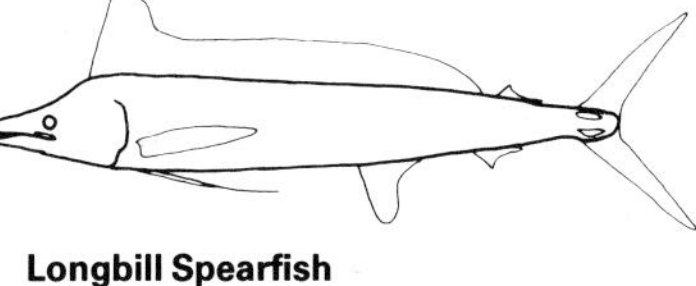

Longbill Spearfish *Tetrapterus pfluegeri*

Bill rounded in cross section;
2 keels on tail stalk;
body slender;
front lobe of 1st dorsal fin not sharply pointed.
Warm eastern and western Atlantic.
2.5 m.

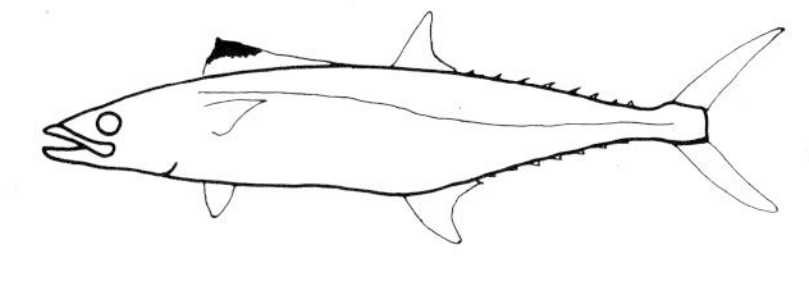

Cero *Scomberomorus regalis*

Orange or yellow streaks and spots;
1st dorsal with black blotch at front;
lateral line dips gently below 2nd dorsal fin.
Western Atlantic.
90 cm.

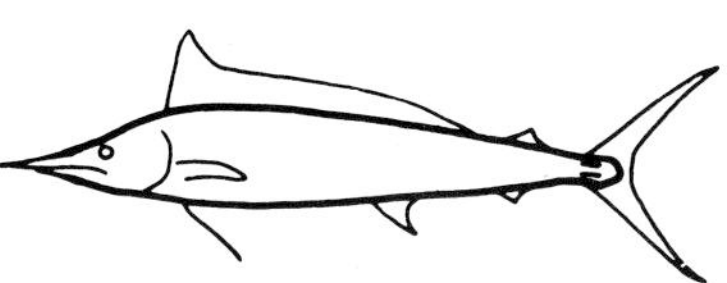

Mediterranean Spearfish *Tetrapterus belone*

Bill rounded in cross section, short;
2 keels on tail stalk;
pectoral fins short.
Mediterranean.
2.4 m.

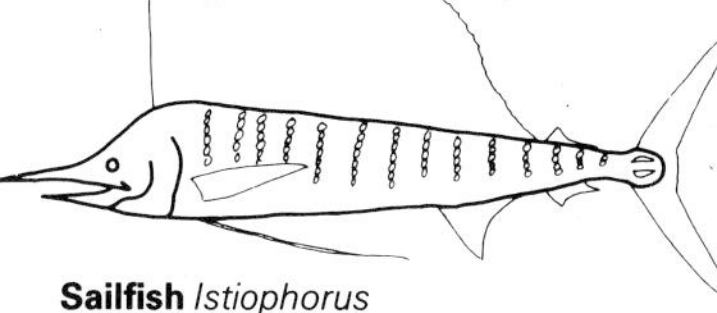

Sailfish *Istiophorus platypterus*

Bill rounded in cross section;
2 keels on tail stalk;
1st dorsal very high along whole length.
Warm eastern and western Atlantic.
3 m.

spots which are present at all ages except in individuals shorter than 5 cm; the 1st dorsal fin has a white band along its lower edge and a dark blotch at the front.

60 cm.

Comes close inshore, the young even into the surf zone, and into low-salinity bays. An important game and commercial fish.

Cero
Scomberomorus regalis (Bloch)

Western Atlantic from Massachusetts south to Brazil; Caribbean.

Closely resembles the Spanish Mackerel (*Scomberomorus maculatus*) in body form except that the 1st dorsal fin is inserted directly over the anal fin and there are scales on the pectoral fin. D XVI-XVIII + 1; A II, 14. There are 8–9 dorsal finlets and 8 anal finlets.

Flanks silvery with dull yellow spots and streaks; there is a dark blotch at the front of the dorsal fin, but there is no lower white band.

90 cm.

King Mackerel; Kingfish
Scomberomorus cavalla (Cuvier)

Western North Atlantic from Gulf of Marine south to Brazil and Gulf of Mexico.

Body resembles the Spanish Mackerel (*Scomberomorus maculatus*) in shape. The main difference is the *lateral line* that dips down sharply below the 2nd dorsal fin; *fins* D XV+I, 15; A II, 15. There are 8–9 dorsal finlets and 8 anal finlets.

No dark blotch on the first dorsal fin, body is a uniform grey colour with no spots in adults. In the young there are darker yellowish spots.

150 cm.

A coastal species which is considered to live further offshore than the Spanish Mackerel and is important to sports fishermen.

Wahoo
Acanthocybium solandri (Cuvier)

Worldwide in warm waters; in western Atlantic south from New Jersey.

Body long bullet-shaped and streamlined; 3 *Keels* on tail stalk; *head* jaw long, snout about half length of head, rear of jaw hidden by bone below eye (preorbital bone); *fins* 23–27 spines in first dorsal fin, 9 dorsal finlets, 9 ventral finlets.

Iridescent blue-green with many vertical bars.

An important game fish which is chiefly found in open waters not far from islands. Solitary but may congregate to feed.

Family:
XIPHIIDAE

Swordfish
Xiphias gladius L.

Eastern Atlantic, in summer Iceland and south; Mediterranean; western Atlantic south from Newfoundland; also worldwide.

Body robust, tapering evenly from nape to tail; 1 large lateral *keel* on each side of tail stalk; *bill* long and flattened; *fins* front dorsal fin short and pointed, no pelvic fin.

Dark brown above often with metallic tints, light brown below.

4.5 m.

From surface waters down to 800 m, solitary and migratory.

Family:
ISTIOPHORIDAE

Mediterranean Spearfish
Tetrapterus belone Rafinesque

Mediterranean.

Body slender, sloping gently from bill to nape; 2 *keels* on tail stalk; *bill* short, round in cross section; *fins* pectoral fins very short, front lobe of 1st dorsal fin has a rounded tip and is slightly greater than body depth.

Sailfish, *Istiophorus platypterus* (Doug Perrine).

Blue-black above, abruptly changing to the silvery colour of the lower flanks and belly.

2.4 m.

In upper waters of the sea, fished commercially between Italy and Sicily.

Blue Marlin
Makaira nigricans Lacépède

Eastern Atlantic from northern Spain south to South Africa; western Atlantic from Massachusetts south to Uruguay; worldwide in warm seas.

Deep, profile rising steeply from eye to 1st dorsal fin; 2 horizontal *keels* on each side of tail stalk; *bill* long, rounded in cross section; *fins* dorsal fin long and low, the front lobe pointed but not as high as body depth.

Bluish or greyish-blue shading gradually to silver belly; faint vertical lines on flanks.

3–4 m.

Usually found in surface water above the thermocline, migratory.

White Marlin
Tetrapterus albidus Poey

Eastern Atlantic south from Portugal and south-western Mediterranean; western Atlantic from Nova Scotia south to Brazil.

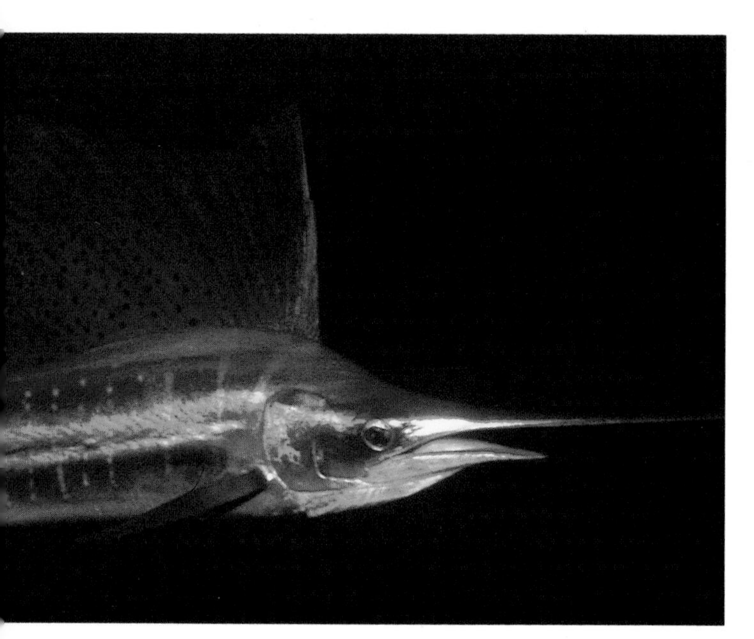

Body rather slender, profile between eye and beginning of 1st dorsal fin moderately steep, 2 keels on each side of tail stalk; *bill* long, round in cross section; *fins* front lobe of 1st dorsal fin rounded, about as high as body depth.

Bluish-green or grey above, silver below; faint vertical lines on flanks.

3 m.

Swims in surface water usually above the thermocline.

Longbill Spearfish
Tetrapterus pfluegeri Robins & de Sylva

Eastern Atlantic from Portugal south; tropical western Atlantic.

Body slender, profile gently sloping from eye to beginning of 1st dorsal fin; 2 lateral *keels* on each side of tail stalk; *bill* long, round in cross section; *fins* pectoral fins long, front lobe of 1st dorsal fin not sharply pointed, a little higher than the body depth.

Blue-black above, silvery below.

2.5 m.

Usually offshore and above the thermocline.

Sailfish
Istiophorus platypterus (Shaw & Nodder)
= *Istiophorus albicans*

Warm eastern Atlantic and south-western Mediterranean; warm western Atlantic from Rhode Island south to Brazil; otherwise, circumtropical.

Body elongate, tapering from neck to tail stalk, profile of nape moderate; 2 *keels* on each side of the tail stalk; *bill* long and round in cross section; *fins* 1st dorsal very high throughout its length, the middle rays being the highest and much greater than the body depth, pectoral fins very long, at least twice the length of the pectoral fins.

Blue-black above shading to silvery below; membrane of dorsal fin is dark blue with rows of darker spots, body with vertical rows of golden spots.

3 m.

Found in open water but often nearer the coast than other species, usually above the thermocline.

Sub-order:

GOBIES

by Peter J. Miller, Department of Zoology, Bristol University

Sub-order:

GOBIOIDEI

Families:

GOBIIDAE, ELEOTRIDIDAE, MICRODESMIDAE

Gobies belong to the sub-order Gobioidei (Gobioidea), a large and very varied array of chiefly tropical and warm-temperate fishes, which, in habitat, are predominantly inshore marine but which also include many estuarine and freshwater species. At least 50 kinds of goby are found in the temperate eastern Atlantic and the Mediterranean, and a further 30 or more are restricted to the more brackish and fresh waters of the Black and Caspian Seas and their tributaries. Gobies in the latter category cannot be dealt with here. In contrast, the much more limited temperate waters of the western Atalntic contain only about 17 species, derived from the numerous gobiid fauna of the Caribbean and Gulf of Mexico, which is much richer in goby species than the tropical eastern Atlantic. All the eastern Atlantic species and most of the western Atlantic species covered in this book belong to the family Gobiidae. However, among the western Atlantic species, there are two representatives of the otherwise tropical Sleeper Gobies (Eleotrididae) and a Wormfish (belonging to the aberrant gobioid family, the Microdesmidae). None of the temperate gobies occur on both sides of the Atlantic.

The temperate Atlantic gobies are mostly small fishes, all less than 30 cm in length, of moderately elongate and subcylindrical or compressed body shape, with a typically rounded, somewhat depressed head, displaying prominent cheeks and dorso-lateral eyes, the latter being separated by a relatively narrow interorbital space. The mouth cleft is oblique, and the snout short, with a tubular front nostril, sometimes bearing a triangular lappet or from one to several narrow tentacles on its free rim, and a generally pore-like rear nostril. There are 2 dorsal fins, the 1st of merely 6 spinous rays in most eastern Atlantic species but 7 in some western Atlantic forms. In the true gobies, the second dorsal base length is always longer than the distance from the end of this fin to the origin of the tail fin, whereas in the sleeper gobies the converse is true. There is a single anal fin, commencing with a flexible spinous ray, and the caudal fin is usually rounded. The pectoral fins are large and, in some species (*Gobius* and *Mauligobius*), the uppermost pectoral rays are more or less free from the fin membrane, so that their branches form a fringe along the upper edge of the fin. The Sleeper Gobies retain the more primitive condition of separate pelvic fins, but an important external characteristic of the true gobies is the fusion of the pelvic fins into a shallow funnel-like disc whose front (anterior) rim is completed by a transverse membrane between the spinous rays of each pelvic fin. This pelvic disc has weak suctorial powers. The anterior membrane may possess conspicuous lateral lobes (as in some *Gobius* species) or carry minute papillae (villi) along its free edge (in some *Pomatoschistus* species). As a secondary modification, the anterior membrane may be lost (*Lebetus* and *Speleogobius* species), or, as in *Gobius auratus*, the fins may be more or less separated again, with the posterior rim of the disc notched and the anterior membrane weak. The tiny *Odondebuenia* and *Vanneaugobius* have the pelvic fins separate, with only a low ridge of membrane between their bases. Wormfish, of

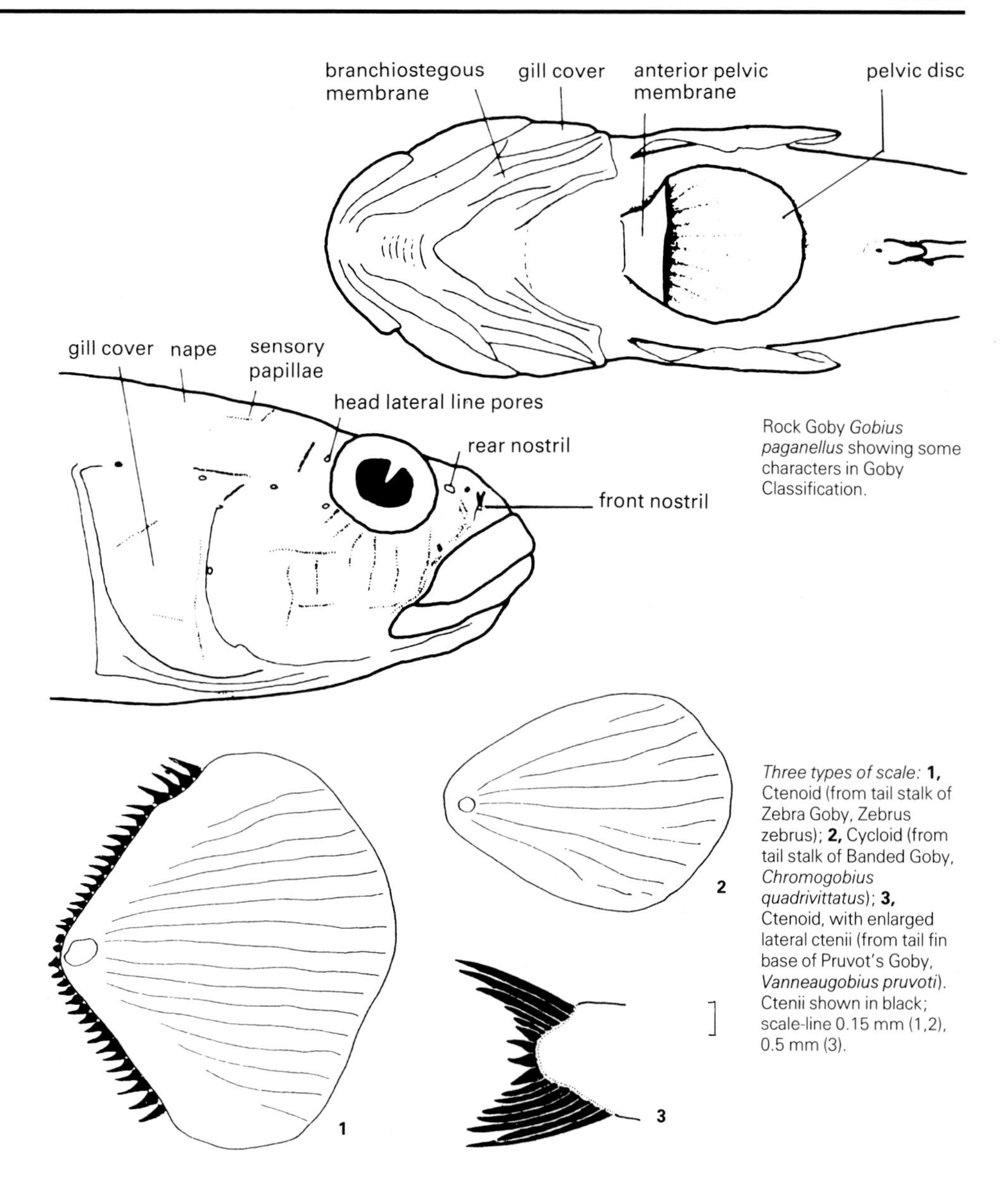

Rock Goby *Gobius paganellus* showing some characters in Goby Classification.

Three types of scale: **1,** Ctenoid (from tail stalk of Zebra Goby, Zebrus zebrus); **2,** Cycloid (from tail stalk of Banded Goby, *Chromogobius quadrivittatus*); **3,** Ctenoid, with enlarged lateral ctenii (from tail fin base of Pruvot's Goby, *Vanneaugobius pruvoti*). Ctenii shown in black; scale-line 0.15 mm (1,2), 0.5 mm (3).

which one species is recorded from the warm temperate western Atlantic, are unusual among gobioids in possessing elongate bodies, with a single long dorsal fin, protruding lower jaw, and reduced opercular opening.

Unlike most fishes, the gobies do not have a lateral-line canal along each side of the body, but canals belonging to this sensory system occur on the head and there are rows of exposed neuromast sensory papillae over the head and body. The precise arrangement of these lines of papillae and the degree of reduction of the head canals are of considerable value in goby classification, but such features are often difficult to examine without chemical treatment of the dead fish or special illumination under a binocular microscope. Their description has therefore been omitted from the following account.

European gobies of marine occurrence fall clearly into 2 groups according to habitat. Most species spend their juvenile and adult life on or near the sea-bed, with merely the postlarval stage in the plankton. Three genera (*Aphia*, *Pseudaphya*,

Crystallogobius) remain nektonic over the entire life-span (apart from breeding activities) and retain gobioid larval features such as transparency and body elongation. No comparable goby occurs in the temperate western Atlantic.

In diet, gobies are typically predatory, feeding chiefly on tiny crustacea, worms, molluscs, and sometimes small fish. Their dentition consists usually of a few rows of fine teeth in each jaw, with similar teeth in the pharynx. In certain species, the lower jaw carries lateral canine teeth, while others may have enlarged median teeth at the rear of the upper jaw row. The western Atlantic Lyre Goby (*Evorthodus*) has a single row of bilobed teeth in the lower jaw.

Gobies breed during spring and summer months. Adult males delimit territories about a nest, which may be excavated under stones, shells, or other rigid objects (tiles, old shoes, etc). After courtship display and conducting the female into the nest, spawning takes place in an inverted position, with the eggs being deposited in a patch on the underside of the nest roof. Each adult female may be expected to produce several batches of eggs during one breeding season. Goby eggs are pear-shaped or more elliptical, not exceeding 3–4 mm in length among the present species, and are attached to the substrate by tiny filaments radiating from the base of each egg. The eggs are guarded and fanned by the male until hatching. The postlarvae (released with yolk sac already absorbed) live in the plankton for a few weeks or months before, as in most species, adopting an existence on the bottom.

Longevity in gobies ranges from one to several years, according to species. The nektonic forms, such as *Aphia* and *Crystallogobius*, have been cited as 'annual' vertebrates, dying immediately after their first and only breeding season at an age of about one year. Smaller species, such as those belonging to *Pomatoschistus* or *Gobiosoma*, become sexually mature after their first winter of life, and survival beyond the second winter is limited. Large, longer-lived European species of *Gobius* and related genera, at least in northern waters, do not mature until at least 2 years old. Among these, maximum ages of 10–12 years have been recorded in the Rock Goby (*Gobius paganellus*) and the Leopard-Spotted Goby (*Thorogobius ephippiatus*), by counting transparent annual rings laid down in the earstones (otoliths).

The following systematic treatment of temperate Atlantic gobies omits some rare species, or those whose specific nomenclature and diagnosis require further investigation. English names are those from FNAM (checklist of fishes of the North-eastern Atlantic, and of the Mediterranean) or AFS (American Fisheries Society). In counts, the number of scales in lateral series refers to those along the lateral mid-line from the pectoral axilla ('armpit') to the base of the tail fin and includes the few extending onto the fin itself. In dealing with fin rays, the last ray in both the 2nd dorsal and anal fins, always split to the base, is counted as one. For all counts, the most frequent value or range of values is given first, followed by extreme range in parentheses when available. Maximum length cited usually refers to total length and therefore includes the tail fin. Most of the gobies mentioned in the text are also shown in diagrammatic lateral outline to indicate certain major features useful for identification. Other details of coloration, etc., are omitted. Except in a few cases, specimens illustrated are adult males.

Detailed accounts of the European gobies will be found in the author's CLOFNAM and FNAM accounts of this family. The western Atlantic gobies are covered in more specialized monographs such as those by Robins & Bohlke, Ginsburg, Dawson, and Birdsong. There is still much to be learned about the systematics, general biology, and distribution of gobies, and the amateur naturalist can help to solve these problems by field observation and judicious collection of specimens. The present author is always glad to receive gobies for identification.

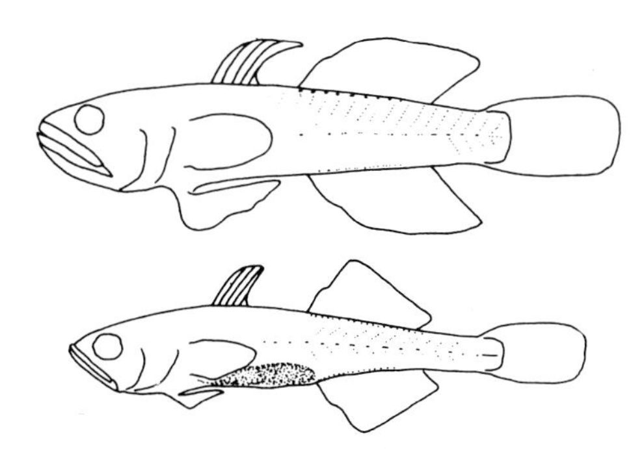

Transparent Goby *Aphia minuta*

5 1st dorsal fin rays;
lateral scales 24–25;
adult male: larger teeth
and fins. (top)
5.8 cm.
Female (bottom): minute teeth;
small pelvic disc.
5.3 cm.

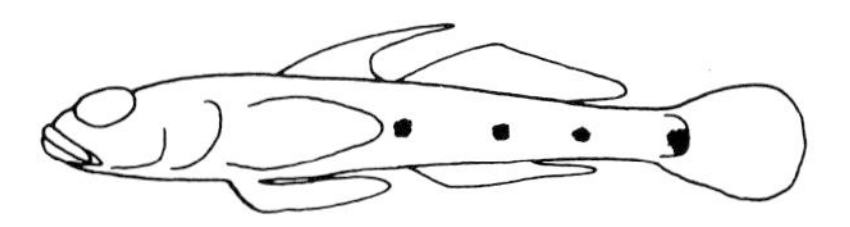

Jeffrey's Goby *Buenia jeffreysii*

Four black spots on lateral mid-line;
nape naked;
lateral scales 25–30.
6 cm.

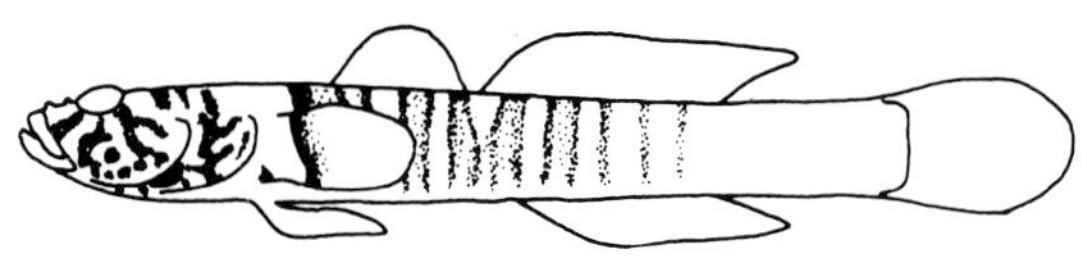

Banded Goby
Chromogobius quadrivittatus

10–14 vertical dark bars, 2 pale saddles;
intense spot in lower angle of gill cover;
edge of pectoral bar straight;
lateral scales 56–72.

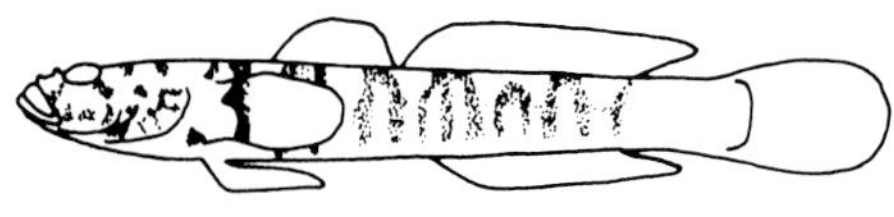

Kolombatovic's Goby
Chromogobius zebratus

5 broad pale saddles;
edge of pectoral bar
acutely bent;
lateral scales 41–52.
5.3 cm.

Liechtenstein's Goby
Corcyrogobius liechtensteini

Dark spot on each side of throat;
lateral scales 27–29 (27–31).
2.5 cm.

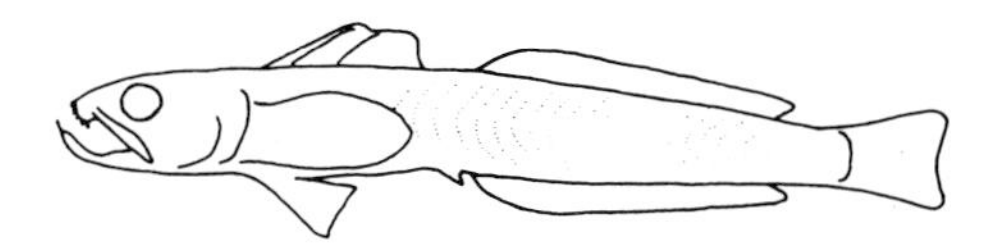

Crystal Goby
Crystallogobius linearis

1st dorsal fin with not more than 2 rays;
no scales;
males with enlarged
canine teeth and deep pelvic disc.
4.7 cm.
Females without teeth, small pelvic disc, and often no 1st dorsal fin.
3.9 cm.

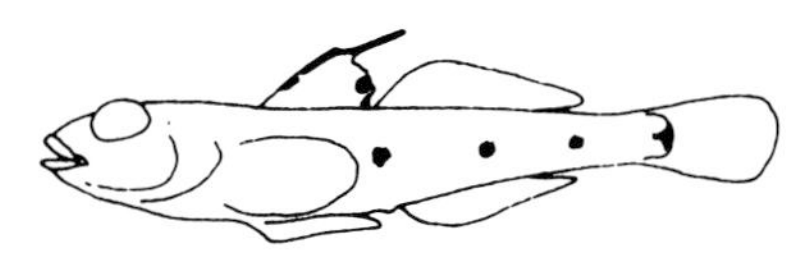

Four-spotted Goby
Deltentosteus quadrimaculatus

Black spots along lateral mid-line;
first dorsal fin with dark markings and spots;
scales on nape;
lateral scales 33–35.
8 cm.

Toothed Goby
Deltentosteus colonianus

As for *Deltentosteus quadrimaculatus* but angle of mouth below rear half of eye.
7 cm.

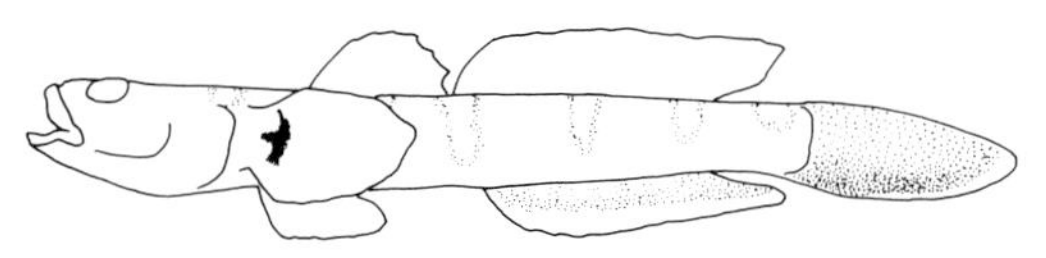

Koch's Goby *Didogobius kochi*

Brownish, with pale saddles;
dark mark on pectoral fin base;
anal fin with broad pale band;
tail fin dark, especially lower edge;
lateral scales 33–37.
5.7 cm.

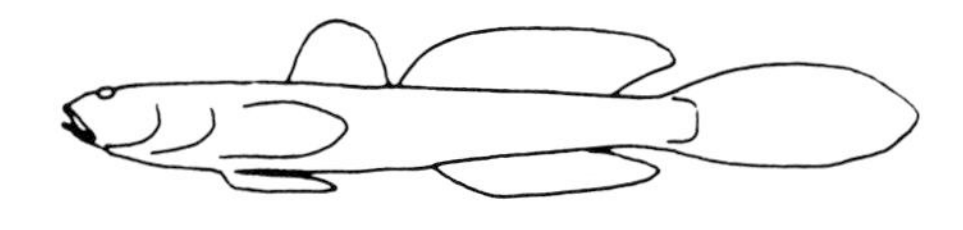

Ben-Tuvia's Goby
Didogobius bentuvii

Pale, somewhat translucent;
eyes relatively small;
tail fin elogate;
lateral scales 65–70.
5 cm.

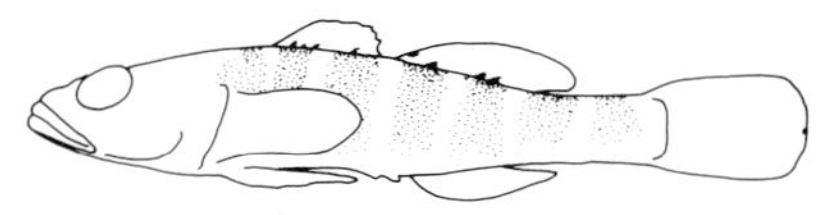

Steinitz's Goby
Gammogobius steinitzi

5–6 broad vertical bands;
dark spots along bases of dorsal fins.
3.8 cm.

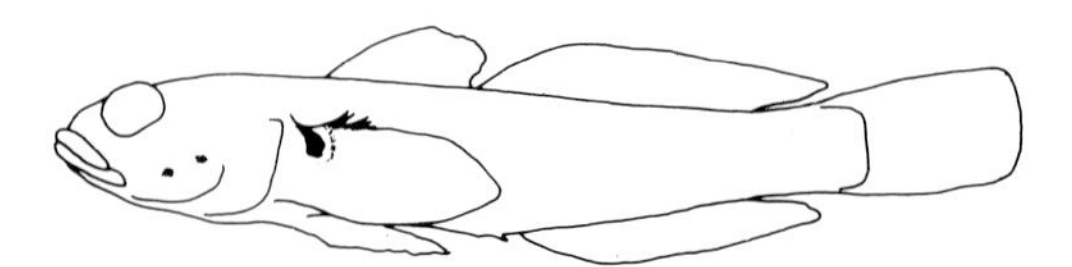

Golden Goby *Gobius auratus*

Golden tint;
2 cheek spots;
pectoral fin base with vertical dark mark;
pelvic disc deeply cleft;
scales in lateral series 44–50.
10 cm.

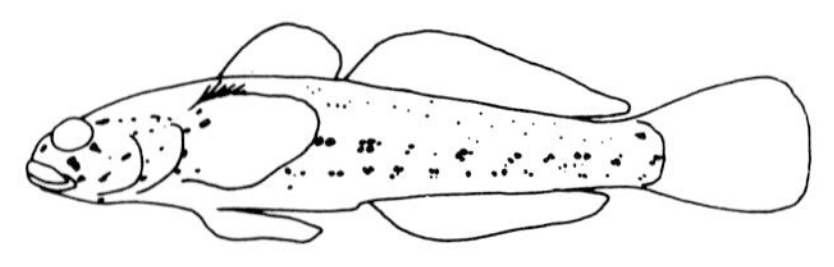

Bucchich's Goby
Gobius bucchichii

Many small spots along head and body;
lateral scales 50–56.
10 cm.

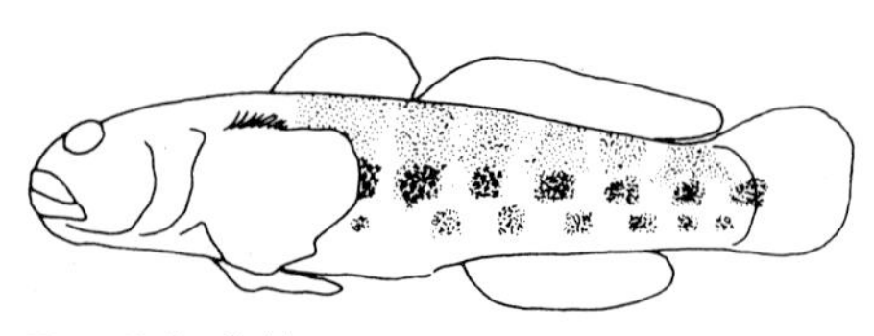

Giant Goby *Gobius cobitis*

Pelvic membrane with lateral lobes;
lateral scales 59–67.
27 cm.

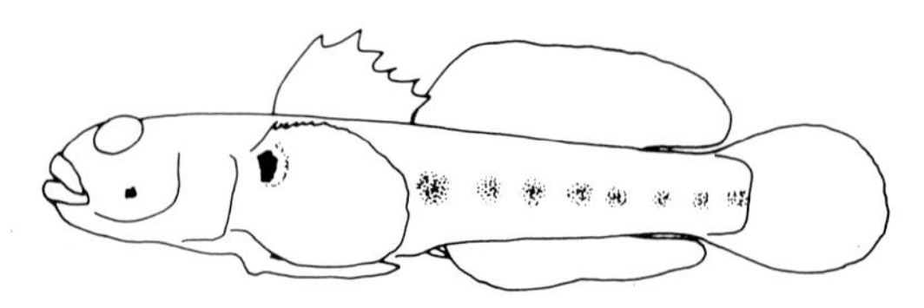

Couch's Goby *Gobius couchi*

Golden flecking;
5 lateral mid-line blotches below 2nd dorsal fin;
pectoral fin base with vertical mark;
1 cheek spot;
scales in lateral series 40–41 (35–45).
9 cm.

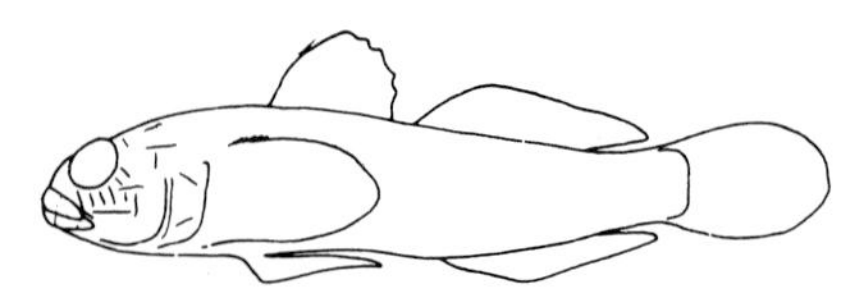

Red-mouthed Goby
Gobius cruentatus

Lips and cheeks with red streaks;
black lines of sensory papillae on head;
lateral scales 52–58.
18 cm.

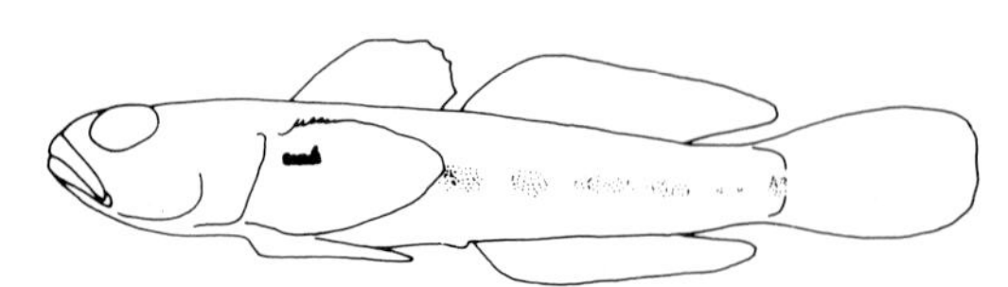

Steven's Goby *Gobius gasteveni*

4 blotches along lateral mid-line below 2nd dorsal fin;
pectoral fin base with dark mark longer than deep;
cheek and gill cover with white spots.
12 cm.

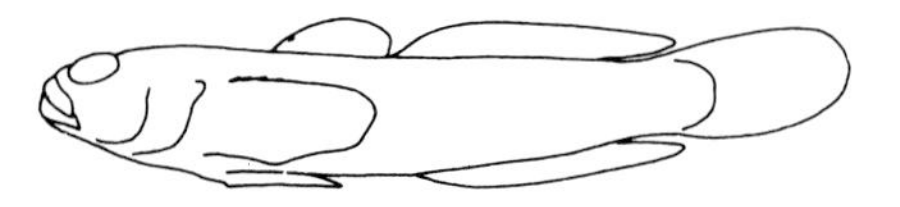

Slender Goby *Gobius geniporus*

Body slender;
pelvic disc anterior membrane reduced or absent;
lateral scales 50–55.
16 cm.

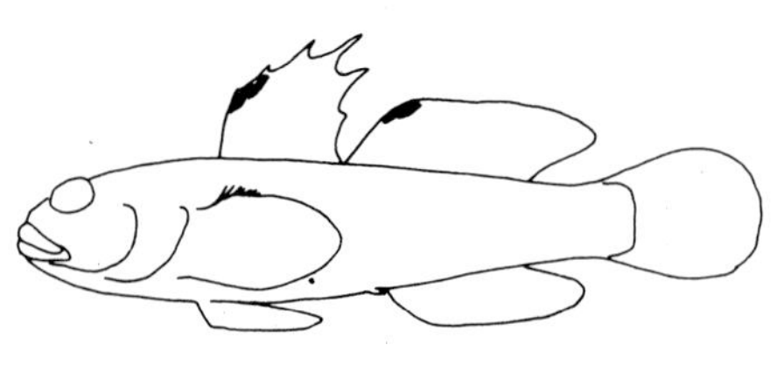

Black Goby *Gobius niger*

Black spots in front corners of dorsal fins;
elongate 1st dorsal rays;
lateral scales 35–41 (32–42).
15 cm.

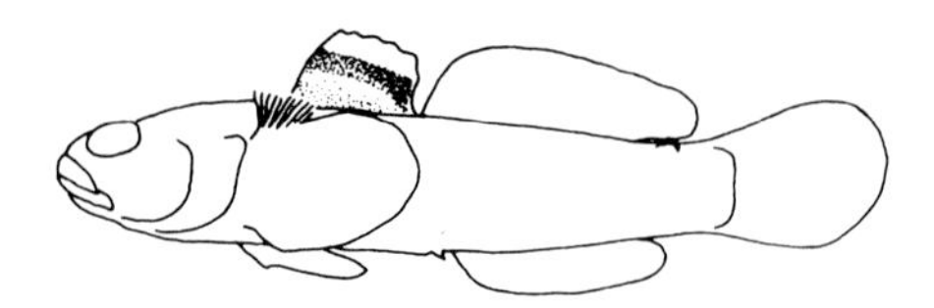

Rock Goby *Gobius paganellus*

Pale to orange band on upper edge of first dorsal fin;
free pectoral rays well developed;
lateral scales 50–55 (46–59).
12 cm.

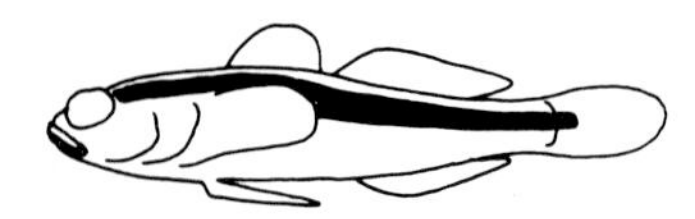

Striped Goby *Gobius vittatus*

Broad dark stripe along head and body;
lateral scales 32–36.
5.8 cm.

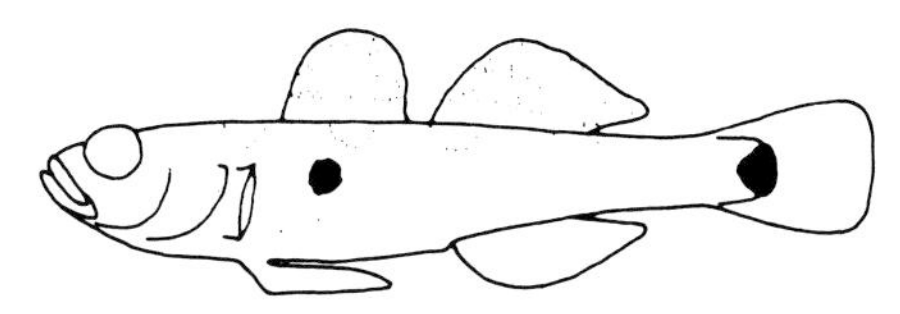

Two-spotted Goby
Gobiusculus flavescens

1st dorsal fin with 7 rays; large caudal spot; males with conspicuous lateral spot below 1st dorsal fin; lateral scales 35–40. 6 cm.

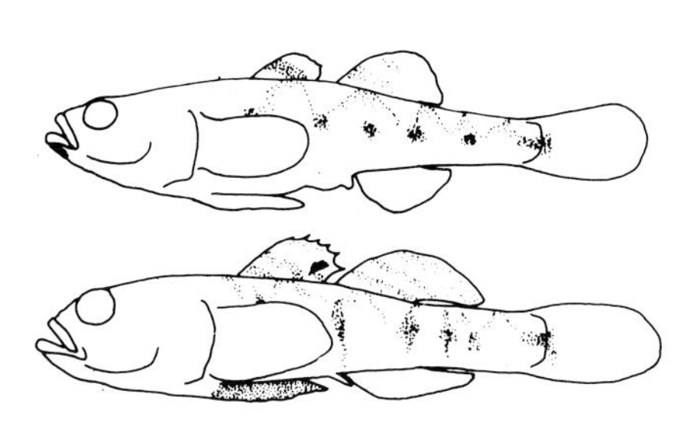

Lagoon Goby
Knipowitschia panizzae

Nape and back scaleless to 2nd dorsal fin base; broad pale saddles; male with lateral bars (above); female with black chin spot (top) and first dorsal dark band; scales in lateral series 32–39. 3.75 cm.

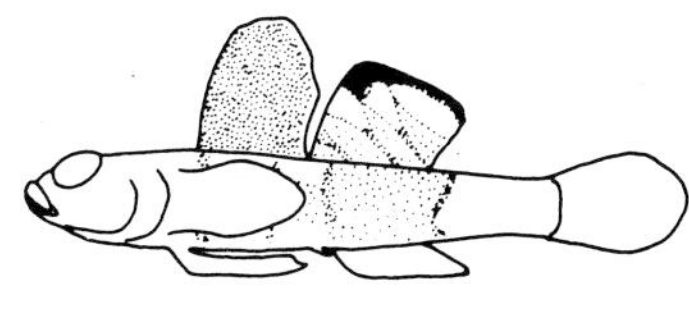

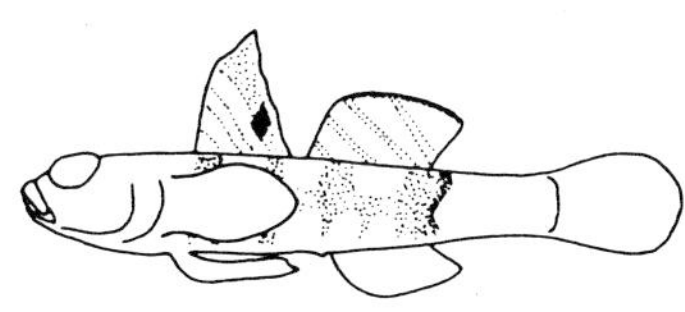

Diminutive Goby
Lebetus scorpioides

No anterior pelvic membrane; lateral scales 25–29; male: large dusky yellow first dorsal fin. 4 cm. Female: purplish-brown lateral markings; dark spot on first dorsal fin. 4 cm.

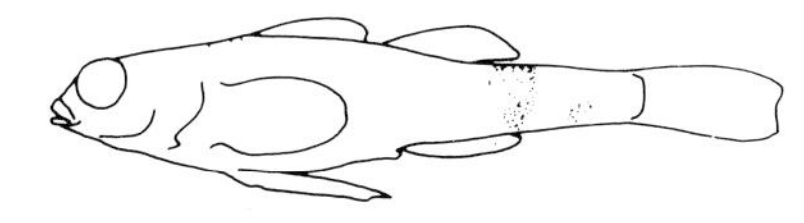

Grotto Goby
Speleogobius trigloides

Reddish; pale band across the tail stalk (like Diminutive Goby, *Lebetus scorpioides*); scales on nape before the first dorsal fin; small mouth and snout. 2.26 cm.

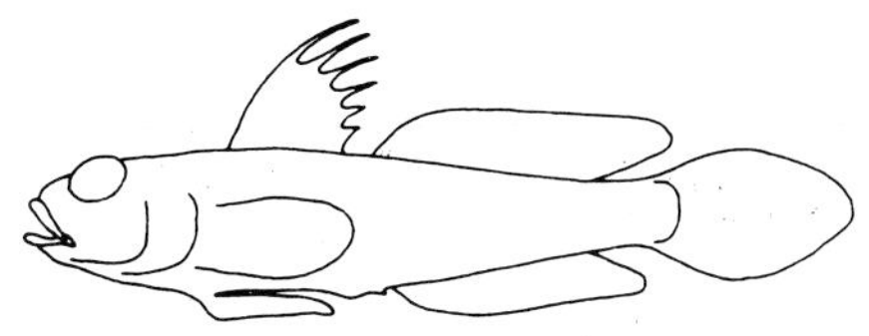

Fries' Goby
Lesueurigobius friesii

Yellow spots; scales on nape; lateral scales 28–29. 10 cm.

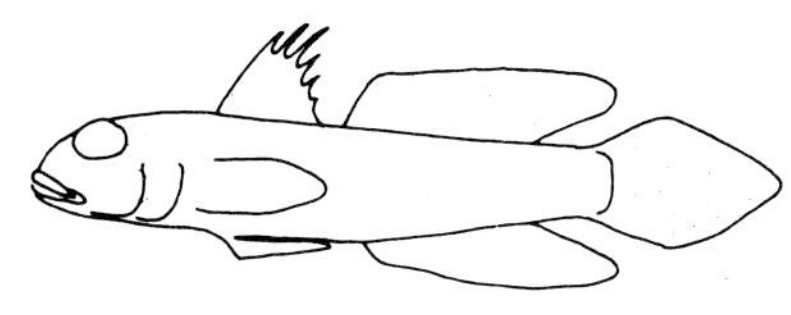

LeSueur's Goby
Lesueurigobius suerii

Blue and yellow markings on head, body and fins; nape naked; lateral scales 26–28. 5 cm.

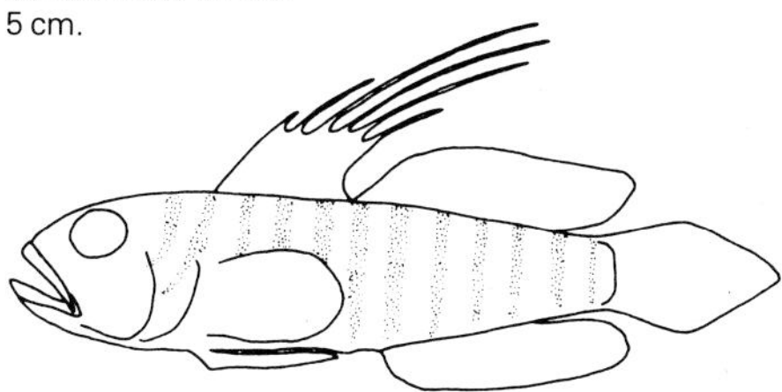

Sanzo's Goby
Lesueurigobius sanzoi

Vertical dark and yellow bands; scales on nape; lateral scales 25–26; 1st dorsal rays very elongate in male. 9.5 cm.

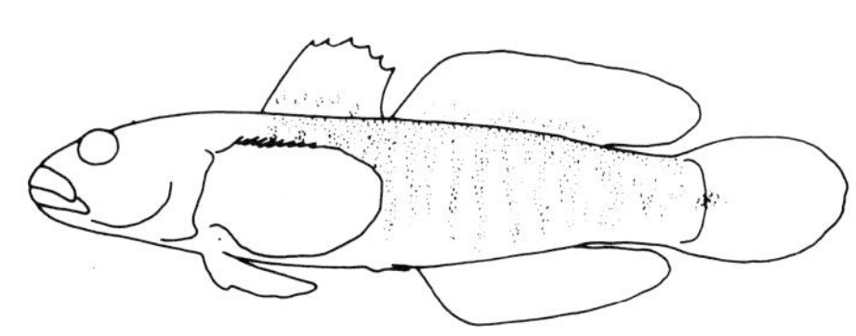

Madeiran Goby
Mauligobius maderensis

Dark, with vertical dark bands; dorsal fins with pale spots near bases; scales on rear cheek; pelvic disc with lateral lobes; free pectoral rays; lateral scales 53–55 (48–57). 15 cm.

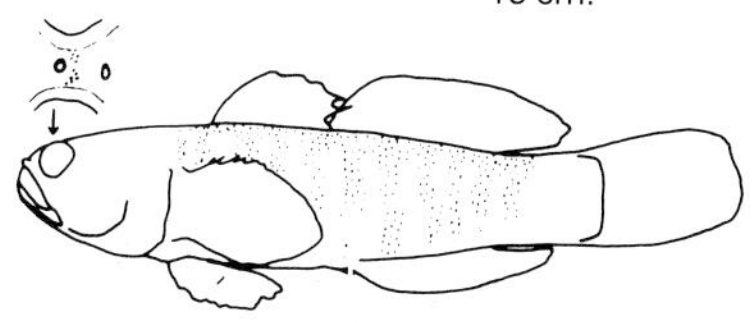

Miller's Goby
Millerigobius macrocephalus

Reddish to grey-brown; vertical brown bars; transverse row of minute papillae between eyes; lateral scales 28–32. 4.35 cm.

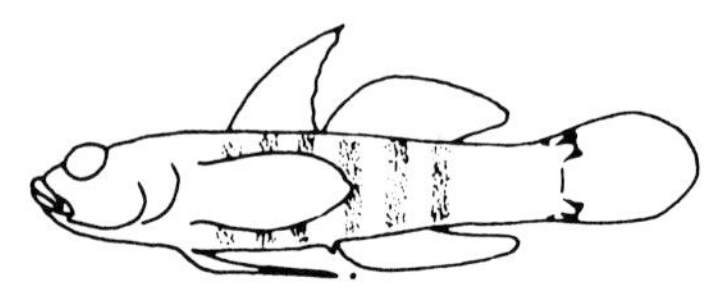

Coralline Goby
Odondebuenia balearica

Pelvic fins separate;
modified scales at tail fin base;
lateral scales 28–30 (24–32).
3.2 cm.

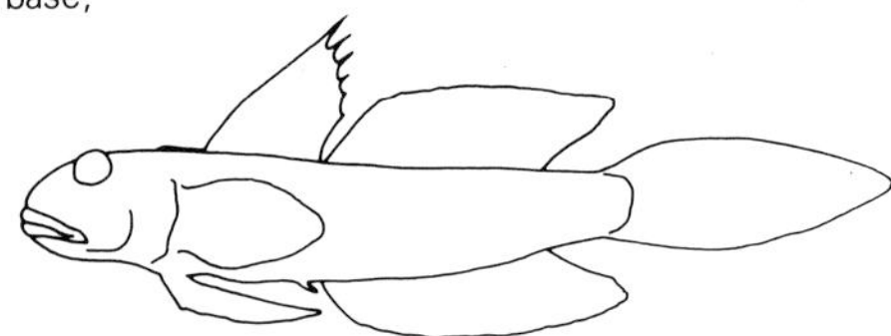

Papuan Goby
Oxyurichthys papuensis

Orange nape ridge;
tail fin lanceolate;
orange flap on upper edge of eye;
lateral scales 75–80.
11 cm.

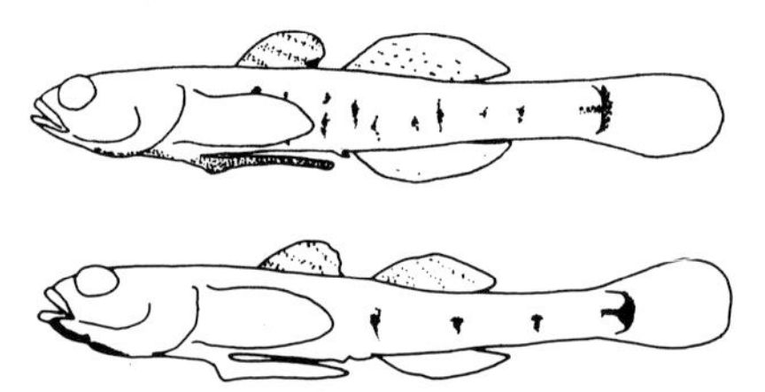

Bath's Goby
Pomatoschistus bathi

Slim;
males with short lateral bars;
females with pale breast, pelvic disc and anal fin, but with dark median band along underside of head;
lateral scales 32–37 (30–38).
3.2 cm.

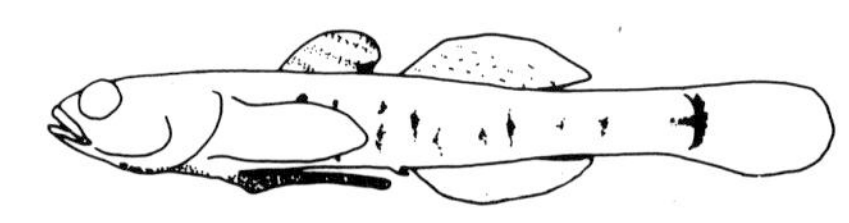

Canestrini's Goby
Pomatoschistus canestrinii

Many small intense black spots;
lateral scales 36–42;
brackish water.
6.7 cm.

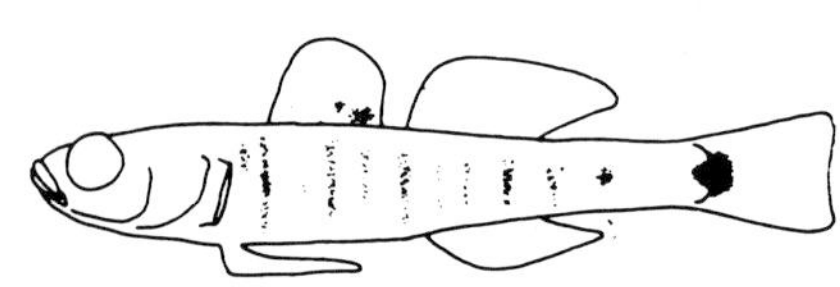

Kner's Goby
Pomatoschistus knerii

Tail fin square-cut;
lateral scales 38–46;
vertical bars in both sexes (more numerous in males), and a large tail spot.
4 cm.

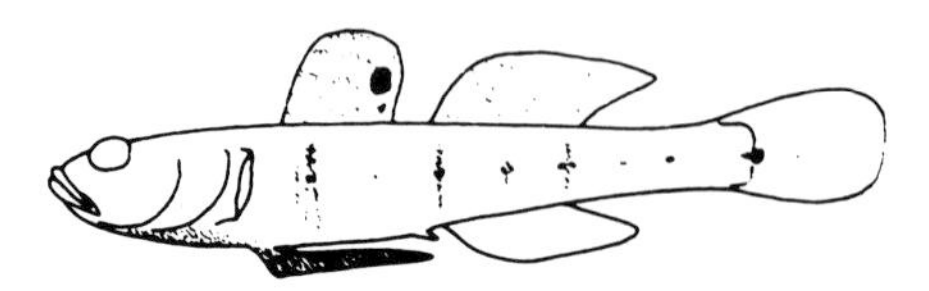

Marbled Goby
Pomatoschistus marmoratus

Anterior pelvic membrane with villi;
lateral scales 40–46 (37–48);
males with about 4 vertical bars, more or less distal first dorsal spot, and dusky breast;
females with dark chin spot.
6.5 cm.

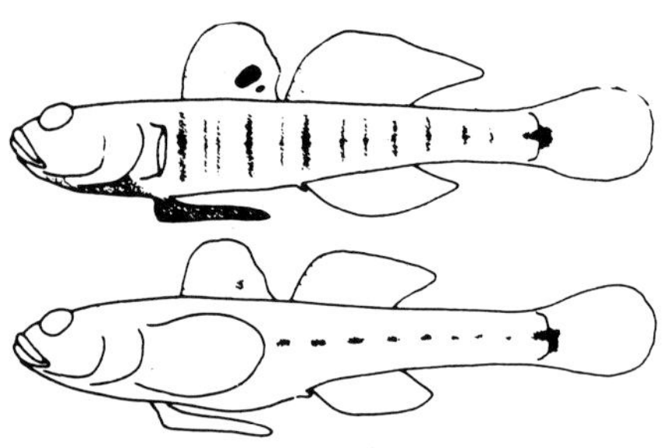

Common Goby
Pomatoschistus microps

Lateral scales 39–52;
males: numerous vertical bars, single 1st dorsal spot (nearer base of fin), and dusky throat and breast.
6.4 cm.
Female: without lateral bars and merely inconspicuous 1st dorsal spot;
breast pale.

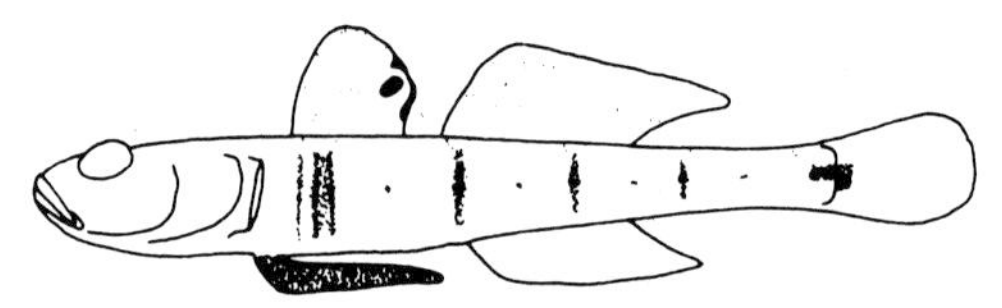

Sand Goby
Pomatoschistus minutus

Anterior pelvic membrane with villi;
lateral scales 58–70;
males with about 4 vertical bars, one distal 1st dorsal spot, and pale breast.
9.5 cm.

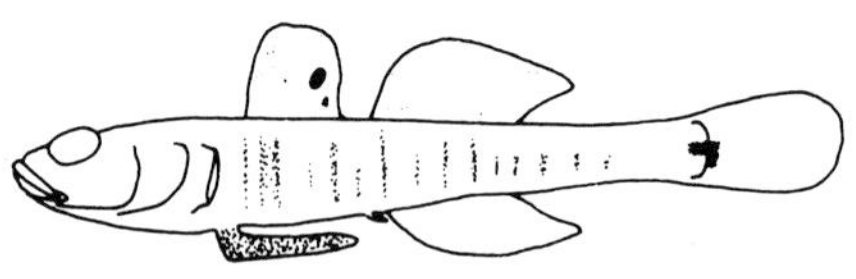

Norwegian Goby
Pomatoschistus norvegicus

Pale;
pelvic villi;
lateral scales 55–60;
males with more numerous vertical stripes.
6.5 cm.

Painted Goby
Pomatoschistus pictus

Dorsal fins with rows of large black spots and rosy bands;
lateral scales 34–43;
dusky breast.
5.7 cm.

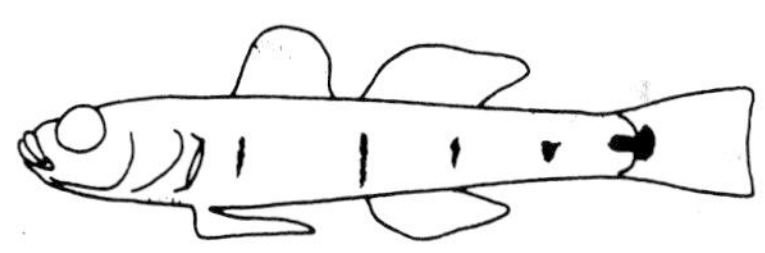

Quagga Goby
Pomatoschistus quagga

Scales in lateral series 35–40; vertical bars in both sexes, but not more than 4 present.
4.6 cm.

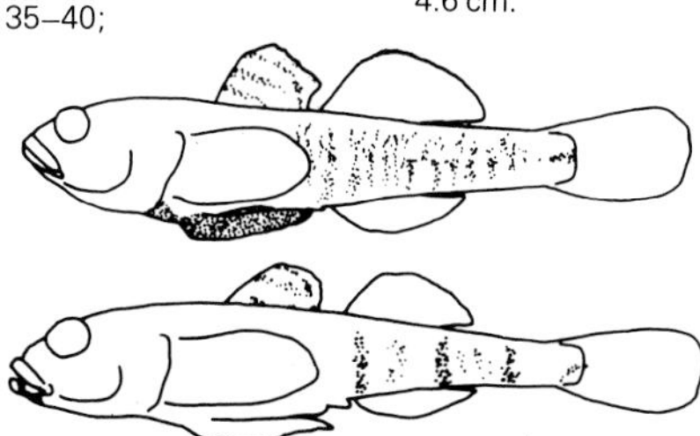

Tortonese's Goby
Pomatoschistus tortonesei

Males: many ill-defined vertical bars, and dark breast and pelvics; females: 3 major vertical bars, bright yellow underside of head, and chin mark.
3.7 cm.

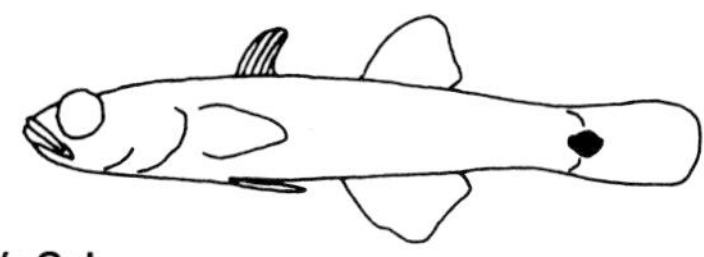

Ferrer's Goby
Pseudaphya ferreri

Large tail spot;
5 1st dorsal fin rays;
lateral scales 25–26.
2.9 cm.

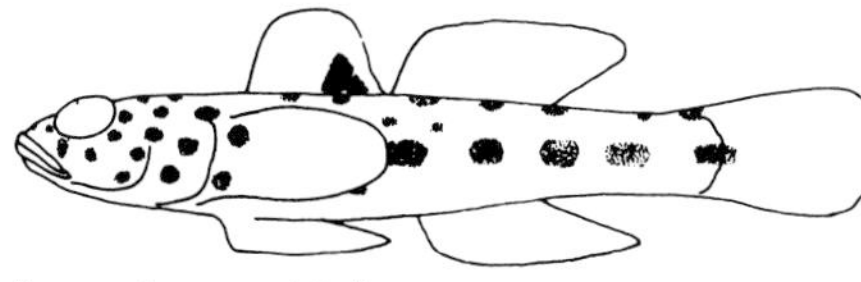

Leopard-spotted Goby
Thorogobius ephippiatus

Large brick-red blotches;
no free pecotral rays;
later scales 36–38 (33–42).
13 cm.

Canary Goby
Vanneaugobius canariensis

Pelvic fins separate;
modified scales at tail fin base;
middle rays of 1st dorsal fin elongate in male;
1st dorsal with pale banding;
pelvics with dark blotch.
4.3 cm.

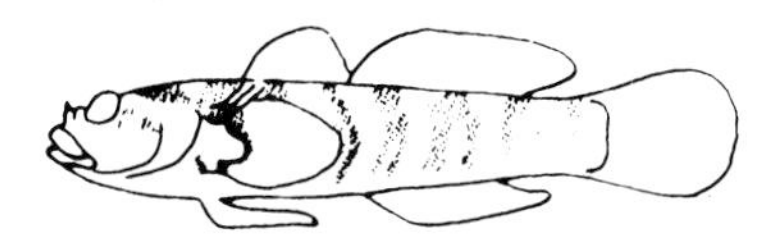

Zebra Goby *Zebrus zebrus*

Free pectoral rays;
lateral bars;
anterior nostril with tentacle;
median snout ridge;
lateral scales 32–34 (29–38).

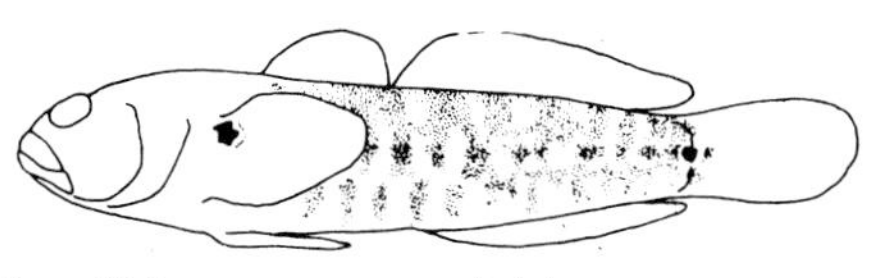

Grass Goby
Zosterisessor ophiocephalus

Pectoral rays free only at tips;
greenish, with irregular dark bars;
small pectoral and tail spots;
lateral scales 59–64 (53–68).
25 cm.

Bridled Goby
Coryphopterus glaucofraenum

Transclucent;
tail fin rounded;
lateral scales 26–27 (25–28).
Western Atlantic.

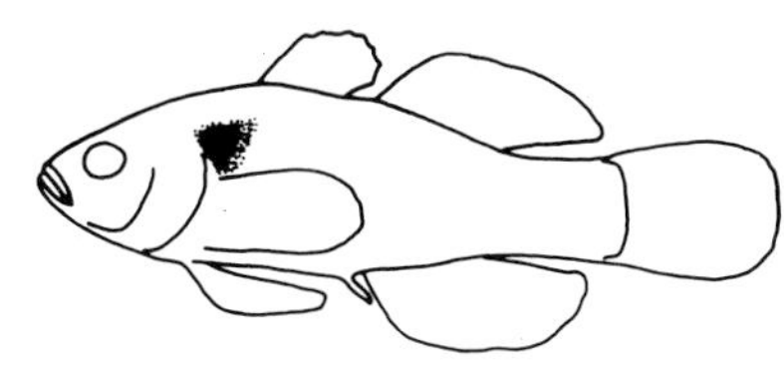

Fat Sleeper *Dormitator maculatus*

Pelvic fins separate;
body deep, compressed;
dark blue shoulder spot;
lateral scales 32–35.
Western Atlantic.
Over 30 cm.

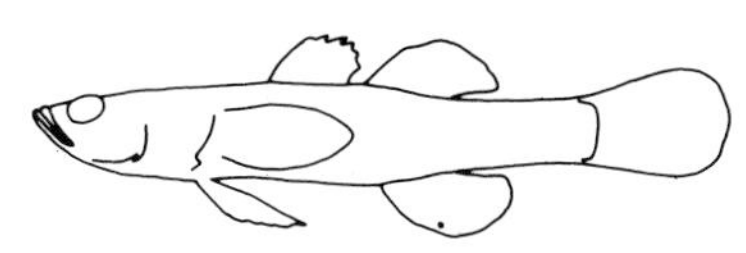

Spinycheek Sleeper
Eleotris pisonis

Pelvic fins separate;
downwardly pointing spine in lower rear corner of cheek;
lateral scales 60–65.
Western Atlantic.
21 cm.

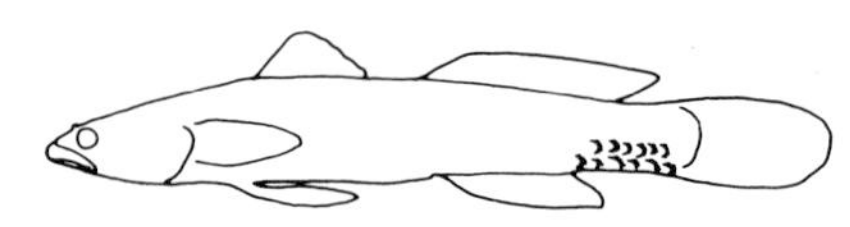

Sponge Goby *Evermannichthys spongicola*

Elongate;
no scales except rows on lower tail stalk;
numerous dark bars;
in large sponges.
Western Atlantic.
2.5 cm.

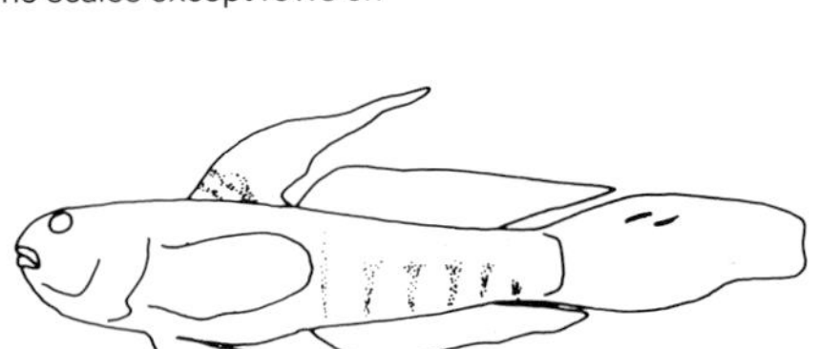

Lyre Goby *Evorthodus lyricus*

Tail fin lanceolate;
nape fully scaled;
several thin vertical bars;
spots on tail;
teeth tips bilobed;
lateral scales 30–35.
Western Atlantic.

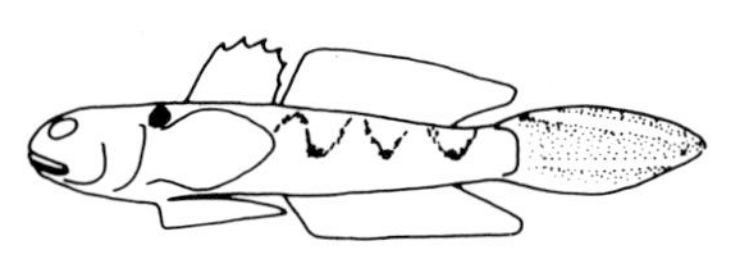

Darter Goby *Gobionellus boleosoma*

V-shaped markings above mid-line;
larger shoulder spot;
male with white streaks on tail fin;
lateral scales 29–33.
Western Atlantic.
6.2 cm.

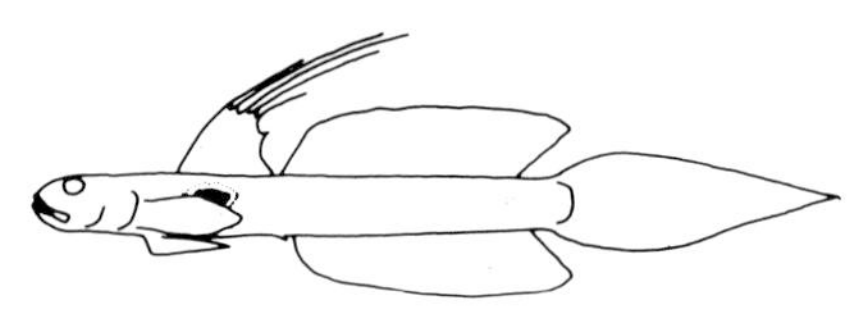

Sharptail Goby *Gobionellus hastatus*

Lateral blotch opposite pectoral fin;
very long tail fin;
lateral scales 76–89;
1st dorsal fin ray very elongate in male.
Western Atlantic.
15 cm (without tail fin).

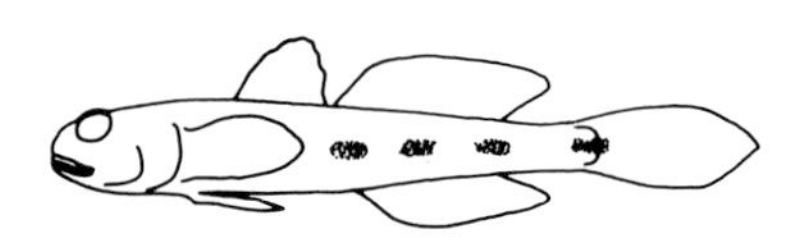

Freshwater Goby *Gobionellus shufeldti*

Mid-line blotches;
lateral scales 34–36.
8.4 cm (males), 7.8 cm
Western Atlantic.
(females).

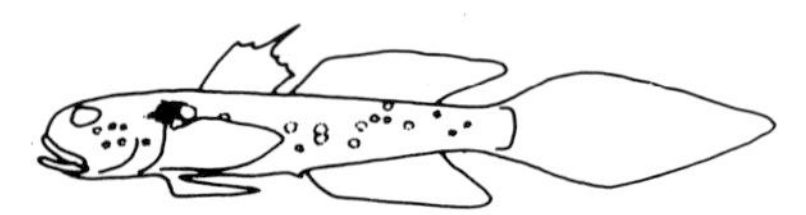

Emerald Goby *Gobionellus smaragdus*

Dark shoulder spot;
brownish rings on head and body;
lateral scales 39–46.
Western Atlantic.
7.6 cm.

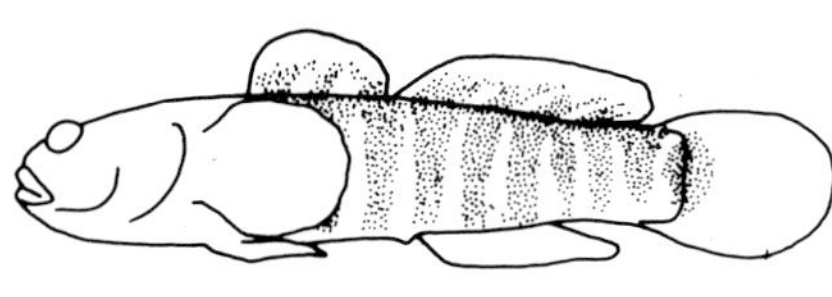

Naked Goby *Gobiosoma bosci*

Dark bars broader than pale interspaces, well defined and no obvious dark mid-line spots;
scales entirely absent;
2 pores over opercle.
Western Atlantic.
6.4 cm.

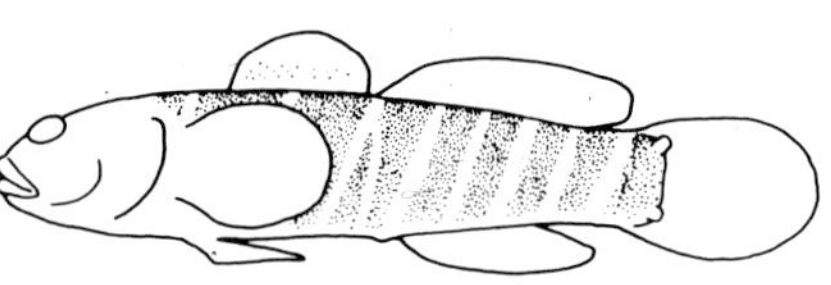

Seaboard Goby *Gobiosoma ginsburgi*

Dark bars broader than pale interspaces and dark mid-line spots;
no scales except for 2 on each side of tail fin base;
2 pores over opercle.
Western Atlantic.
5.2 cm.

Pink Wormfish *Microdesmus longipinnis*

Very elongate;
continuous dorsal fin;
pelvics small, separate.
Western Atlantic.
26 cm.

Green Goby *Microgobius thalassinus*

Green, 1st dorsal with reddish spots;
female with dark markings on 1st dorsal fin;
lateral scales 43–50, mostly cycloid.
Western Atlantic.
4 cm.

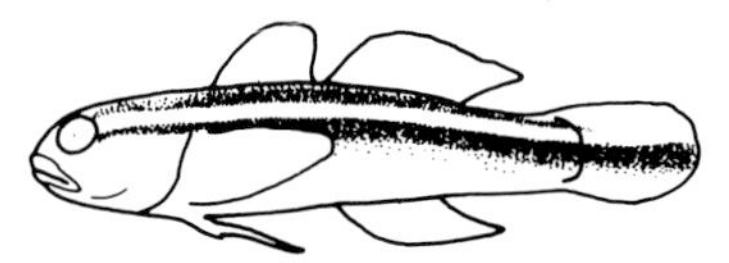

Yellow-prow Goby *Elacatinus xanthiprorus*

Tail rounded; no scales;
bright yellow and black stripes along head and body; on or in sponges.
Western Atlantic.
4.0 cm.

Diagnostic features of species not illustrated.

Roule's Goby *Gobius roulei*

No scales on nape;
elongate 1st dorsal rays;
lateral scales 33–34.
8 cm.

Knipowitchia caucasica

Similar to Lagoon Goby (*Knipowitschia panizzae*) but with scales forward to the origin of 2nd dorsal fin, and males have a longer dark bar behind root of pectoral fin.
5 cm.

Guillet's Goby *Lebetus guilleti*

Similar to Diminutive Goby (*Lebetus scorpioides*) but 2nd dorsal fin with 1/7–9 and anal fin 1/5–6 rays.
2.4 cm.

Maul's Goby *Lesueurigobius heterofasciatus*

Narrow vertical dark bands, varying in intensity;
no scales on nape;
lateral scales 28.
4.4 cm.

Lozano's Goby *Pomatoschistus lozanoi*

Paler than Sand Goby, often more speckled, but differing chiefly in pattern of head sensory papillae.

Large-scaled Goby *Thorogobius macrolepis*

Pale spots on head;
no free pectoral rays;
lateral scales 27–28.
6 cm.

Pruvot's Goby *Vanneaugobius pruvoti*

Separate pelvics;
modified scales at tail fin base;
dark spot on rear part of 1st dorsal fin;
lateral scales 28–30;
offshore.
3.9 cm.

Transparent Goby

Aphia minuta (Risso, 1810)
= *Aphya pellucida* (Nardo, 1824)

Mediterranean and Black Sea; eastern Atlantic from Morocco to Trondheim, Norway; western Baltic.

Body laterally compressed; *eyes* lateral with the orbits well separated; *fins* pelvic membrane and disc complete, larger in males; no free pectoral rays; *teeth* adult males with large canine-like teeth; *nostrils* rim of front nostril without process; *scales* cycloid, those on body easily lost; scales do not extend onto nape; scales in lateral series 19–25 (18–25). *Fin rays* 1D V (IV-VII); 2D I, 12 (11–13); A I, 13–14 (11–15).

Body transparent, with minute dots along bases of median fins and on head.

5.8 cm (males), 5.3 cm (females).

Pelagic, inshore and estuarine, surface to 70–80 m over sand, mud, eel-grass, etc.; feeds on zooplankton. Breeding season May (Adriatic), June-August (Oslofjord). Adults subsequently die at about 1 year old.

Jeffrey's Goby

Buenia jeffreysii (Günther, 1867)

Eastern Atlantic from Portugal to Foldenfjord, Norway, Faroes, and south-west Iceland; western Mediterranean, at Banyuls.

Fins pelvic anterior membrane and disc complete; edge of anterior membrane lacking villi and lobes; no free pectoral rays; 1st dorsal fin elongate in males; *nostrils* no process on rim of front nostril; *scales* none on nape; 25–30 in lateral series. *Fin rays* 1D VI (V-VI); 2D I, 8–10; A I, 7–8.

A coarse dark reticulate pattern flecked with rusty dots and interrupted by paler dorsal; saddles opposite, large single black spots on lateral mid-line.

6 cm.

Usually found offshore, typically over 10 m and known even from deep water (330 m) beyond the continental shelf. Found on coarse shell to offshore muddy grounds. Breeds March-August (British Isles).

A related Mediterranean species, *Buenia affinis* Iljin, 1930, little known, may differ in somewhat more (36) scales in lateral series.

Banded Goby

Chromogobius quadrivittatus (Steindachner, 1863)
= *Relictogobius kryzhanovskii* Ptchelina, 1939

Mediterranean and Black Sea.

Fins pelvic anterior membrane complete, with small lateral lobes; pelvic disc short, with a straight rear edge and rounded corners (rounded truncate); pectoral fin with at best rudimentary free rays, usually indistinguishable; all scales cycloid; *nostrils* front nostril without process on rim, rear nostril a short tube; *teeth* lateral canines in lower jaw, enlarged rear canines in middle of upper jaw; *scales* no scales on nape; 56–72 in lateral series. *Fin rays* 1D VI; 2D I, 10 (8–11); A I, 9 (7–9).

Pale brown with 10–14 vertical dark bars across flanks; pale saddle at origin and end of 2nd dorsal fin; broad pale band across nape and root of pectoral fin, with its posterior edge sharply demarcated by straight black band across pectoral base; head, cheeks and gill cover with convolute pattern, intense black spot in lower front corner of gill cover.

6.5 cm.

Inshore and intertidal.

Kolombatovic's Goby

Chromogobius zebratus (Kolombatovic, 1891)
= *Gobius depressus* Kolombatovic, 1891

Mediterranean.

Similar elongate body form to the Banded Goby (*Chromogobius quadrivittatus*), but original scales ctenoid and only 41–52 in lateral series.

Five broad pale saddles across back; cheek with simpler pattern not extending onto branchiostegous membrane and no black gill cover spot; edge of pectoral band bent.

5.3 cm.

Habitat intertidal to offshore, probably on coralline grounds.

Liechtenstein's Goby
Corcyrogobius liechtensteini
(Kolombatovic, 1891)

Adriatic.

Fins pelvic anterior membrane and disc complete except for emargination of rear edge, and no lateral lobes or villi on anterior membrane; no free pectoral rays; 1st dorsal fin elongate in males, reaching to rear half of 2nd dorsal fin-base; *nostrils* front nostril without process on rim; *scales* 27–29 (27–31) in lateral series. *Fin rays* 1D VI; 2D I, 9: A I, 8 (7–9).

An intense black spot on the underside of the head on the branchiostegous membrane of each side.

2.5 cm.

Habitat probably offshore, on coralline grounds.

Crystal Goby
Crystallogobius linearis (Von Düben, 1845)
= *Cr. nilssonii* (Von Düben & Koren, 1846)

Mediterranean; eastern Atlantic from Gibraltar to Lofoten Islands, Norway, and Faroes.

Body laterally compressed; *eyes* lateral with a wide space between orbits; *fins* pelvic disc a deep funnel-like structure in males, but both pelvic disc and 1st dorsal fin vestigial or lacking in females; no free pectoral rays; *jaws and teeth* lower jaw of male with prominent canine teeth; jaws in female smaller and toothless; *scales* absent. *Fin rays* 1D II-III (male), 0 (females); 2D I, 18–20; A I, 20–21.

Transparent, with dots of pigment on chin and along bases of medium fins.

4.7 cm (male), 3.9 cm (female).

Nektonic, but more offshore in distribution than Transparent Goby (*Aphia minuta*), in depths of even 400 m, over bottoms of dead shells, sand or mud. Males live on sea-bed during breeding season (May-August in Oslofjord), guarding eggs deposited in empty tubes of large worms like *Chaetopterus*. Feeds on plankton.

Four-spotted Goby
Deltentosteus quadrimaculatus
(Valenciennes, 1837)

Mediterranean, and adjoining eastern Atlantic, from Mauretania to southern Bay of Biscay.

Fins pelvic anterior membrane and disc complete, the anterior membrane with neither villi or lateral lobes; no free pectoral fin rays; second ray of 1st dorsal fin elongate, reaching to front half of 2nd dorsal fin when depressed; *scales* nape and breast completely scaled; 33–35 scales in lateral series. *Fin rays* 1D VI; 2D I, 8–9; A I, 8–9; pectorals 17–18.

Fawn with reticulate pattern and 4 large black spots along lateral mid-line, opposite paler saddles across back; 1st dorsal fin with conspicuous black spot at rear end and black streak along anterior edge, including elongate 2nd ray; dusky pelvic and anal fins.

8 cm.

Inshore on sand or muddy sand, to 90 m. Breeding season in spring (Mediterranean).

Toothed Goby
Deltentosteus colonianus (Risso, 1826)
= *D. lichtensteinii* (Steindachner, 1883)

Western Mediterranean and Adriatic; eastern Atlantic along south-west Portugal.

Similar to Four-spotted Goby (*D. quadrimaculatus*), but longer jaw (angle of mouth below rear half of eye) and more rays in 2nd dorsal fin (I, 10–11), anal fin (I, 10–11) and pectorals (18–19).

7 cm.

Inshore, including sea-grass meadows, to about 120 m. Biology otherwise unknown.

Koch's Goby
Didogobius kochi Van Tassell, 1988

Eastern Atlantic, at the Canary Islands (Gran Canaria).

Body elongate; *eyes* moderate, widely separated; *fins* tail fin lanceolate; pelvic anterior membrane and disc complete; pectoral fins lacking upper free rays; *nostrils* anterior nostril tubular, without process from rim; *teeth* lower jaw with lateral canines; upper jaw with large median rear teeth;

scales all cycloid, 33–37 in lateral series. *Fin rays* 1D VI; 2D I, 12; A I, 11.

Brownish, with a few thin pale saddles; base of pectoral fin with vertical dark bar; anal fin with broad pale edge; tail fin dark, black towards lower edge.

5.7 cm.

Inshore rocky outcrops on sandy bottom. Biology otherwise unknown.

Ben-Tuvia's Goby
Didogobius bentuvii Miller, 1966

Eastern Mediterranean, off Israel.

As Koch's Goby (*D. kochi*), but *eyes* small and widely separated, and *scales* all cycloid, 65–70 in lateral series. *Fin rays* 1D VI; 2D I, 14; A I, 12.

Pale, somewhat translucent in life; preserved, pale fawn with numerous melanophores over head and body.

3.65 cm.

The only specimen on record was trawled on muddy sand in about 37 m.

Steinitz's Goby
Gammogobius steinitzi Bath, 1971.

Mediterranean (near Marseilles).

Fins pelvic anterior membrane and disc complete; *nostrils* front nostril without process on rim; *scales* ctenoid, 29–37 in lateral series; nape naked. *Fin rays* 1D VI; 2D I, 10; A I, 9.

Brownish, with 5–6 dark vertical bands, broader than pale interspaces; small dark spots along bases of dorsal fins.

3.8 cm.

Habitat inshore, in submarine grottoes, 2–15 m. Biology unknown.

Bellotti's Goby
Gobius ater Bellotti, 1888
= *Gobius balearicus* Lozano y Rey, 1919.

Mediterranean.

Fins pelvic anterior membrane without prominent lateral lobes; pelvic disc rounded; pectoral fins with well-developed upper free rays; 1st dorsal fin rays not elongate; *nostrils* anterior nostril with divided process from rim; *scales* nape scaled, cheek naked, 38–40 in lateral series. *Fin rays* 1D VI; 2D I, 12–14; A I, 11.

Brown, with dark pectoral fin lobe; 1st dorsal fin with narrow pale edge.

7.1 cm.

Inshore, and in lagoons, among seagrass.

Golden Goby
Gobius auratus Risso, 1810

Mediterranean; eastern Atlantic, Canaries to northern Spain.

Fins pelvic anterior membrane reduced, pelvic disc deeply cleft (right); pectoral fin upper rays free; 1st dorsal fin not elongate; *nostrils* front nostril with a triangular flap; *scales* nape scaled, 44–50 in lateral series. *Fin rays* 1D VI; 2D I, 14–15 (14–16); A I, 14.

Body with gold flecks and mottling; blotches along sides, 5 below 2nd dorsal fin; pectoral fin base with dark mark deeper than long; usually 2 cheek spots.

10 cm.

Inshore, among and under stones.

Bucchich's Goby
Gobius bucchichii Steindachner, 1870

Mediterranean and Black Sea; eastern Atlantic, Morocco to Algarve coast of Portugal.

Fins pelvic anterior membrane and disc complete, former without lateral lobes; upper pectoral fin rays free; *nostrils* front nostril with simple elongate process (right), sometimes forked; *scales* nape scaled, 50–

Golden Goby, *Gobius auratus* (Andrea Ghisotti).

56 in lateral series. *Fin rays* 1D VI; 2D I, 14 (13–14); A I, 13 (12–14).

Fawn or darker brown, with longitudinal rows of numerous small dark spots along head and body, usually more noticeable on lateral mid-line and below.

10 cm.

Occurs on inshore sandy and muddy grounds, near Snakelocks anemones (*Anemonia sulcata*), among whose tentacles the goby seeks refuge when alarmed. Breeds June to August (Crimea, Black Sea).

Giant Goby
Gobius cobitis Pallas, 1811
= *Gobius capito* Valenciennes, 1837

Mediterranean and Black Sea; eastern Atlantic, from Morocco to western English Channel; also Gulf of Suez (probably via Suez Canal).

Fins pelvic anterior membrane with conspicuous lateral lobes (below right); pelvic disc short and rounded; free upper pectoral fin rays; *nostrils* front nostril with long process from rim usually with several divisions; *scales* nape scaled; 59–67 in lateral series. *Fin rays* 1D VI; 2D I, 13 (13–14); A I, 11 (10–12).

Brownish-olive 'pepper and salt' speckling with dark mottling and rounded blotches along and below lateral line, the latter especially in smaller individuals. Ripe males dark, their median fins with narrow white edges.

27 cm, the largest Mediterranean-Atlantic goby.

Found on rocky and weedy ground in shallows to about 10 m; in Atlantic, occurs in high-tide pools on sheltered rocky shores, sometimes in brackish conditions. Breeds from March to May (Naples), May to early July (Varna, Black Sea). May be sold in Mediterranean fish-markets.

Couch's Goby
Gobius couchi Miller & El-Tawil, 1974

Eastern Atlantic (southern Cornwall, south and west Ireland).

Fins pelvic anterior membrane somewhat reduced, pelvic disc rear edge rounded to truncate; pectoral fin upper rays free; 1st dorsal fin not elongate; *nostrils* front nostril with a triangular flap; *scales* nape scaled, 40–41 (35–45) in lateral series; *fin rays* 1D VI; 2D I, 13 (12–14); A I, 12 (11–13).

Golden flecks and mottling below mid-line; blotches along sides, 5 below 2nd dorsal fin; pectoral fin base with dark mark deeper than long; usually 1 cheek spot.

9.0 cm.

Inshore and intertidal, under stones on sheltered muddy sand. Breeds in spring (western English Channel).

Bucchich's Goby, *Gobius bucchichii* (Andy Purcell).

Couch's Goby, *Gobius couchi* (B. Picton).

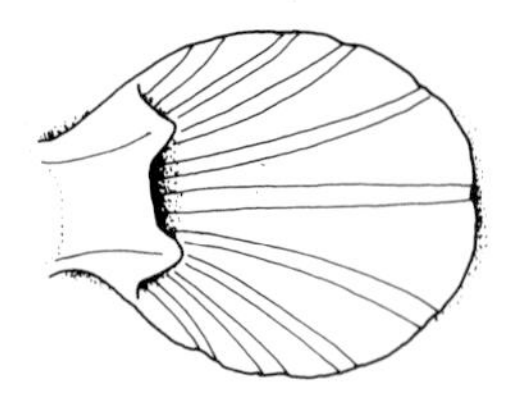

Red-mouthed Goby
Gobius cruentatus Gmelin, 1789
= *Gobius strictus* Fage, 1907 (probably a juvenile of this species)

Mediterranean; eastern Atlantic, Morocco to south-west Ireland (Lough Hyne).

Red-mouthed Goby, *Gobius cruentatus* (B. Picton).

Fins pelvic anterior membrane little too well developed, without lateral lobes; pelvic disc with somewhat cleft rear margin; pectoral fin with free uppermost rays; *nostrils* anterior nostril with simple process from rim; *scales* present on rear part of cheek as well as nape and upper gill cover; 52–58 in lateral series; *fin rays* 1D VI; 2D I, 14: A I, 12–13.

Reddish-brown with lateral blotches; lips and cheeks display vivid red markings and rows of sensory papillae on the head are black.

18 cm.

Found inshore, 15–40 m, on rocky ground, sand and sea-grass meadows.

Sarato's Goby

Gobius fallax Sarato, 1889

Mediterranean; eastern Atlantic at Canaries.

Similar to the Golden Goby (*Gobius auratus*), but body coloration with rows of small dots. *Scales* 39–48 in lateral series; *fin rays* 1D VI; 2D I, 14–16; A I, 13–15.

9 cm.

Found inshore, to 20 m, on sand near rocky ground and in grottoes. This goby may intergrade with the Golden Goby.

Steven's Goby

Gobius gasteveni Miller, 1974

Eastern Atlantic, in western English Channel and around Madeira and the Canaries.

Fins pelvic anterior membrane somewhat reduced, pelvic disc rear edge rounded to notched; pectoral fin upper rays free; 1st dorsal fin not elongate; *nostrils* front nostril with a thin tentacle; *scales* nape scaled, 40–45 (37–45) in lateral series; *fin rays* 1D VI; 2D I, 14 (13–15); A I, 13 (12–14).

Greyish to fawn; blotches along sides; 4 below 2nd dorsal fin; pectoral fin base with dark mark longer than deep; cheek and gill-cover with white spots.

12 cm.

Offshore, 35–100 m, on muddy sand and coarser grounds.

Slender Goby
Gobius geniporus Valenciennes, 1837
= *Gobius arenae* Bath, 1972 (juvenile).

Mediterranean.

Body slender, maximum depth about one-seventh of total length; *fins* pelvic anterior membrane reduced or absent, and lacking lateral lobes; pelvic disc rear edge truncate; *nostrils* front nostril with lappet or simple tentacle from rim; *scales* 50–55 in lateral series. *Fin rays* 1D VI; 2D I, 12–14; A I, 11–14.

Brownish, with rectangular lateral mid-line blotches; breeding males dusky.

16 cm.

Inshore, on sand or mud near sea-grass. Breeds from April to May (Taranto).

Gobius luteus Kolombatovic, 1891

Mediterranean.

Also resembling the Golden Goby (*Gobius auratus*), but canary yellow in colour, lacking cheek and mid-line spots and blotches. *Scales* 43–45. *Fin rays* 1D VI; 2D I, 14; A I, 13.

At least 6.5 cm.

Found in deeper inshore waters, 15–80 m, over rocks, with algae and gorgonians. Another species whose status with respect to the Golden Goby requires study.

Black Goby
Gobius niger L.
= *Gobius jozo* L.

Mediterranean and Black Sea; eastern Atlantic from Cape Blanco, Mauretania, and the Canaries, to Trondheim, Norway, and the Baltic Sea.

Fins pelvic anterior membrane without lateral lobes; pelvic disc (right) rounded (not emarginate); pectoral fins with short free upper rays; 1st dorsal fin elongate, especially in male, reaching to at least middle of 2nd dorsal fin when depressed; *nostrils* front nostril with simple flap on rim; *scales* the nape is scaled but to a variable extent; *fin rays* 1D VI (V-VII); 2D I, 12–13 (11–13); A I, 11–12 (10–13). Scales in lateral series 35–41 (32–42).

Pale brownish, with darker lateral blotches and dots, to dusky in breeding males; each dorsal fin with dark spot in upper anterior corner; branchiostegous membrane dark.

15 cm.

Found in coastal waters down to 50–75 m, on sandy or muddy ground, often in sea-grass meadows and, being tolerant of brackish water, penetrating estuaries and lagoons. Breeds from March-May (Naples), May-August (Baltic). Often sold in Mediterranean fish markets.

Black Goby, *Gobius niger* (B. Picton).

Rock Goby, *Gobius paganellus* (B. Picton).

Rock Goby
Gobius paganellus Linnaeus, 1758

Mediterranean and Black Sea; eastern Atlantic from Senegal to western Scotland, including Canaries, Madeira and Azores; also reported from Gulf of Aqaba, probably as migrant through Suez Canal.

Fins pelvic anterior membrane well developed, with at best only small lateral lobes; pectoral fin free rays very conspicuous (left), reaching origin of 1st dorsal fin in smaller individuals; *nostrils* front nostril with tentacle from rim divided into several processes in larger fish; *scales* nape scaled, and cheek sometimes scaled in upper rear corner; 50–55 (46–59) in lateral series. *Fin rays* 1D VI; 2D I, 13–14 (12–15); A I, 11–12 (10–13).

Fawn, with darker mottling and lateral blotches, becoming dark; upper margin of 1st dorsal fin with pale horizontal band and rear corner with dark spot in small fish; breeding males are deep purplish-brown, with 1st dorsal fin-band yellow or orange.

12 cm.

Inshore rocky shallows in Mediterranean; along Atlantic coasts, common under stones in pools on sheltered rocky shores with much weed cover. Breeding season January to June (Naples), April to June (Isle of Man).

Roule's Goby
Gobius roulei De Buen, 1928

Western Mediterranean; eastern Atlantic off south-west Portugal.

Similar to the Black Goby, *Gobius niger*, with elongate 1st dorsal fin, but no scales on nape. *Scales* 33–34 in lateral series. *Fin rays* 1D VI; 2D I, 12; A I, 11.

Coloration unknown in life, but no dark blotches in upper anterior corners of dorsal fins.

8 cm.

Inshore in Mediterranean, but from 320–385 m off Portugal. Rare.

Striped Goby
Gobius vittatus Vinciguerra, 1883

Mediterranean.

Fins pelvic anterior membrane vestigial; pelvic disc deeply cleft; pectoral free rays little developed; *nostrils* front nostril with lappet on rim; *scales* nape scaled, 32–36 in lateral series. *Fin rays* 1D VI; 2D I, 11–13; A I, 11–13.

Olive to yellowish, with broad black stripe along head and body, from snout to root of tail fin.

5.8 cm.

Offshore, 20–42 (85) m, on coralline grounds.

Two-spotted Goby
Gobiusculus flavescens (Fabricius, 1779)
= *Gobius ruthensparri* Valenciennes, 1837

Eastern Atlantic from north-west Spain to Vesteralen, Norway, including the Faroes and western Baltic, but not the south-eastern North Sea.

Body somewhat laterally compressed, *eyes* lateral with a wide space between the orbits; *fins* pelvic anterior membrane and

Two-spotted Goby, *Gobiusculus flavescens* (B. Picton).

disc complete, without villi along edge of former; no free pectoral rays; *nostrils* front nostril simple and tubular, without process on rim; *scales* none on nape; 35–40 in lateral series. *Fin rays* 1D VII (VI-VIII); 2D I, 9–10; A I, 9–10.

Reddish to olive-brown, with dark reticulate pattern and several pale saddle-like markings on nape and back; many small alternate dark and pale bluish marks along lateral mid-line; dorsal fins banded red; a large black spot, partly edged with yellow, at base of tail fin. Males have another large intense dark spot on lateral mid-line below 1st dorsal fin.

6 cm.

Inshore, above sea-bed, in groups about weed-grown structures and over beds of oarweed or eel-grass, down to 20 m; also in shore pools among weed. Breeds April to August (western English Channel).

Lagoon Goby
Knipowitschia panizzae (Verga, 1841)

Adriatic Sea.

Fins pelvic disc complete; pelvic anterior membrane with smooth edge; pectoral fins without free rays; tail fin rounded; *nostrils* front nostril rim without process; *scales* nape and anterior back naked to 2nd dorsal base; 32–39 in lateral series. *Fin rays* 1D VI (V-VI); 2D I, 7–8 (7–9); A I, 7–8.

Fawn or olive, with broad pale saddle-like markings across back; adult male with lateral bars, dark head and breast, and dark spot in rear corner of 1st dorsal fin; female with black chin spot and dark band across 1st dorsal fin.

3.75 cm.

Brackish estuaries and lagoons. Breeds April to August (Venezia). The related Black Sea species, *Knipowitschia caucasica*, may occur along Aegean coasts; it has scales forward to the origin of 2nd dorsal fin, and males have a longer dark bar behind root of pectoral fin.

Diminutive Goby
Lebetus scorpioides (Collett, 1874)
= *L. orca* (Collett, 1874)

Eastern Atlantic, from the northern Bay of Biscay to Norway (Hemnefjord), the Faroes, and south-west Iceland.

Fins pelvic disc with a concave rear edge

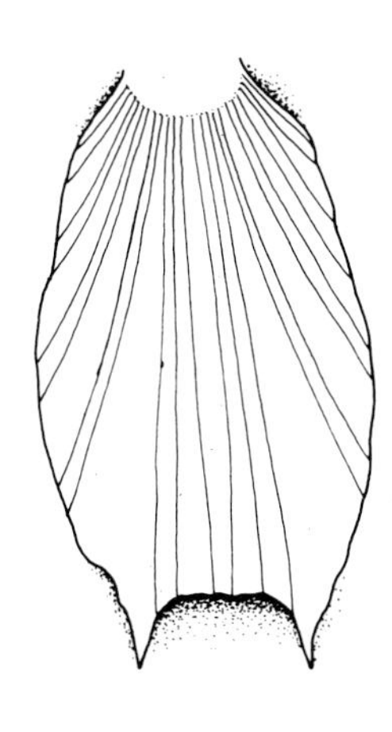

and lacking an anterior membrane (left); pectoral fins without free rays; 1st dorsal fin high, and especially voluminous in adult males, when it may reach to at least the middle of the 2nd dorsal fin base; *nostrils* front and rear nostrils tubular, but without rim processes; *scales* nape without scales; 26–27 (25–29) in lateral series. *Fin rays* 1D VI (VI-VII); 2D I, 9–10; A I, 7–8.

Vertical bars on body and broad, well-demarcated pale band across tail stalk; males yellowish to dusky grey, underside of head and breast reddish orange; 1st dorsal fin dusky yellow edged with white, 2nd dorsal fin with intense black edge and oblique yellow and white bands; females pale brown, with bars purplish brown, and 1st dorsal fin displaying oblique yellow to orange-red bands with black spot at rear end, while 2nd dorsal fin has narrow dark edge and thin oblique orange-red bands.

3.9 cm.

Offshore, to 375 m, mostly on coarse ground, with coralline deposits, but also in muddy areas. Breeds March to August (Isle of Man).

Guillet's Goby (*Lebetus guilleti*) (Le Danois, 1913), with 2D I/7–9 and A I/5–6, is even smaller (to 2.4 cm), from inshore coarse grounds, to 29 m, in the western Mediterranean and eastern Atlantic (Portugal to Kattegat and Belt Seas). Another tiny species (to 2.26 cm), the Grotto Goby (*Speleogobius trigloides* Zander & Jelinek, 1976), from inshore waters of the Adriatic, has a similar pale band across the tail stalk, but has scales before the 1st dorsal fin, small mouth and snout, and minute lateral-line canals around upper border of the eye.

Fries' Goby

Lesueurigobius friesii (Malm, 1874)
= *Gobius macrolepis* auct. (not Kolombatovic, 1891)

Mediterranean, into Sea of Marmora; eastern Atlantic from Portugal to southern Norway.

Fins pelvic disc complete, anterior membrane without villi or lateral lobes; no free pectoral rays; 1st dorsal fin rays elongate; tail fin lanceolate; *scales* cover nape; 28–29 in lateral series. *Fin rays* 1D VI; 2D I, 13–16; A I, 12–15.

Pale fawn with numerous golden yellow dots over nape, body, dorsal and tail fins.

10 cm.

Burrowing in muddy sand or mud, 10–130 m, often trawled with *Nephrops norvegicus*. Breeds from late May to August (western Scotland).

LeSueur's Goby

Lesueurigobius suerii (Risso, 1810)
= *L. lesuerii* (Valenciennes, 1837)

Mediterranean; eastern Atlantic, Canaries and Morocco.

Fins pelvic disc complete, anterior membrane without villi or lateral lobes; no free pectoral rays; 1st dorsal fin rays somewhat elongate, especially in males; tail fin lanceolate; *scales* nape without scales; 26–28 in lateral series. *Fin rays* 1D VI; 2D I, 13–14; A I, 13–14.

Body with vertical markings of blue and yellow; head and gill cover exhibit oblique yellow bands on blue background; yellow lines on dorsal and tail fins.

5 cm.

Inshore, but young may occur to 230 m. Breeds in summer and autumn (Mediterranean).

Sanzo's Goby

Lesueurigobius sanzoi (De Buen, 1918)

Western Mediterranean (Alboran Sea); eastern Atlantic from Mauretania to Portugal.

Fins pelvic disc complete, anterior membrane without villi or lateral lobes; no free pectoral rays; 1st dorsal fin rays very elongate, especially in males, reaching to end of 2nd dorsal fin or root of tail fin when depressed; tail fin lanceolate; *scales* nape with scales; 25–26 in lateral series. *Fin rays* 1D VI; 2D I, 15; A I, 16–17.

Vertical dark brown and yellow bands across flanks.

9.5 cm.

Found offshore, on muddy sand or mud, 47–100 m.

Maul's Goby (*Lesueurigobius heterofasciatus*) (Malm, 1971), 4.4 cm, with narrow vertical dark bands on the flanks, varying in intensity, and no scales on the nape, has been found off Madeira and Morocco.

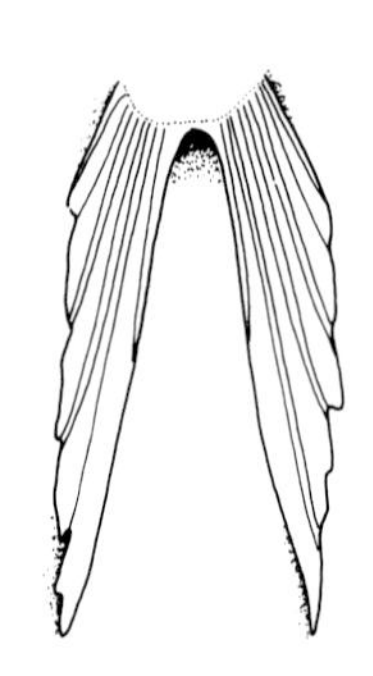

Madeiran Goby
Mauligobius maderensis (Valenciennes, 1837)

Eastern Atlantic islands of Madeira, the Salvages, and the Canaries.

Fins pelvic anterior membrane with lateral lobes; pelvic disc rounded; pectoral fins with well-developed free upper rays; *nostrils* front nostril with simple or branched tentacle on rim; *scales* nape and rear cheek scaled; 53–55 (48–57) in lateral series. *Fin rays* 1D VI; 2D I, 13 (13–14); A I, 11 (11–12).

Dark, with about a dozen vertical dark bands more or less evident; head with pale spots in smaller fish; dorsal fins with rows of pale spots nearer bases and narrow light edge.

15 cm.

Found in intertidal rock pools.

Miller's Goby
Millerigobius macrocephalus (Kolombatovic, 1891)

Mediterranean.

Head transverse row of minute papillae across space between eyes; *fins* pelvic anterior membrane and disc complete, former without villi; pectoral fins with uppermost rays free at tips; *nostrils* front nostril tubular, to over upper lip, but without process on rim; *scales* ctenoid, 28–32 in lateral series; nape naked. *Fin rays* 1D VI (V-VI); 2D I, 10–11; A I, 9–10.

Reddish to pale grey-brown, with about a dozen dark brown vertical bands especially noticeable in smaller fish.

4.35 cm.

Habitat inshore shallows and lagoons, about and beneath stones.

Coralline Goby
Odondebuenia balearica (Pellegrin & Fage, 1907)

Mediterranean.

Fins pelvic fins (above right) virtually separate, without anterior membrane and only a low ridge between the bases of the innermost rays of the fins; pectoral fins without free rays; 1st dorsal fin elongate, reaching to at least middle of 2nd dorsal fin in males; *scales* none on nape; uppermost and lowermost scale on each side of tail fin is enlarged, with very long lateral ctenii; 28–30 (24–32) in lateral series. *Fin rays* 1D VI; 2D I, 10 (9–10); A I, 9 (9–10).

Male blood-red, with narrow vertical blue bands across flanks; females without vertical bars.

3.2 cm.

Found offshore, 25–70 m, coralline ground and also *Cladophora* turf.

Papuan Goby
Oxyurichthys papuensis (Valenciennes, 1837)

A common Indo-Pacific species now established along the eastern Mediterranean Levant coast after migration via Suez Canal.

Head nape ridge along mid-line before 1st dorsal fin; *fins* pelvic disc complete, anterior membrane without villi and lateral lobes; pectoral fins without free upper rays; 1st dorsal fin elongate; tail fin lanceolate; *eye* flap on upper edge; *nostrils* front nostril with a short tube, lacking rim process; *scales* nape scaled; 75–80 in lateral series. *Fin rays* 1D VI; 2D I, 12; A I, 13.

Greyish-blue, with irregular yellow band along lateral mid-line; head streaked yellow and purple; eye process and nape ridge orange.

Over 11 cm.

Found on offshore muddy grounds, in 35–45 m.

Bath's Goby
Pomatoschistus bathi Miller, 1982

Mediterranean and Sea of Marmora.

Fins pelvic disc complete, anterior membrane with smooth edge; no free pectoral rays; tail fin rounded; *nostril* front nostril rim without flap; *scales* none on back before middle of 1st dorsal fin or on breast; 32–37 (30–38) in lateral series. *Fin rays* 1D VI (V-VI), 2D I, 8 (7–9), A I, 8 (6–9).

Fawn with darker reticulation. Males with several short vertical dark marks on flanks, and dark pelvic disc, breast and underside of head; females with pale breast and pelvic disc, but a conspicuous dark band along the mid-line of the underside of the head.

3.2 cm.

Inshore sand and gravel, to 12 m, sometimes in brackish water.

Canestrini's Goby

Pomatoschistus canestrinii (Ninni, 1883)

Mediterranean (Adriatic Sea).

Fins pelvic disc complete, anterior membrane with straight to wavy rear edge; no free pectoral rays; tail fin rounded; *nostril* front nostril rim without flap; *scales* absent on back before 2nd dorsal fin; 36–42 in lateral series. *Fin rays* 1D VI, 2D I, 8–9, A I, 8–9.

Fawn to greyish, with many scattered tiny intense black spots. Males with 4 vertical dark bars on flanks and black spot on rear of 1st dorsal fin.

6.7 cm.

Found in brackish waters of estuaries and lagoons.

Kner's Goby

Pomatoschistus knerii (Steindachner, 1861)

Western Mediterranean and Adriatic.

Fins pelvic disc complete, anterior membrane with more or less straight free edge; no free pectoral rays; tail fin with slightly concave rear edge; *nostril* front nostril rim without flap; *scales* absent from nape and breast; 41–44 (38–46) in lateral series. *Fin rays* 1D VI (VI-VII), 2D I, 10 (9–11), A I, 9–10 (8–10).

Reddish-orange with saddles, vertical bars on flanks and large spot at base of tail fin in both sexes. Males with bars darker and more numerous, and dark blotch in rear corner of 1st dorsal fin.

4 cm.

Inshore, possibly swimming above seabed.

Marbled Goby

Pomatoschistus marmoratus (Risso, 1810)
(= *P. microps leopardinus* (Nordmann, 1840)

Mediterranean and Black Sea; eastern Atlantic (Iberian Peninsula); also in Suez Canal, and introduced to Lake Qarun, Egypt.

Fins pelvic disc complete, anterior membrane edged with tiny villi; no free pectoral rays; tail fin rounded; *nostril* front nostril rim without flap; *scales* absent from nape but present on rear part of breast; 40–46 (37–48) in lateral series. *Fin rays* 1D VI (V-VII), 2D I, 9 (8–10), A I, 9 (8–10).

Sandy with darker reticulation and small dorsal paler saddles. Males with 4 vertical dark bars on flanks, dark breast, and conspicuous blue-black spot on 1st dorsal fin, in distal part of membrane between rays V and VI. Females with dark chin spot.

6.5 cm.

Inshore sandy deposits, usually to about 20 m, and entering brackish or saline waters. Breeds in spring and summer (Mediterranean).

Common Goby

Pomatoschistus microps (Krøyer, 1838)

Western Mediterranean; eastern Atlantic, from Morocco to Norway and most of Baltic Sea.

Fins pelvic disc complete, anterior membrane edged without distinct villi; no free pectoral rays; tail fin rounded; *nostril* front nostril rim without flap; *scales* absent from nape and back to about interdorsal space; none on breast; 39–52 in lateral series. *Fin rays* 1D VI (V-VII), 2D I, 8–9 (8–11), A I, 8–9 (7–10).

Greyish to fawn, with saddles and coarse reticulation. Males with up to 10 vertical dark bars on flanks, dark pelvic, breast and underside of head (tinged orange), and conspicuous blue-black spot on 1st dorsal fin, in proximal part of membrane between rays V and VI.

6.4 cm.

Inshore, in estuaries as well as saltmarsh and high shore pools. Breeds April-August (Isle of Man, Baltic Sea).

Sand Goby

Pomatoschistus minutus (Pallas, 1770)

Mediterranean and Black Sea; eastern Atlantic from Spain to Tromso, Norway, and Baltic Sea.

Sand Goby, *Pomatoschistus minutus* (David Maitland).

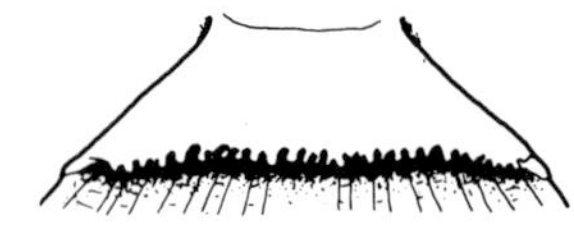

Fins pelvic disc complete, anterior membrane edged with tiny villi (left); no free pectoral rays; tail fin rounded; *nostril* front nostril rim without flap; *scales* present on rear part of nape; 55–75 in lateral series. *Fin rays* 1D VI (VI-VII), 2D I, 10–12, A I, 9–12.

Sandy with ferruginous specks, fine reticulation, and small dorsal paler saddles. Males with conspicuous dark blue spot on 1st dorsal fin, in distal part of membrane between rays V and VI; dorsal fins otherwise with reddish-brown bands; four main vertical dark bars on flanks; pelvic disc dark; breast pale in Atlantic populations.

9.5 cm.

Inshore sand and muddy sand deposits, usually to about 20 m, sometimes even to 60–70 m; juveniles in lower estuaries. Breeds from February to May (Plymouth).

Two other smaller species, related to the Sandy Goby, are the Mediterranean and Atlantic Norwegian Goby (*Pomatoschistus norvegicus* (Collett, 1903)) (6.5 cm), offshore from 18–325 m, with P 17 (16–18), usually paler coloration, but males have more, thinner vertical bars, and dark breasts; and Lozano's Goby (*Pomatoschistus lozanoi* (De Buen, 1923)) (8.0 cm), occurring in the eastern Atlantic from Portugal to the North Sea, very similar to the Sand Goby but paler, often distinctly speckled, P 18–21, usually inshore and in larger estuaries. Identification of these two species is based on examination of head sensory papillae patterns and counting of vertebrae, and needs specialist reference material.

Painted Goby

Pomatoschistus pictus (Malm, 1865)

Mediterranean (Adriatic Sea and Sea of Marmora); eastern Atlantic, from Canaries to Norway (Trondheimfjord).

Fins pelvic disc complete, anterior membrane without distinct villi; no free pectoral rays; tail fin rounded; *nostril* front nostril rim without flap; *scales* absent from nape and back before origin of 1st D; 34–43 in lateral series. *Fin rays* 1D VI (V-VII), 2D I, 9 (7–10), A I, 8–9.

Fawn with coarse dark reticulation and large pale dorsal saddles reaching lateral mid-line on each side; four 'double-spots' along lateral mid-line. In both sexes, dorsal fins with rows of large black spots, surmounted by rosy oblique banding.

5.7 cm.

Inshore, low water to 50–55 m, on gravel or sand, but sometimes found in shore pools. Breeds from early April to July (Isle of Man).

Quagga Goby

Pomatoschistus quagga (Heckel, 1840)

Western Mediterranean and Adriatic.

Fins pelvic disc complete, anterior membrane lacking villi; no free pectoral rays; tail fin with truncate or slightly concave rear edge; *nostril* front nostril rim without flap; *scales* absent from nape and back to end of 1st D, and from breast; 35–40 in lateral series. *Fin rays* 1D VI (VI-VII), 2D I, 9, A I, 9 (8–9).

Yellowish, with coarse reticulation between pale saddles; males with 4, females with 3, vertical bars on flanks, none under 1st dorsal fin in latter sex.

4.6 cm.

Inshore, possibly at least partly midwater in habits.

Tortonese's Goby
Pomatoschistus tortonesei (Miller, 1968)

Mediterranean (Sicily and Tripolitania).

Fins pelvic disc complete, anterior membrane without distinct villi; no free pectoral rays; tail fin rounded; *nostril* front nostril rim without flap; *scales* absent from nape, back to middle of 1st dorsal base, and breast; 32–34 (30–36) in lateral series. *Fin rays* 1D VI (V-VI), 2D I, 7 (6–8), A I, 7 (6–8).

Fawn, with darker reticulation; males with many ill-defined vertical dark bars on flanks, and dark breast and pelvic disc, suffused yellowish; females with 3 major dark bars, underside of head bright yellow, and chin mark.

3.7 cm.

Lagoons, in shallows on sand near seagrass; breeding in May at Farwah Lagoon, Libya.

Ferrer's Goby
Pseudaphya ferreri (O. de Buen & Fage, 1908)
= *Ps. pelagica* De Buen, 1931

Western Mediterranean and Adriatic.

Body compressed; *fins* pelvic disc and membrane complete; no free pectoral rays; tail fin rounded; *nostrils* front nostril without flap; *teeth* similar in both sexes; *scales* ctenoid, absent from nape; 25–26 scales in lateral series. *Fin rays* 1D V; 2D I, 8; A I, 10.

Body transparent with rosy dots; reddish line along bases of median fins; large triangular or rhomboidal dark spot at base of caudal fin.

2.9 cm.

Inshore, on sandy ground, especially from December to February.

Leopard-spotted Goby
Thorogobius ephippiatus (Lowe, 1839)
= *Gobius forsteri* Corbin, 1958

Mediterranean; Eastern Atlantic, Canaries and Azores to Skagerrak.

Fins pelvic disc and anterior membrane complete, without lateral lobes; no free rays in pectoral fin; tail fin rounded; *nostrils* front nostril without process from

Painted Goby, *Pomatoschistus pictus* (Jim Greenfield).

Leopard-spotted Goby, *Thorogobius ephippiatus* (B. Picton).

rim; *scales* none on nape; 36–38 (33–42) in lateral series. *Fin rays* 1D VI (VI-VII); 2D I, 11 (10–12); A I, 10.

Pale fawn or sandy, with orange to brick-red blotches (dark after preservation) over head and body, including lateral mid-line; black spot in rear corner of first dorsal fin.

13 cm.

Coastal, in or near crevices associated with vertical rock faces, from low water of spring tides to 40 m; rarely in deep shore pools. Breeds from May-July (Plymouth, Connemara).

Large-scaled Goby
Thorogobius macrolepis (Kolombatovic, 1891)

Western Mediterranean (near Marseilles) and Adriatic (Split).

Similar to preceding, but pelvic disc slightly emarginate and anterior membrane absent or rudimentary. *Scales* in lateral series 27–28. *Fin rays* 1D VI; 2D I, 11; A I, 10.

Coloration in life not recorded; head with many small pale spots instead of large brick-red blotches seen in the Leopard-Spotted goby (*Thorogobius ephippiatus*).

Canary Goby
Vanneaugobius canariensis Van Tassell, Miller & Brito, 1988

Canary Islands and Guinea.

Fins pelvic fins separate except for low membrane between bases; 1st dorsal rays elongate in male; no free pectoral rays; tail fin rounded; *nostril* front nostril without process on rim; *scales* absent from nape; uppermost and lowest scales at base of tail fin enlarged, with very long lateral ctenii; 28–30 in lateral series. *Fin rays* 1D VI (VI-VII), 2D I, 11, A I, 10.

Reddish-brown, with coarse dark reticulation and lateral blotches, head dark; 1st dorsal fin with pale band across middle; pelvics with central dark blotch.

4.3 cm.

Inshore, under loose rocks on sand, 9–45 m.

Pruvot's Goby (*Vanneaugobius pruvoti*) (Fage, 1907), length 3.9 cm, occurs in deeper water on coarse grounds, including coralline beds, at 60–163 m; it differs from the Canary Goby in possessing a dark spot on the rear part of 1st dorsal fin; A I, 10.

Zebra Goby
Zebrus zebrus (Risso, 1826)

Mediterranean.

Fins pelvic disc complete, but anterior membrane without lobes; pectoral fin with well developed free rays; tail fin rounded; *nostril* front nostril with tentacle on rim as long as nostril tube; *scales* none on nape; 32–34 (29–38) in lateral series. *Fin rays* 1D VI (V-VI), 2D I, 11, A I, 9 (7–10).

Brownish-olive, with several vertical dark bands across flanks; pale band across anterior nape behind eyes, continuing very obliquely backwards to upper gill cover; pectoral base with dark band, having deeply concave front edge; 1st dorsal fin with 2 dark bands, upper edge reddish.

5.5 cm.

Inshore, to 3 m; lagoons and intertidal pools; under stones, among algae and sea-grass. Breeds in early July (Messina).

Grass Goby
Zosterisessor ophiocephalus (Pallas, 1811)
= *Gobius lota* Valenciennes, 1837

Mediterranean and Black Sea.

Fins pelvic disc and anterior membrane complete but with no lateral lobes; uppermost pectoral rays free only at tips; tail fin rounded; *nostril* front nostril rim without process; *scales* present on nape; 59–64 (53–68) in lateral series. *Fin rays* 1D VI (V-VII), 2D I, 14–15 (13–16); A I, 14–15 (12–16).

Yellowish-olive to green, with many irregular vertical bars; blotches along lateral mid-line, with 1 at base of tail more intense, and sometimes small dark spot at upper end of pectoral fin-base.

25 cm.

Brackish water, in estuaries and lagoons, on mud and eel-grass beds; especially common in Venetian lagoon. Breeds from April to May (Black Sea). Sold in fish-markets.

Bridled Goby
Coryphopterus glaucofraenum Gill, 1863

Western Atlantic, North Carolina and Bermuda to Bahamas, Antilles, and Brazil.

Fins pelvic anterior membrane and pelvic disc complete; pectoral fins without free upper rays; 1st dorsal fin rays not elongate; tail fin rounded; *nostrils* front nostril without process from rim; *scales* nape naked; 26–27 (25–28) in lateral series. *Fin rays* 1D VI; 2D I, 9 (8–9); A I, 9.

Translucent, yellowish to golden, with variable darker internal pigmentation; black spot above opercle, and spots or bar at base of tail fin.

7.5 cm.

Inshore, from shallows to 25 m, chiefly around 6 m, from turbid embayments to clear open coral sand.

Fat Sleeper
Dormitator maculatus (Bloch, 1790)

Western Atlantic from North Carolina to Brazil, and Gulf of Mexico; also reported from New Jersey.

Body deep, somewhat compressed; *head* cheek without spine; *fins* pelvic fins entirely separate, without intervening membrane between bases; pectoral fins without free upper rays; 1st dorsal fin rays not elongate; 2nd dorsal base shorter than distance from it to origin of tail fin; tail fin rounded; *nostrils* front nostril without process from rim; *scales* nape with large scales, onto snout; 32–36 scales in lateral series. *Fin rays* 1D VII; 2D I, 9: A I, 10.

Dark brown to bluish, with dark blue-black spot at upper end of pectoral fin base; dorsal and anal fins reddish.

Over 30 cm.

Brackish waters of saltmarsh and pools in upper estuaries; also in fresh and inshore marine habitats; young may occur among water hyacinth root masses.

Spinycheek Sleeper
Eleotris pisonis (Gmelin, 1801)

Western Atlantic from South Carolina and Bermuda to Brazil, and Gulf of Mexico.

Head a downwardly-pointing spine in lower rear corner of cheek; *fins* pelvic fins entirely separate, without intervening membrane between bases; pectoral fins without free upper rays; 1st dorsal fin rays not elongate; 2nd dorsal base shorter than distance from it to origin of tail fin; tail fin rounded; *nostrils* front nostril tubular, without process from rim; *scales* nape scaled; 53–65 ctenoid scales in lateral series. *Fin rays* 1D VI; 2D I, 8; A I, 8.

Dark brown, sometimes back pale, with dark spot at upper end of pectoral fin base.

21 cm.

Brackish waters of saltmarsh and pools in estuaries, and also in fresh water; young more marine.

Sponge Goby

Evermannichthys spongicola (Radcliffe, 1917)

Western Atlantic, North Carolina (off Beaufort) to Florida and Gulf of Mexico.

Body elongate; *fins* pelvic anterior membrane and pelvic disc complete; pectoral fins without free upper rays; dorsal fins low, widely separated; tail fin rounded; *nostrils* front nostril without process from rim; *scales* absent, except for 3 rows of scales, with large ctenii, along lower edge of tail stalk between anal and tail fins. *Fin rays* 1D VI-VII (V-VII); 2D I, 12 (11–13); A I, 9 (8–10).

About 17 dark bars, and scales with dark edging.

2.5 cm.

Shelf waters; lives within large sponges, including the loggerhead (*Speciospongia vesparia*).

Lyre Goby

Evorthodus lyricus (Girard, 1858)
= *Mugilostoma gobio* Hildebrand & Schroeder, 1928

Chesapeake Bay to Gulf of Mexico and Surinam.

Fins pelvic anterior membrane and pelvic disc complete; pectoral fins without free upper rays; 1st dorsal fin rays elongate in male; tail fin lanceolate; *nostrils* front nostril without process from rim; *teeth* tips bilobed; *scales* nape fully scaled; 30–35 in lateral series. *Fin rays* 1D VI; 2D I, 10; A I, 11.

Body brownish-grey, with 5–6 vertical bars more or less evident; elongate black spots on upper tail fin.

7.7 cm.

Shallows of muddy estuaries, saltmarsh creeks and pools; breeding season May to August (Georgia), eggs lining burrows.

Darter Goby

Gobionellus boleosoma (Jordan & Gilbert, 1882)
= *Ctenogobius stigmaticus* Smith, 1907

Western Atlantic, New Jersey to Panama and Brazil; also reported from Massachusetts (Dartmouth).

Fins pelvic anterior membrane and pelvic disc complete; pectoral fins without free upper rays; 1st dorsal fin ray III elongate in males; tail fin moderately lanceolate; *nostrils* front nostril without process from rim; *scales* nape naked; 29–33 in lateral series. *Fin rays* 1D VI; 2D I, 10 (9–11); A I, 11 (10–12).

5 diffuse lateral mid-line blotches, produced dorsally into V-shaped marks; large dark shoulder spot over pectoral fin base; tail fin of males with 2 bands, orange and yellow, directed towards tip.

6.2 cm.

Inshore shallows and in estuaries, on muddy ground, including turtle-grass meadows. Breeds from May to October or November (Beaufort, NC).

Sharptail Goby

Gobionellus hastatus Girard, 1858

North Carolina; Gulf of Mexico (Campeche).

Fins pelvic anterior membrane and pelvic disc complete; pectoral fins without free upper rays; 1st dorsal fin ray III elongate in males; tail fin very long, lanceolate, with elongate pointed tip; *nostrils* front nostril without process from rim; *scales* nape with small cycloid scales; 75–89 (73–92) in lateral series. *Fin rays* 1D VI; 2D I, 13; A I, 14.

Body with large oval brown spot opposite pectoral fin; 1st dorsal fin with a few small dark marks on proximal part of first ray.

15 cm (excluding tail fin), total over 22 cm.

Estuaries and inshore waters.

A southern form, the Highfin Goby, *Gobionellus oceanicus* (Pallas, 1770), with 60–76 scales in lateral series, is reported off North Carolina; recent work suggests that the Sharptail Goby may be merely a variant of this species.

Freshwater Goby
Gobionellus shufeldti (Jordan & Eigenmann, 1886)

Western Atlantic, South Carolina (Georgetown); Gulf of Mexico (Louisiana and Texas).

Fins pelvic anterior membrane and pelvic disc complete; pectoral fins without free upper rays; 1st dorsal fin not elongate; tail fin rhomboidal; *nostrils* front nostril without process from rim; *scales* nape mostly scaleless, except before 1st dorsal origin; 35–40 in lateral series. *Fin rays* 1D VI; 2D I, 11 (10–12); A I, 12 (11–13).

5 darker brown blotches along lateral mid-line, without shoulder spot and V-shaped markings.

8.4 cm (male), 7.8 (female).

Saltmarsh and upper estuarine waters.

Emerald Goby
Gobionellus smaragdus (Valenciennes, 1837)

Western Atlantic, South Carolina (Charleston); Florida, Cuba and Brazil.

Fins pelvic anterior membrane and pelvic disc complete; pectoral fins without free upper rays; 1st dorsal fin ray III moderately elongate in both sexes; tail fin lanceolate; *nostrils* front nostril without process from rim; *scales* rear nape scaled; 39–46 in lateral series. *Fin rays* 1D VI; 2D I, 10; A I, 11.

Diffuse lateral mid-line blotches, not forming V-shaped marks; head and anterior body with numerous small brownish rings; large dark shoulder spot over pectoral fin base.

Over 7.6 cm.

Inshore.

Naked Goby
Gobiosoma bosci (Lacépède, 1800)

Western Atlantic from Long Island, New York, to Florida; Gulf of Mexico to Campeche, Mexico.

Canal pores 2 pores over opercle; *fins* pelvic anterior membrane and pelvic disc complete, short; pectoral fins without free upper rays; 1st dorsal fin not elongate; tail fin rounded; *nostrils* front nostril without process from rim; *scales* Absent. *Fin rays* D VII (VI-VIII); 2D I, 13 (12–14); A I, 10 (9–11).

9–10 sharply defined dark bars, broader than intervening pale bars, without conspicuous dark spots along lateral mid-line; sometimes uniformly pale or almost black.

6.4 cm.

Inshore shallows, typically under empty shells in oyster and clam beds; can tolerate fresh and fully salt water but prefers reduced salinity. Breeds from April to September (Beaufort, NC), May to August (Rhode Island), with postlarvae migrating to upper reaches of estuaries.

A related species, the Code Goby (*Gobiosoma robustum*) (Ginsburg, 1933), lacking canal pores above opercle and all scales, common in the Gulf of Mexico and Florida, is reported from Maryland (Patuxent River).

Seaboard Goby
Gobiosoma ginsburgi Hildebrand & Schroeder, 1928

Western Atlantic, Massachusetts to South Carolina (May River).

Canal pores 2 pores over opercle; *fins* pelvic anterior membrane and pelvic disc complete; pectoral fins without free upper rays; 1st dorsal fin not elongate; tail fin rounded; *nostrils* front nostril without process from rim; *scales* absent except for 2 scales on each side of tail fin base; *fin rays* 1D VII; 2D I, 11 (11–12); A I, 10 (9–11).

Dark bars, broader than intervening pale bars, and dark spots along lateral mid-line.

5.2 cm.

Found in coastal waters, from shallows to greater depths than *Gobiosoma bosci*, but including silty sheltered inshore areas with substrate shelter; tolerant of low oxygen levels and can adapt to brackish water. Breeds from April to September (Beaufort, NC), May to August (Rhode Island).

Pink Wormfish
Microdesmus longipinnis (Weymouth, 1910)

Western Atlantic, South Carolina (Charleston County) to Texas and Cayman Islands.

Body very elongate; *head* eyes small; mouth curved; opercular opening very narrow; *fins* dorsal fin single, long; pelvic fins separate, with only minute spine and 3 unbranched soft rays; pectoral fins short, without free upper rays; tail fin rounded; *scales* minute, not overlapping. *Fin rays* D XX-XXI (IXX-XXII) + 50; A I, 42–43 (40–46).

Tan with brownish stippling, paler below; pink when stressed.

26 cm.

Burrows in sand and mud of estuaries and inshore areas, from shallows to about 15 m but may also swim near surface at night.

Green Goby
Microgobius thalassinus (Jordan & Gilbert, 1883)

Western Atlantic, Chesapeake Bay to Texas, excluding south-east Florida and Keys.

Head low median ape ridge in females; *fins* pelvic anterior membrane and pelvic disc complete; pectoral fins without free upper rays; 1st dorsal fin rays slightly elongate in males; tail fin moderately lanceolate; *nostrils* front nostril without process from rim; *tongue* bilobed; *scales* mostly cycloid; nape and back to middle of 1st dorsal base naked; 43–50 in lateral series. *Fin rays* 1D VI; 2D I, 15 (14–16); A I, 15 (14–16).

Bright greenish, with vertical bars anteriorly, head with blue and green lines, 1st dorsal fin with reddish-brown spots, anal fin dark, and pelvic disc orange-yellow in male; female with dark markings on 1st dorsal fin.

5 cm.

Tide pools and inshore brackish and marine shallows, usually to 6 m, on sandy and muddy ground, including oyster beds in association with the sponge *Microciona*. Breeds from July to September (Beaufort, North Carolina).

The related Clown Goby (*Microgobius gulosus* (Girard, 1858)), Florida to Gulf of Mexico, has been reported from Maryland (Patuxent River); it may be distinguished from the Green Goby by more drab coloration, with many dark spots and blotches, absence of nape ridge in both sexes, and elongate 1st dorsal rays in males.

Yellow-prow Goby
Elacatinus xanthiprorus Böhlke & Robins, 1968

North Carolina to Central America.

Fins pelvic anterior membrane and pelvic disc complete; pectoral fins without free upper rays; 1st dorsal fin rays not elongate; tail fin rounded. *Nostrils* front nostril without process from rim. *Scales* absent. *Fin rays* 1D VII; 2D I, 11 (10–12); A I, 10.

Bright yellow stripe between upper and lower black stripes, from head to tail fin; snout and upper eye yellow.

4.0 cm.

Inshore, on or in tubular sponges.

Family:
CALLIONYMIDAE

The body is rather long and the head is flattened with the eyes set almost on top of the head. There is a strong branched spine at the angle of the front gill cover and this is important in identifying the species. The pelvic fin is inserted in front of the pectoral fin.

These are bottom-living fishes in shallow water. The gill opening is situated high up on the body and plumes of mud may be blown out of it, presumably as part of the feeding process. The adult males often have strikingly elongated dorsal fin rays and brilliant colouring.

Callionymus Reticulatus Valenciennes

Eastern Atlantic.

Body slender; *head* and front part of body flattened from above; *eye* large (diameter greater than the space between the eyes); *mouth* small; *jaws* upper jaw slightly longer than lower; *spines* 3 spines present on a small bone situated on the front gill cover; *fins* 2 dorsal fins, the 1st small with 3 rays; 2nd fin longer. In the male the last ray of the 2nd dorsal fin is branched and taller than the rest. Tail fin rounded with no elongate rays. 1D III; 2D 8–9; A 8.

Greyish-yellow with large pale spots and small dark spots. 1st dorsal fin is entirely black in the female and immature males, and with a black and white spot in the adult males.

Up to 8 cm.

Found on sand and mud at 15–150 m and probably much deeper. Breeding takes place in the spring and summer.

Callionymus phaeton Valenciennes

Mediterranean and adjacent Atlantic.

Body slender; *head* flattened from above; *snout* pointed; *eye* very large; *mouth* reaches back to the front margin of the eye; *lips* fleshy; *spines* 2 spines on the bone on the front gill cover; *fins* 2 dorsal fins. All the rays of the 2nd dorsal and tail fins are branched. In the adult male the last ray of the 2nd dorsal fin and the centre ray of the tail fin are elongate as is also the last ray of the anal fin, which is also branched. The female and immature fish have no elongate rays and the 2nd dorsal fin is shorter. 1D IV; 2D 8–9; A 8–9.

The colour of this fish is very variable, usually orange or pinkish with greenish spots on the back and sides and pinkish silver on the undersurface. The dorsal fins are reddish. The 1st dorsal fin has a black spot between the 3rd and 4th rays. There are faint darker spots on the dorsal fins and the anal fin has a darker lower edge. The lower half of the tail fin has dark spots. Pectoral and pelvic fins are reddish.

Females up to 12 cm; Males up to 18 cm.

A fairly deep-water fish found at 100–500 m on mud or sand bottoms. They feed on

Callionymus reticulatus

3 front gill cover spines; 1st dorsal fin short with 3 rays.
Eastern Atlantic.

Adult male: last ray of dorsal fin branched and elongate:
1st dorsal fin with black and white spot.

Female and immature male: 1st dorsal fin entirely black.
8 cm.

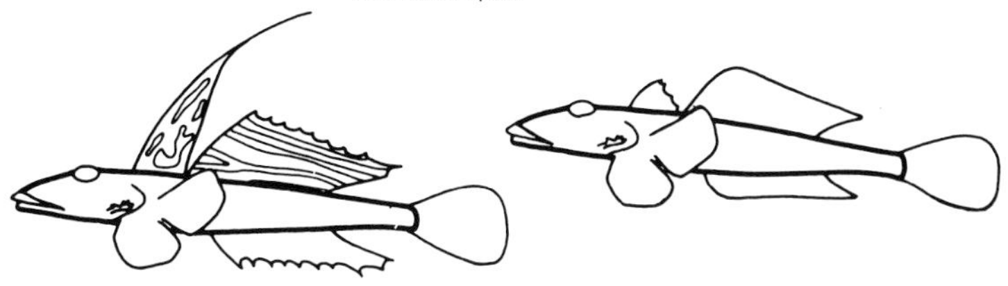

Dragonet
Callionymus lyra

4 front gill cover spines.
Eastern Atlantic;
Mediterranean.

Adult male: very elongate 1st ray of 1st dorsal fin; yellow and blue dorsal fins.
30 cm.

Female and immature male: uniform fin colour; 6 greenish or brownish blotches along sides.
20 cm.

Callionymus risso

3 front gill cover spines; bluish spots on body; 10 rays in 2nd dorsal fin.
Mediterranean.

Adult male: dark spots and wavy blue lines on dorsal fins.
10 cm.
Females 8 cm.

Males 14 cm.

Spotted Dragonet
Callionymus maculatus

Dorsal fins with alternating dark and pearl spots;
4 front gill cover spines.
Eastern Atlantic;
Mediterranean.

Females 11 cm.

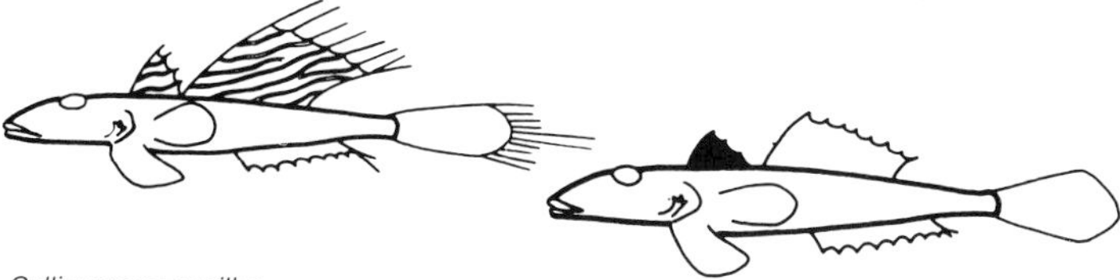

Callionymus pusillus

3 front gill cover spines.

Adult male: 1st dorsal fin shorter than 2nd; wavy bluish lines along the fins and blue spots and stripes along body.
Mediterranean.
15 cm.

Female and immature male: 1st dorsal fin black: colour similar to adult male but paler.
10 cm.

small crustacea and worms, etc., which they find on the bottom.

Dragonet
Callionymus lyra L.

Mediterranean; eastern Atlantic from Norway and Iceland to Senegal; North Sea; English Channel; west Baltic.

Body slender; *head* and front half of body flattened from above. The head appears triangular when seen from above; *snout* long; *jaws* upper jaw longer than the lower; *lips* fleshy; *spines* there are 4 spines present on a bone on the front gill cover; *skin* without scales; *fins* in the adult male the 2 dorsal fins are large and the 1st spine of the 1st fin is greatly elongated and reaches beyond the base of the tail fin. The anal fin is also large. In the female and immature male the 2nd dorsal and anal fins are smaller than in the adult male. The 1st dorsal is very much smaller and is lower than the 2nd dorsal. The pectoral and pelvic fins are joined by a small membrane. 1D IV; 2D 9; A 9.

Colour extremely variable and may depend on the individual, the locality and the season. There is a considerable difference between the male and female. Male fish are usually yellowish or brownish with bluish striping and spotting on the head and body. The 1st dorsal fin is yellow with blue blotches and the 2nd dorsal is yellow with longitudinal blue stripes. The females and young males are predominantly yellowish-brown with 6 greenish or brownish blotches along the sides and 3 dark saddles across the back. The fins are uniform in colour.

Females up to 20 cm. Males up to 30 cm.

These fish are very common in some areas and may be found from very shallow water to 50 m and sometimes below 100 m. They usually live on sand or mud in which they sometimes bury themselves. They feed on small molluscs, worms and crustacea which they find buried in the sand.

Spawning occurs from January through to the end of August, but the month varies from one locality to another. The fish move

Dragonet, *Callionymus lyra* (Jo Jamieson).

into shallow water prior to spawning, where the male displays his large fins to the female. The female, if she accepts the male, swims to his side and spawning takes place in mid-water with the fish swimming side by side nearly vertically upwards towards the surface.

Callionymus risso Le Sueur

Mediterranean.

This species is very similar to *Callionymus maculatus* except that there are 3 spines on the front gill cover bone and always 10 rays in the 2nd dorsal fin. Mature males have the 1st spine of the 1st dorsal fin elongated and the last ray of the 2nd dorsal fin branched.

Adult females and immature males are orange-brown on the back and sides and off-white below. There are bluish spots and 4 clearly defined orange-red saddles on the back. Adult males are similar in colour but the dorsal fins have rows of dark spots and bluish wavy lines.

Males up to 10 cm. Females up to 8 cm.

These fish are easily mistaken for *Callionymus maculatus* and may well be more common than records suggest. They are normally found at 20–40 m on sand or shell bottoms.

Spotted Dragonet
Callionymus maculatus Rafinesque

Eastern Atlantic from Norway to North Africa; Mediterranean.

Body slender; *head* somewhat flattened from above; *spines* there are 4 spines present on a small bone situated on the front gill cover; *skin* without scales; *fins* in the male the 2 dorsal fins are enlarged; the 1st ray of the 1st dorsal fin reaches to the end of the base of the 2nd dorsal. The 1st dorsal is separated from the 2nd. In the female the fins are shorter and the bases of the 2nd dorsal and anal fins are larger. 1D IV; 2D 9–10; A 8–9.

The colour of both the males and the females are similar. The ground colour is brownish-yellow darker on the back and lighter on the belly. There are longitudinal rows of dark and pearly spots. Male fish have four dark spots interspaced with pearl spots on each membrane between the rays of the 2nd dorsal fin. The 1st dorsal and tail fin are also spotted. The female fish have 2 rows of dark spots on the 2nd dorsal fin. The spots are less distinct in the female and immature male than in the adult male.

Females up to 11 cm. Males up to 14 cm.

Found from shallow water to below 300 m, usually on sandy bottoms. They are frequently found on the large offshore sandbanks. They are carnivorous and feed on small bottom-living animals, worms, crustacea, etc. In the Atlantic they spawn from April to June and in Mediterranean from January to May.

Callionymus pusillus Delaroche

Mediterranean.

Body slender; *snout* rather long and rounded; *mouth* small; *lips*; fleshy; *skin* without scales; *spines* 3 spines present on a bone situated on the front gill cover; *fins* in the male the 2nd dorsal fin is higher than the 1st and all the rays are elongated beyond the margin of the membrane. The last ray of the anal fin and all the rays of the tail fin are similarly elongated and the 2 central rays of the tail fin are longer than the others. The female fish has none of the elongated rays and the 2nd dorsal and anal fins are more equal in size. 1D IV; 2D 7; A9.

The ground colour is greyish, sandy or brownish. In the adult male there are bluish spots and stripes on the back and sides of the body and on the head. The fins are yellow with wavy bluish lines outlined in dark brown. The lower edge of the tail and anal fins are blackish. The female is similar to the male but less bright, also the 1st dorsal fin is black.

Male up to 15 cm. Female up to 10 cm.

These fish live in very shallow water on sand. They may bury themselves in the sand and can even travel through it to emerge a little distance further off. Breeding occurs during summer. The male displays to the female with his large fins and circles her in gradually decreasing circles. The male and female swim close together and spawning occurs in an almost vertical position when swimming towards the surface.

Family:
BLENIIDAE

The **Blennies** generally live in very shallow water, even between tide marks. They superficially resemble the gobies but do not have the pectoral fins united into a sucker. When alarmed they often take refuge in holes or crevices amongst rocks and this may explain why some Blennies (e.g., *Blennius rouxi*) are quite often seen by divers but are rare in museum collections.

They are elongate fishes without scales and with a somewhat slimy body. There is a single dorsal fin which is usually divided into the hard and soft rayed portions by a notch. The pelvic fins are reduced and contain only 1–3 rays. The hard rays are flexible and jointed at the base. Many species have little 'tentacles' on the head and these are particularly useful in identifying blennies from photographs.

Butterfly Blenny
Blennius ocellaris L.

Mediterranean; north-east Atlantic from Senegal to northern Scotland.

Body deep and laterally flattened; *teeth* there are 2 prominent canine teeth at the back of each jaw preceeded by smaller rounded teeth; *tentacles* there are 2 rather squat branched tentacles above each eye and two very small ones on each side of the 1st dorsal ray; *fins* the dorsal rays are very long, the 1st being the longest. There is a notch separating the two parts of the dorsal fin. D X-XII, 14–16; A I-II, 15–16.

Greenish-brown with 5–7 darker vertical bands on the flanks extending onto the dorsal fin. There is a characteristic white spot surrounded by a pale-bluish halo on the 6th and 7th rays of the dorsal fin.

Up to 20 cm.

Red-speckled Blenny
Parablennius sanguinolentus Pallas

Mediterranean; north-east Atlantic from Senegal to the English Channel.

Body heavy with a paunchy stomach. The front part of the body is laterally flattened; *profile* rounded; *jaws* rather small; *teeth* there are two rather small canine teeth in each jaw behind files of massive conical teeth; *tentacles* there is a minute tentacle above each eye about ⅓ diameter of the eye; *fins* the dorsal fin is of about the same height throughout its length. D XII-XIII; A 22–24.

Colour variable, usually dull olive. There are often red spots on the tail and pelvic fin rays and often a black spot on the 1st and 2nd dorsal rays. There is often a dark margin to the anal fin.

Red-speckled Blenny, *Parablennius sanguinolentus* (Andy Purcell).

Up to 15 cm, sometimes 20 cm.

Lives in shallow water amongst rocks where there are plenty of crevices. They are voracious carnivores using their strong teeth to crush up molluscs and echinoderms. They breed in early summer laying their eggs in empty mollusc shells, etc.

Tompot Blenny
Parablennius gattorugine Brünnich

Mediterranean; north-east Atlantic from West Africa to northern Scotland; not in North Sea.

Body rather elongate and laterally flattened behind; *eyes* set high on the head; *profile* rounded; *mouth* large extending back to the centre of the eye; *teeth* there is a single row of minute sharp teeth in each

Tompot Blenny (top), *Parablennius gattorugine* (Andrea Ghisotti).

Tompot Blenny (bottom), *Parablennius gattorugine* (B. Picton).

jaw. No canines; *tentacles* there is a large much-branched tentacle; above each eye. There is also a minute fringed tentacle on the nostril beneath each eye; *fins* the dorsal fin is continuous but the last rays are the longest. D XII-XIV, 17–20; A I, 20–21.

Ground colour yellow-brown, olive-brown or reddish-brown. There are about 7 darker vertical bars on the flanks extending onto the dorsal fin. Eye is reddish-brown.

Up to 20 cm, sometimes 30 cm.

In the northern part of its range it is found in the kelp (*Laminaria*) zone but in southern England and in the Mediterranean it may live 1 m or less beneath the surface. They favour places where there are medium-sized stones. They are very common in the eastern English Channel.

Spawns in early spring. The male guards the eggs, which are laid in crevices between rocks. The male also keeps them oxygenated by fanning a current of water over them. After a month the eggs hatch. The larvae are free-swimming but probably settle on the bottom by mid-summer.

Horned Blenny
Parablennius tentacularis Brünnich

Mediterranean; north-east Atlantic from Senegal to Portugal.

Body similar to Tompot Blenny (*Parablennius gattorugine*) but the body is somewhat longer and less massive. The most obvious point of difference is in the *tentacles* above each eye, which are branched on the rear edge only. On each of the front nostrils there is a small tubular tentacle with a tiny tongue-like projection on the rear part of its tip. *Fins* D XII-XIV, 18–23; A II, 21–24.

Colour variable. The most common form is chestnut brown and pale ochre arranged in alternating blotches; the body is speckled with darker brown. The iris is reddish with blue spots. There is often a black spot on the 2nd dorsal ray.

Up to 15 cm.

Generally lives on sandy bottoms where there are some stones at depths of 6–30 m. Also found in brackish water on mud.

Parablennius zvonimiri Kolombatovic

Mediterranean.

Body rather flattened laterally; *eye* set high in the head; *profile* almost vertical in front of the eye; *mouth* small, not extending past the hind edge of the eye. Lips fleshy, particularly the upper; *teeth* 2 large canines in each jaw behind a single file of small conical teeth; *tentacles* above each eye there is a prominent tentacle resembling a stag's antlers and on each nostril there are smaller horned tentacles; *fins* dorsal is obviously divided into two halves. D XII, 17; A I, 19.

Grey or red-brown with darker marbling. There are red spots and stripes on the sides of the head and on the lips.

Up to 6 cm.

Uncommon or rare. They live in water of moderate depth where there is a rich growth of encrusting red algae.

Parablennius cristatus L.

Both sides of tropical Atlantic; rare in Mediterranean.

Profile steep and rounded; *tentacles* there is a fringe of filaments running along the mid-line of the back from between the eyes to the dorsal fins, and a tuft of 7 filaments above each eye; *fins* there is a small notch in the dorsal fin at the 12th ray. D XI-XII, 13–16; A I-II, 15–18.

Olive-green with a dark spot between the 1st and 2nd dorsal rays.

Up to 10 cm.

Rare. In shallow water, probably lives amongst rocks.

Peacock Blenny
Lipophrys pavo Risso

Mediterranean.

Body rather long and laterally compressed; *eye* very small; *profile* steep and rounded. In the mature male there is a large hump above the eye; *mouth* very small, set low in the head; *teeth* a single file of blunt teeth in each jaw. At the rear of each file there is a single canine tooth; *tentacles* very small, one above each eye; *fins* the dorsal fin is continuous, not obviously divided into two parts. D XII, 21–23; A 24–46.

The ground colour is yellow-green with darker vertical bands. The front ones are outlined in iridescent blue. At the rear the blue pattern breaks down into scattered spots. There is always a black spot edged with iridescent blue behind the eye. In the female the black spot may be bordered with

Peacock Blenny, *Lipophrys pavo* (Andy Purcell).

Striped Blenny, *Parablennius rouxi* (Andy Purcell).

pinkish white. In the mature male the hump has patches of golden brown.

Up to 10 cm.

Found in very shallow water, usually on mud and sand near rocks. Very tolerant of extremes of temperature and salinity and may enter water that is nearly fresh.

Breeds in summer. They deposit their eggs amongst stones or on empty shells.

Zebra Blenny
Lipophrys basilicus Valenciennes

Mediterranean.

Resembles the female of the Peacock Blenny (*Lipophrys pavo*) but differs in having violet vertical bars on the flanks and a maze-like pattern of dark markings outlined in white on the sides of the head.

Up to 12 cm.

Very rare, shallow coastal water.

Aidablennius sphynx Valenciennes

Mediterranean.

Body long and laterally compressed; *profile* very steep; *eye* small and set high in the head; *mouth* very small, set low in the head; *teeth* 2 canine-like teeth in each jaw behind a single file of small sharp teeth; *tentacles* there is a conspicuous unbranched tentacle above each eye; *fins* the dorsal fin is clearly divided into two parts. D XII, 16–17; A 18–20.

Ground colour yellowish-brown with irregular dark vertical bands outlined in whitish blue; behind the eye there is a blue-grey spot which is outlined in red.

Up to 8 cm.

Generally lives in very shallow water where the bottom is rocky but the seaweed is not thick. Rests with its high dorsal fin folded over to one side, but when alarmed the fin becomes erect.

Spawns in early summer in rock cavities, shells, etc. The courtship is typical of Blennies. The male displays in front of any passing female, following and butting her until she enters the nest. A single male may mate with a succession of females. The eggs are attached to the nest wall with sticky filaments.

Striped Blenny
Parablennius rouxi Cocco

Mediterranean.

Body long; *eyes* large; *profile* almost vertical; *mouth* very small; *tentacles* above each eye there is a tentacle made up of 5 or 6 filaments. There is also a very small tentacle on the lower nostrils; *fins* dorsal fin not obviously divided. D XI, 24; A II, 24.

Pearly white with a black longitudinal stripe running from the eye to the tail. The fins are colourless except that there is sometimes a black spot between the 1st and 2nd dorsal ray and the anal fin has a black margin.

Up to 7 cm.

Normally lives in depths between 1 and 7 m where there are rocks, some detritus and possibly some sand. They are often seen with their heads looking out of tiny holes in the rocks. Considered to be quite a rare fish, although divers often see it, perhaps because of its very conspicuous coloration.

Shanny
Lipophrys pholis (L.)

North-east Atlantic south to Portugal.

Body distance of snout to hind edge of eye equals distance of hind edge of eye to the 1st dorsal ray; *profile* steep but not vertical; *mouth* rather large; *teeth* 2 canine

Shanny, *Lipophrys pholis* (Andy Purcell).

Lipophrys trigloides (Evan Jones).

teeth in each jaw behind a single row of small sharp teeth; *tentacles* no tentacle over eye; *fins* a shallow notch separates the 2 parts of the dorsal fin. D XI-XIII, 18–20; A I, 17–19.

Colour is blotched greenish- or yellowish -brown with an indistinct spot between the 1st and 2nd parts of the dorsal fin. During courtship and nesting the male becomes intense black with white lips.

Up to 16 cm.

Common amongst rocks and stones in shallow water, often in the intertidal. Also found on sandy or muddy shores where they will hide beneath fronds of seaweed. The Shanny feeds on almost anything including barnacles and algae. Spawning is in early summer. The eggs are usually laid on the underside of rocky crevices and are guarded by the male, which also fans them to keep them oxygenated. A long-lived fish, a large bleny of 16 cm may be 16 years old.

Lipophrys trigloides Valenciennes

Mediterranean.

Resembles the Shanny (*Lipophrys pholis*) except that the distance between the tip of the snout and the hind edge of the eye is greater than the distance of the hind edge of the eye to the first dorsal ray. D XII, 16; A 17–18.

Up to 9 cm.

Lipophrys canevai Vinciguerra

Mediterranean.

Head rounded; *eye* very small, set high in the head; *mouth* set low in the head; *teeth* 2 canines in each jaw, followed by a single row of conical teeth; *tentacles* none on head; *fins* the 2 parts of the dorsal fin are divided by a deep notch. D XIII, 16; A II, 15–16.

Cheeks and throat are golden-brown speckled with red spots. The body colour is dark brown covered with an irregular network of lighter lines arranged in a more or less longitudinal fashion.

Up to 7 cm.

Uncommon, lives amongst rocks in water so shallow that the fish is exposed by the waves.

Black-headed Blenny
Lipophrys nigriceps Vinciguerra

Mediterranean.

Body elongate; *eyes* large, set high in the head; *profile* rather steep; *mouth* set very

Lipophrys canevai (top), male (Evan Jones).

Lipophry canevai (middle) (Evan Jones).

Black-headed Blenny, *Lipophrys nigriceps* (Evan Jones).

low; *appendages* none on head; *teeth* 2 well developed canines in each jaw; *fins* no predorsal fin (see *Tripterygion* spp.) The 2 parts of the dorsal fin are conspicuously divided by a deep notch. D XII, 15; A I, 15.

Up to 4 cm.

Greatly resembles *Tripterygion minor* in general colour but differs in having no predorsal fin. Lives in the same habitat as *Tripterygion minor*.

Montagu's Blenny

Coryphoblennius galerita L.
= *Blennius montagui* Fleming

North-east Atlantic from Guinea to south Ireland; rare in Mediterranean.

Body long; *profile* gently sloping; *mouth* large; *teeth* 2 well-developed canines in the lower jaw; *tentacles* there is a characteristic crest of fringed filaments running between the eyes. From the centre line of this is a row of 3–7 simple filaments running back along the mid-line of the head; *fins* there is a conspicuous notch between the two parts of the dorsal fin. D XII-XVIII, 5–18; A 17–18.

Olive-brown often with darker vertical bands. In young fish there are bluish-white spots on the head and body. In the males the crest is orange and longer than in the females. The body colour is also darker.

Up to 8 cm.

Chiefly lives in rock pools and rarely extends down to the *Laminaria* seaweed zone. They live in pools where there is little algal growth and mostly feed on the arms of barnacles.

Breeds in mid-summer in rock pools. The male guards the eggs which are attached to the ceiling of a rock crevice with adhesive filaments.

Lipophrys dalmatinus (Steindachner and Kolombatovic).

Mediterranean.

Body long and laterally compressed; *profile* almost vertical; *tentacles* there are no tentacles above the eye but there is a pair of exceedingly small simple tentacles above the lower nostrils; *jaws* set very low; *teeth* 20–22 small almost conical teeth in each jaw. On each side of the lower jaw at the back is a larger tooth; *fins* the spiny rays of the dorsal fin are shorter than the soft rays. D XII, 15–16; A II, 18–19.

Montagu's Blenny, *Coryphoblennius galerita* (Peter Scoones).

Olive-green above, greenish-gold below. There are 8–11 darker brown bands on the back which may be outlined in silver. In some specimens the dark bands are divided along a line running from the gill cover to the tail. On the top of the head is a dark area which is divided from the greenish-gold cheeks by a wavy line.

Up to 4 cm.

Before the advent of divers this little fish was thought to be very rare as it can only be caught in a hand net. However, it may, in reality, be quite common. It lives on rocky bottoms where there is little weed from just beneath the surface to a maximum of 2 m.

Lipophrys adriaticus (Steindachner and Kolombatovic)

North-west Mediterranean, Adriatic and Black Sea.

Head rounded, rather steep profile, slight crest in some males; *tentacles* none above eyes; *teeth* canine-like teeth in both jaws. D XII, 15; A II, 16–17; pectorals 12.

Row of large white spots on each side of the dorsal fin and 5–6 dark vertical bars joining to form an uneven band along the flanks, black spots at the beginning of the dorsal fin. Spawning males have a yellow head and a black crown.

5 cm.

In water less than 1 m on sheltered rocky

coasts. Often on steep walls or sheltering in mollusc burrows. Only active in sunlight.

Parablennius incognitus (Bath)

West coast of Spain, Mediterranean.

Head profile steep, tentacles, simple one above each eye; *teeth* canine-like teeth in both jaws; *fins* dorsal fin with distinct notch between spiny and soft parts. D XII–17; A II–19; pectoral 14.

7–9 dark vertical bars, no spot on dorsal fin or on head.

6 cm.

Amongst seaweeds in water usually 50 cm to 1 m deep. Males entice females to resident hole for spawning; sometimes several females spawn into the same hole. Males fight off rivals by threatening yawns, tail-beating and biting.

Parablennius pilicornis (Cuvier)

Western Mediterranean, Spain and north Africa.

Head rather shallow profile; *tentacles* short tuft above eye; *teeth* canine-like teeth present. D XII–21; A II–23; pectoral 14.

Irregular spotty pattern with no black spot on dorsal fin or behind eye.

11 cm.

Down to 6 m on steep rocky slopes exposed to wave action.

Parablennius incognitus (Evan Jones).

Seaweed Blenny
Parablennius marmoreus (Poey)

Western Atlantic, New York south to Venezuela.

Head front profile rather steep, cluster of tentacles above the eyes, but none on the nape between the eyes and dorsal fin. D XII 17–18; A II 19–20.

8 cm.

Dark mottled above, paler below; often with a small spot between the 1st and 2nd rays of the dorsal fin.

Striped Blenny
Chasmodes bosquianus (Lacépède)

Western Atlantic, New York south to Florida.

Head profile pointed, lower jaw large, reaching back to centre of eye, no tentacles on head; *teeth* 18–19 slender recurved teeth in lower jaw, no canine-like teeth. D XI, 17–19; A II, 18–19.

Dorsal and anal fins striped; dorsal fin of males blue at the front, females are mottled.

8 cm.

Common on sea-grass meadows.

Feather Blenny
Hypsoblenius hentzi (Lesueur)

Western Atlantic, New Jersey south to Yucatan.

Head profile rounded and rather steep, pronounced feather-like branched tentacles above each eye; *teeth* no canine-like teeth. D XII 16, A II 16.

Brownish mottled with small dark spots on the gill cover and above the pectoral fin, a faint dark stripe along middle of dorsal spiny fin.

10 cm.

Usually on soft muddy bottoms but also on sea-grass meadows and oyster beds.

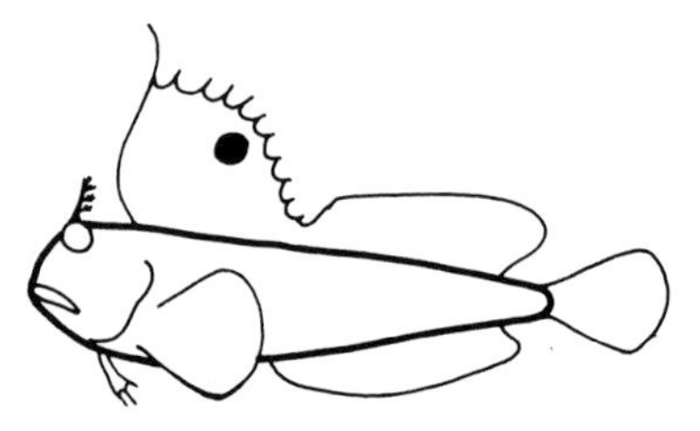

Butterfly Blenny
Blennius ocellaris

High dorsal fin with large eye spot;
Eastern Atlantic;
Mediterranean.
20 cm.

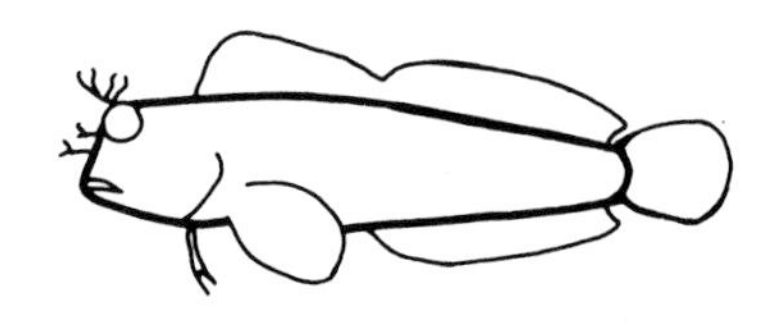

Parablennius zvonimiri

Staghorn-like appendage over eye;
smaller appendages on each nostril.
Mediterranean.
6 cm.

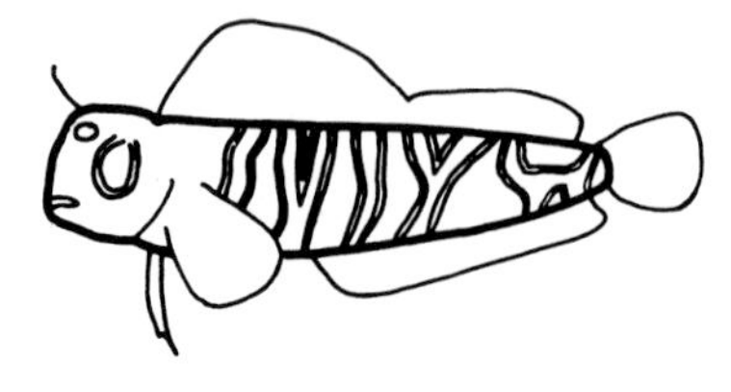

Aidablennius sphynx

Red-bordered spot on gill cover;
simple appendage over eye.
Mediterranean.
8 cm.

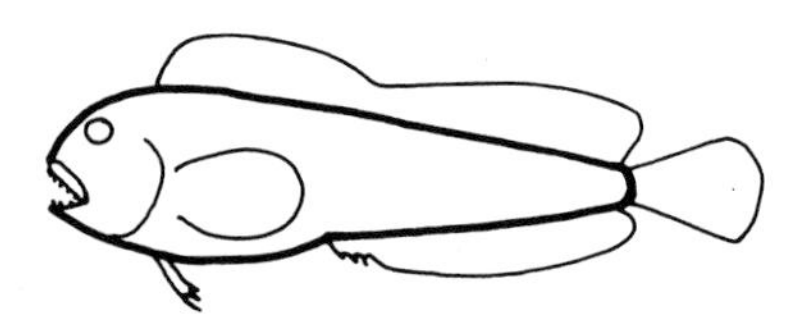

Red-speckled Blenny
Parablennius sanguinolentus

Body heavy and paunchy;
massive conical teeth.
Eastern Atlantic;
Mediterranean.
15 cm.

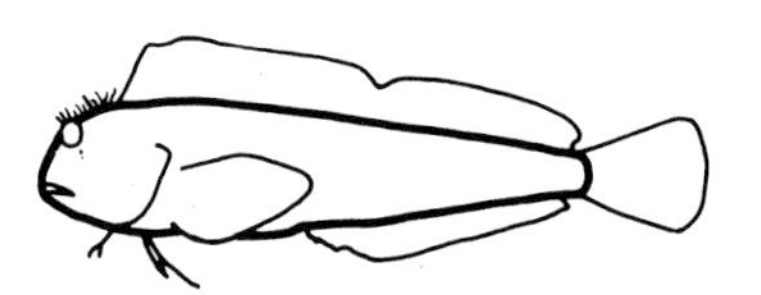

Parablennius cristatus

Tuft of 7 filaments over each eye;
line of filaments along mid-line of back from eyes to dorsal fin;
Eastern Atlantic;
Mediterranean.
10 cm.

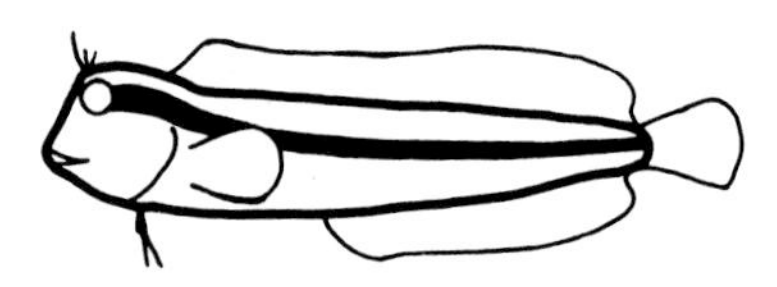

Striped Blenny
Parablennius rouxi

Black stripe on flanks.
Mediterranean.
7 cm.

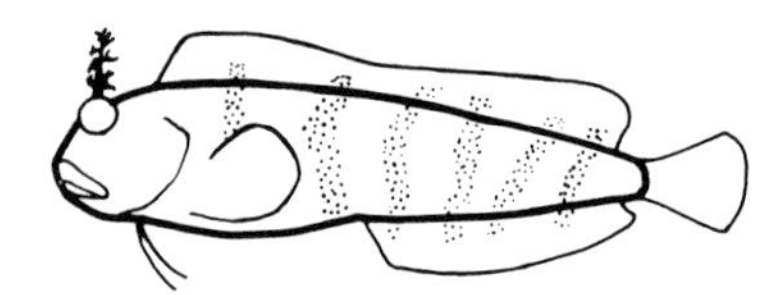

Tompot Blenny
Parablennius gattorugine

Large branched tentacles above eye.
Eastern Atlantic;
Mediterranean.
20 cm.

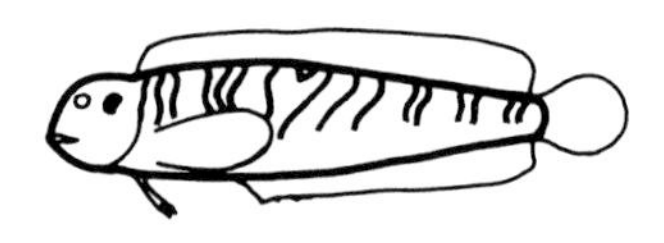

Peacock Blenny
Lipophrys pavo

Black spot with blue margin behind eye;
dark bands on flanks outlined in blue.
Mediterranean.
10 cm.

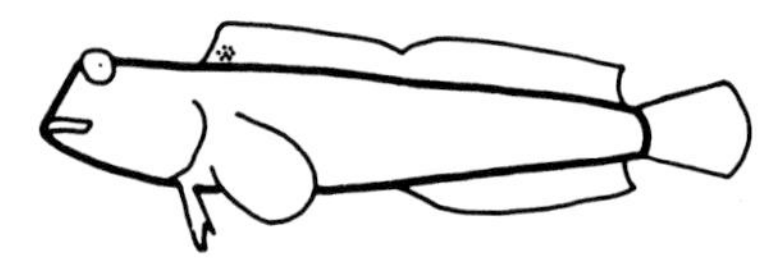

Shanny *Lipophrys pholis*

No appendages on head.
Eastern Atlantic.
16 cm.

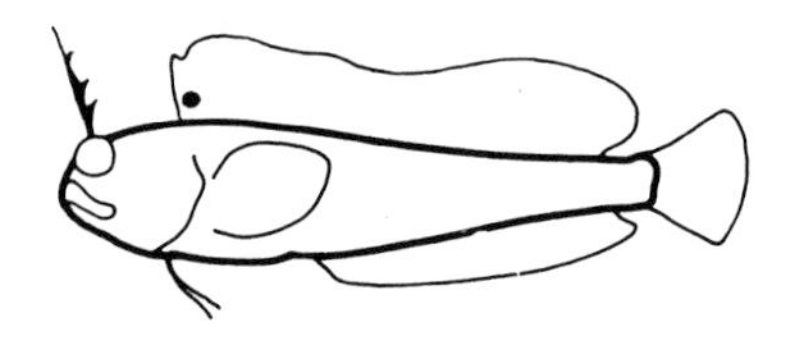

Horned Blenny
Parablennius tentacularis

Tentacle above eye, fringed on rear edge.
Eastern Atlantic;
Mediterranean.
15 cm.

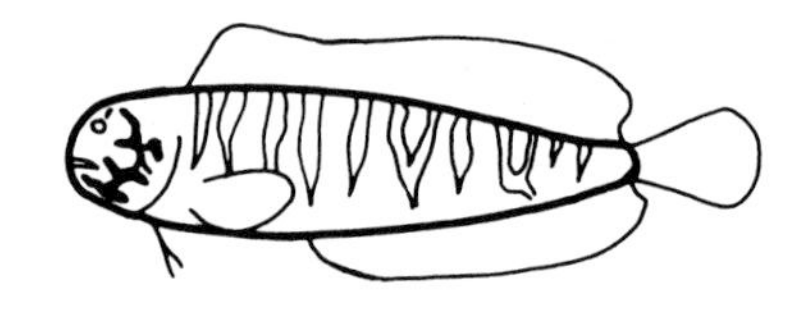

Zebra Blenny *Lipophrys basilicus*

Maze-like pattern on cheeks.
Mediterranean.
12 cm.

Lipophrys trigloides

No appendages on head.
Mediterranean.
9 cm.

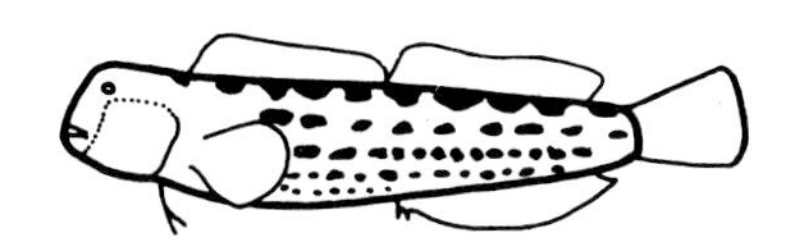

Lipophrys canevai

Golden cheeks.
Mediterranean.
7 cm.

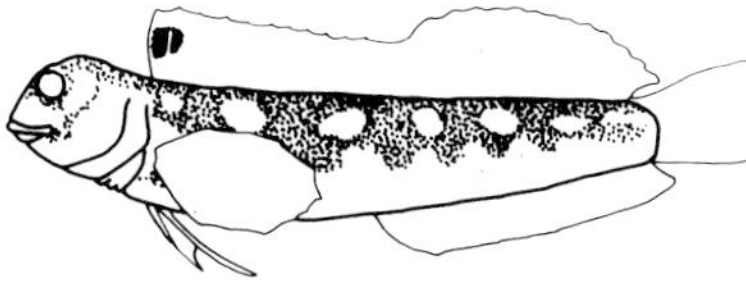

Lipophrys adriaticus

Row of large white spots at base of dorsal fin; dark spot at front of dorsal fin.
Mediterranean.
5 cm.

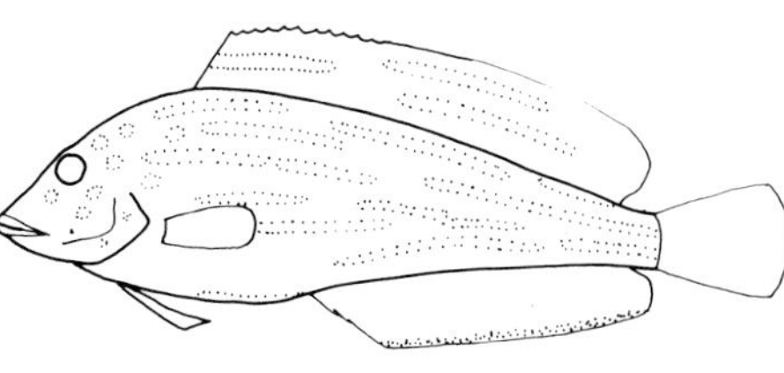

Striped Blenny
Chasmodes bosquianus

No canine-like teeth; dorsal and anal fins striped.
Western Atlantic.
8 cm.

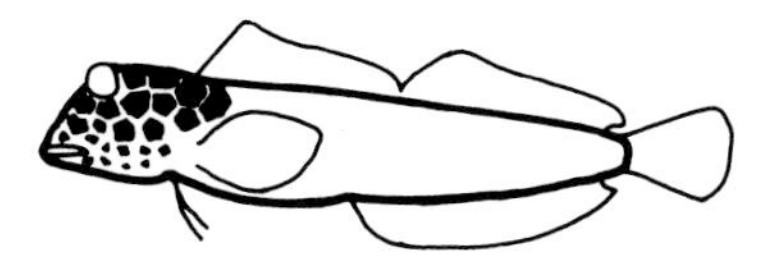

Black-headed Blenny
Lipophrys nigriceps

2 dorsal fins;
red body;
no appendages on head.
Mediterranean.
4 cm.

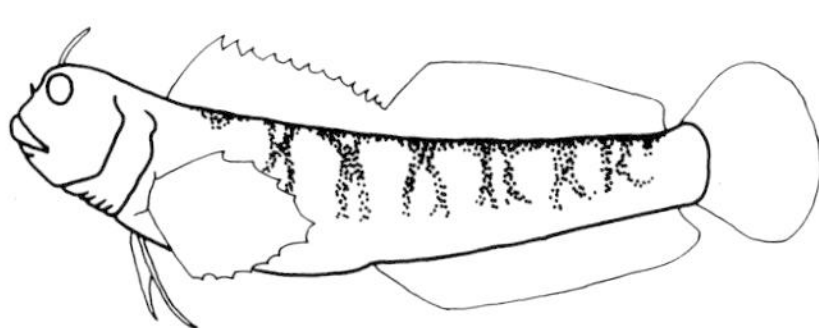

Parablennius incognitus

Simple tentacle above eye;
7–9 dark vertical bars.
Mediterranean; eastern Atlantic.
6 cm.

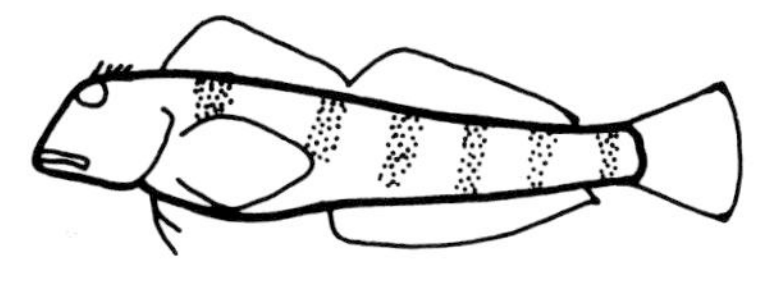

Montagu's Blenny
Coryphoblennius galerita

Fringe of branched appendages above eyes;
row of simple appendages on mid-line of head.
Eastern Atlantic;
Mediterranean.
8 cm.

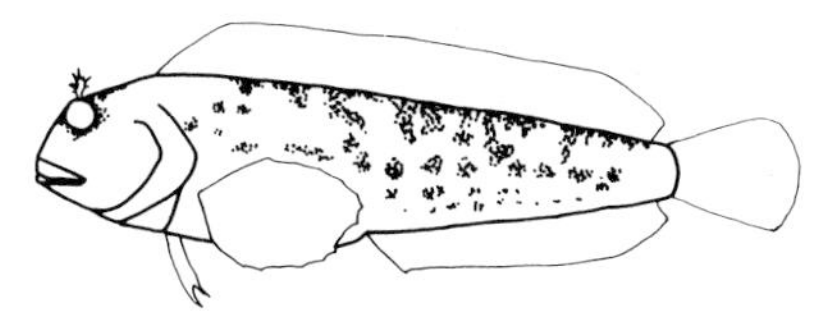

Parablennius pilicornis

Tuft of short tentacles above eye;
irregular spotty pattern;
14 pectoral fin rays.
Eastern Atlantic.
11 cm.

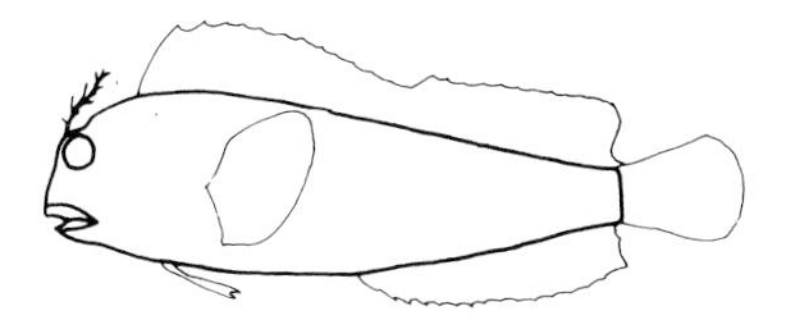

Feather Blenny
Hypsoblenius hentzi

Tentacles above eyes branched and feather-like;
No canine teeth;
mottled with small dark spots.
Western Atlantic.
10 cm.

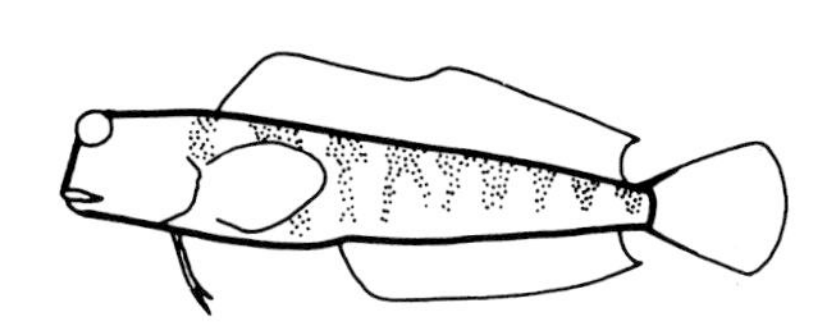

Lipophrys dalmatinus

Dorsal fin profile;
no appendages over eyes.
Mediterranean.
4 cm.

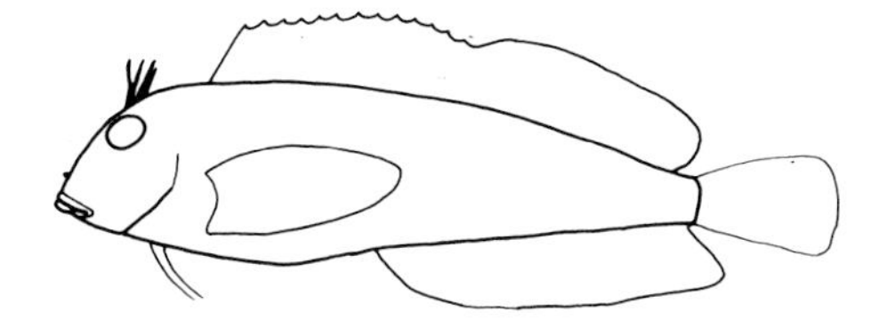

Seaweed Blenny
Parablennius marmoreus

Cluster of tentacles above eyes, none on nape;
Dark spot between 1st and 2nd dorsal spines.
Western Atlantic.
8 cm.

Family:
ANARHICHADIDAE

The **Wolf Fish** or **Cat Fish** are blenny-like fishes with no pelvic fins, a long dorsal fin and a distinct tail fin. There are large canine-like teeth in the front of the jaws and flattened grinding teeth at the sides.

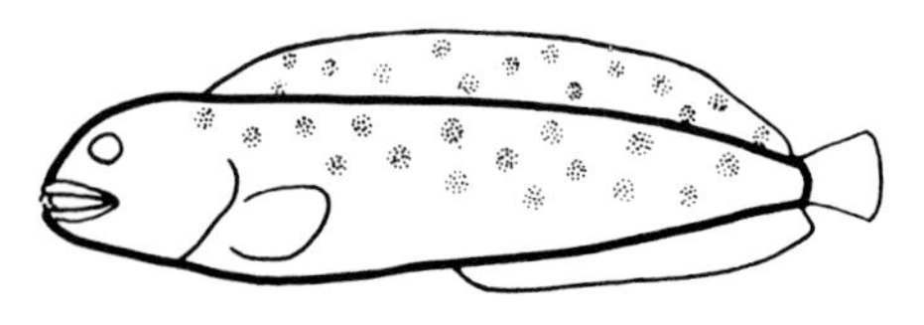

Spotted Catfish; Lesser Catfish *Anarhichas minor* Olafsen

Large conical teeth; spots scattered over back and flanks.
Eastern and western Atlantic.
200 cm.

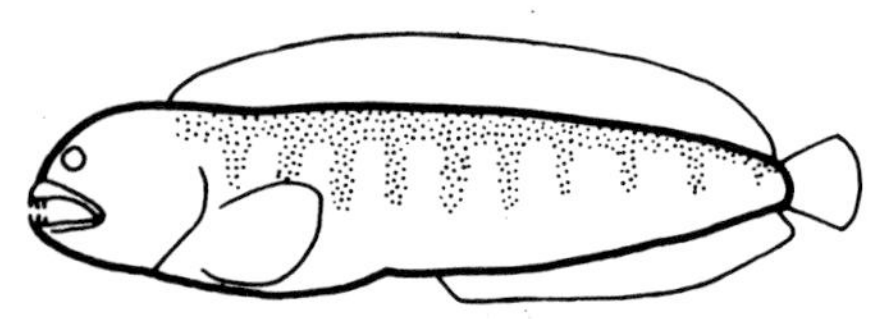

Wolf Fish; Cat Fish *Anarhichas lupus*

Large curved teeth in front of jaws, flattened teeth at sides; dark vertical stripes on the back and flanks.
Eastern and western Atlantic.
120 cm.

Wolf Fish; Cat Fish
Anarhichas lupus L.

Eastern Atlantic south to the English Channel, North Sea, Iceland, Scandinavia, Greenland; also western Atlantic south to New Jersey.

Body deepest behind the head, tapering to the tail; *head* large; *teeth* curved in the front of the jaws; large, flattened teeth at the sides; *fins* 1 dorsal fin, and 1 anal fin which terminates nearer the tail than does the dorsal fin. No pelvic fins. D 69–79; A 42–48.

Greyish, brownish-red or greenish with darker vertical stripes which continue onto the dorsal fin.

Up to 120 cm.

Usually found at 100–300 m but young fish may be seen in much shallower water. Spawning occurs in winter in water between 40–200 m. The eggs are laid in clumps on the sea-bed amongst weeds and stones. The adult fish feed on small echinoderms, molluscs, crustacea, etc.

Spotted Catfish; Lesser Catfish
Anarhichas minor Olafsen

Northern-east Atlantic south to Scotland; Iceland; Greenland; Barents Sea; White Sea; also western Atlantic south to Maine.

Body deepest behind the head tapering strongly to the tail; *head* large; *snout* rounded; *teeth* large, curved teeth in the front of the jaws and rounded or pointed teeth on the palate and the sides of the jaws; *fins* 1 dorsal fin, and 1 anal fin which terminates nearer the tail than the dorsal fin. No pelvic fins. D 74–80; A 45–47.

Brownish with distinct darker spots irregularly scattered over the back and sides.

Up to 200 cm.

Found at 25–460 m but most commonly between 100–250 m. They live mainly on mud or sand. During the spring there is an offshore migration into deeper water for spawning which occurs at around 250 m. The eggs are laid in large clumps. The adult fish feed on small molluscs, echinoderms, crustacea and fish.

Wolf Fish, *Anarhichas lupus* (Jim Greenfield).

Wolf Fish, *Anarhichas lupus* (Gilbert van Rijckevorsel).

Family:
CLINIDAE

Elongate fishes with tiny scales embedded in the skin. The pelvic fins contains only one or two rays.

Cristiceps argentatus (Risso)

Mediterranean.

Body long with a sharp snout; *mouth* small and terminal with fleshy lips; *fins* there are 2 dorsal fins. 1st dorsal is small and triangular set very far forward at the level of the front gill cover. 2nd dorsal very long and of uniform height. The outer rays of the tail are fused in pairs, the inner rays are single and widely spaced. 1D III, 2D XXV-XXVIII, 3–4; A II, 18–20.

The ground colour can be almost any shade of black, olive, violet or brown. The flanks have rows of lighter blotches and the dorsal and anal fins have regular dark vertical bands.

Up to 10 cm.

Lives near rocks especially where there is sea-grass (*Posidonia*). They rarely swim but instead crawl over the bottom on their pectoral fins. Feed on the small animals coating the fronds of sea-grass and algae. The eggs are laid in early summer in about 1 m of water. The eggs are provided with tufts of silky filaments and these are used to attach the eggs to algal fronds. The larvae do not have a free-living stage but immediately settle on the bottom.

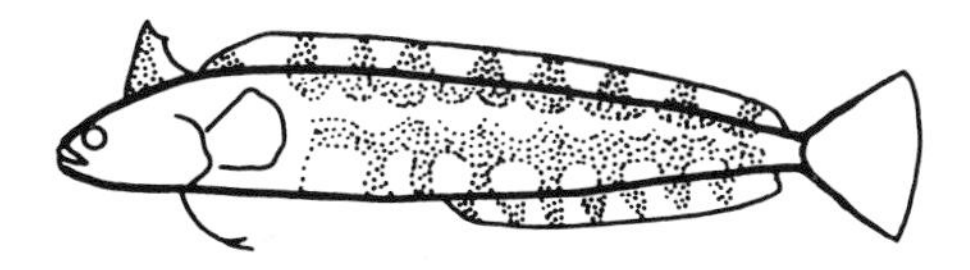

Cristiceps argentatus

2 dorsal fins the 1st between eye and gill cover.
10 cm.

Family:
TRIPTERYGII

Small blenny-like fish with three dorsal fins, the first of which can be erected independently of the other two. The black head and red or yellow body of the courting males is very characteristic.

Blackfaced Blenny
Tripterygion tripteronotus (Risso)
= *Tripterygion nasus* Risso

Mediterranean.

Body scaly; *head* profile rather steep, lips not protruding; *fins* dorsal fin in territorial males has prolonged rays in 2nd dorsal fin. D III; 2D XV-XVII; 3D 11–13; A II, 23–26.

Territorial males have black heads and red bodies, the black head mask in territorial males extends onto the pectoral fin. There is never a dark bar at the base of the tail. Females and non-territorial males are grey with 5 broad vertical bars.

7 cm.

On rocks down to 6 m, but prefer shallower than 3 m.

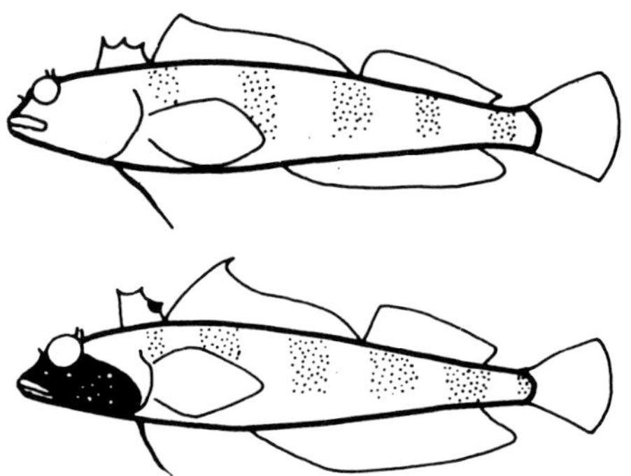

Blackfaced Blenny
Tripterygion tripteronotus

3 dorsal fins;
head profile steep;
males with red body and black head;
pectoral fins black in males.
Mediterranean.
7 cm.

Blackfaced Blenny
Tripterygion delaisi (Cadenat & Blache)
= *Tripterygion xanthosoma* (Zauder & Heymer)

Mediterranean; eastern Atlantic coast from Morocco as far north as southern England.

Body scaly; *head* profile rather steep, lips

Blackfaced Blenny, *Tripterygion melanurus* (Andy Purcell).

Blackfaced Blenny, *Tripterygion delaisi* (Andrea Ghisotti).

not protruding; *fins* dorsal fin in territorial males has prolonged rays in 2nd dorsal fin. 1D III; 2D XVI-XVIII; 3D 11–14; A II, 24–28.

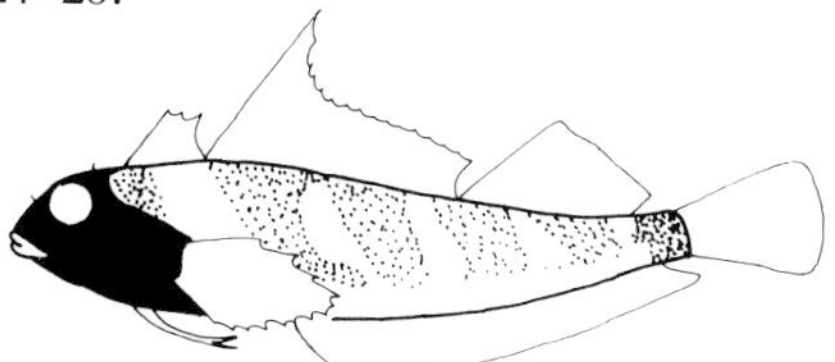

Blackfaced Blenny
Tripterygion delaisi =

3 dorsal fins;
head profile steep;
males yellow with black head;
pectoral fins not black in males.
Eastern Atlantic; Mediterranean.
7 cm.

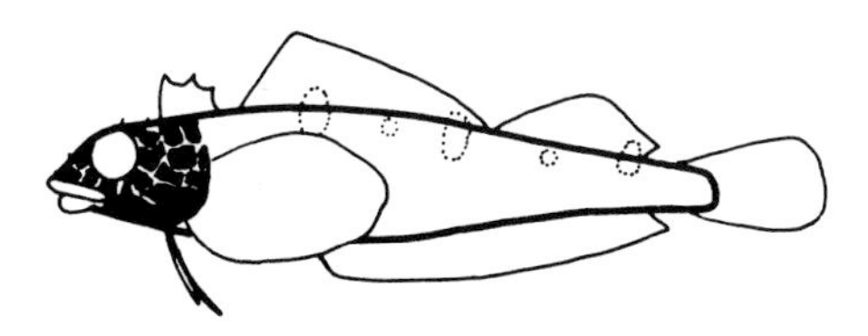

Blackfaced Blenny
Tripterygion melanurus

Head profile pointed;
lips protruding;
always red, males with black head;
typically in caves.
Mediterranean.
5 cm.

Territorial males have black heads and yellow bodies, the black head mask not extending onto the pectoral fins; females and non-territorial males are grey with 5 broad vertical bars, the last forming a distinct dark blotch at the base of the tail.

7 cm.

On rocks from 3 to 40 m, but prefers 6–12 m in shaded places; in the Atlantic also frequents shallow well-lit areas.

Blackfaced Blenny
Tripterygion melanurus Guichenot

Mediterranean.

Body scaly; *head* profile rather pointed and the lips protrude; *fins* rays of 2nd dorsal fin extended in territorial males. 1D III; 2D XIV-XVI; 3D 10–13; A II, 22–25.

Body always red, head marbled black in females and non-territorial males; in territorial males it is black. Some varieties have a black spot on the tail stalk.

5 cm.

Characteristic inhabitant of sea caves, where it clings to the walls and roof.

Family:
CRYPTACANTHODIDAE

Wrymouth
Cryptacanthodes maculatus Storer

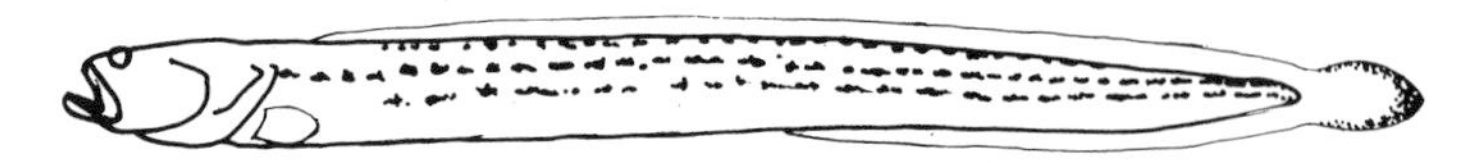

Western Atlantic from southern Labrador to New Jersey.

Body long and eel-like; *head* flattened, jaw strongly oblique, the lower jaw protruding, mucous-secreting pits on sides of head; *teeth* strong and conical; *fins* no pelvic fins, dorsal fin continuous and low with 73–77 spiny rays, tail fin small and pointed.

Brown with 3 rows of brown spots along the upper sides; head, dorsal and anal fins spotted.

90 cm.

Found on muddy bottoms from the intertidal down to 110 m. It burrows extensive tunnel systems 3–8 cm beneath the surface and 5 cm in diameter. Each burrow is inhabited by one fish.

Wrymouth
Cryptacanthodes maculatus

Long eel-like body;
mouth oblique;
no pelvic fins
3 rows of dark spots.
Western Atlantic.
90 cm.

Family:
STICHAEIDAE

Coldwater fish of the northern hemisphere, mostly in the north Pacific, but also in the north Atlantic. Tend to live in moderately shallow water. Body elongate or eel-like, but moderately robust and covered with small scales; dorsal fin is very long, composed almost entirely of spiny rays; fleshy tentacles on head or body.

Atlantic Warbonnet; Yarrell's Blenny
Chirolophis ascanii

Western and southern coasts of Iceland, Norway and Britain; central east coast of England; western English Channel.

Body rather deep behind the head and tapers to the tail fin; *snout* rounded; *tentacles* there is a large, branched tentacle above each eye and a smaller tentacle at the rear edge of each nostril. There are small filaments present on the top of the head; *scales* very small, not present on the head; *fins* dorsal fin long. It originates behind the head and extends to the tail. The first few spines may be extended into short fila-

Atlantic Warbonnet, *Chirolophis ascanii* (David Maitland).

ments. These are particularly apparent in adult males. Pelvic fins are small (shorter than dorsal fin height). D L-LXV; A I, 35–40.

Yellowish or greenish-brown with darker vertical bands. The eye is circled with dark brown and there is also a dark stripe which runs from the eye to the corner of the mouth.

Up to 25 cm.

Usually found at 40–50 m, occasionally down to 400 m, amongst rocks. Very little is known about this fish except that it feeds on molluscs, worms, etc.

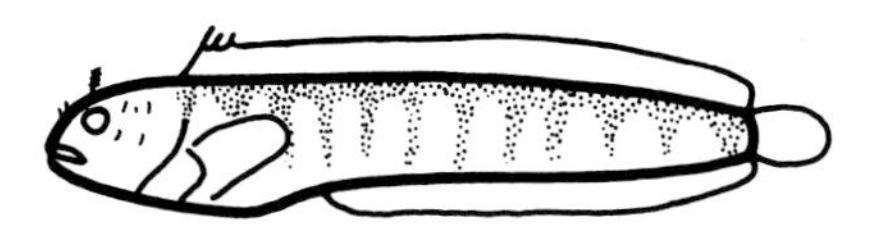

Atlantic Warbonnet; Yarrell's Blenny *Chirolophis ascanii*

Branched tentacles over each eye;
dorsal fin long, only spiny rays;
red-brown markings in life, dark rings round eyes extending down onto cheeks.
Eastern Atlantic.
20 cm.

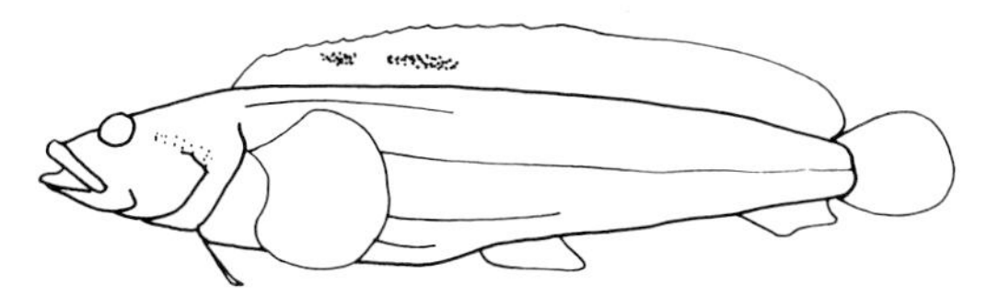

Fourline Snakeblenny *Eumesogrammus praecisus*

Dorsal fin long, only spiny rays;
western Atlantic.
1 complete and 3 partial lateral lines;
1 or 2 dark spots with light edge on dorsal fin.
Western Atlantic.
18 cm.

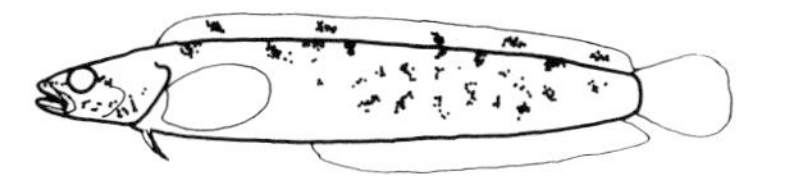

Arctic Shanny *Stichaeus punctatus*

dorsal fin long, only spiny rays;
1 incomplete lateral line above complete lateral line;
5 black blotches along dorsal fin;
brownish or bright scarlet.
Western Atlantic.
15 cm.

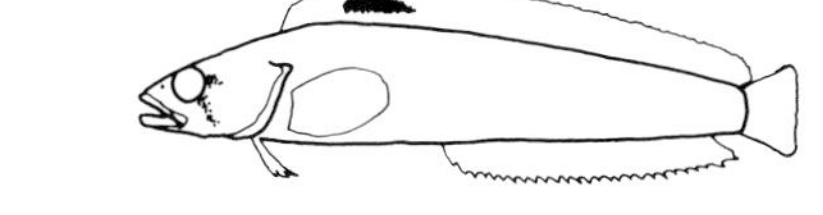

Radiated Shanny *Ulvaria subbifurcata*

dorsal fin long only spiny rays;
lateral line complete but branched;
dark blotch towards front of dorsal fin.
Western Atlantic;
16 cm.

Fourline Snakeblenny
Eumesogrammus praecisus (Kroyer)

Cold north-west Atlantic south to Gulf of St Lawrence.

Body stout and only moderately elongate; *lateral lines* one complete lateral line, with two incomplete lateral lines below and one incomplete lateral line above; *head* pointed with many pores, lower jaw projecting; *teeth* small and conical; *fins* dorsal long, composed only of 47–49 spiny rays, dorsal fin attached to tail fin by membrane, pectoral fins large and rounded.

Brownish above, yellowish below. There is a dark spot ringed with white on the dorsal fin, often with a similar spot in front. Fins dark with light edges.

18 cm.

On muddy bottoms in 30–70 m at about 0°C water temperature.

Arctic Shanny
Stichaeus punctatus (Fabricius)

Circumpolar, in the western Atlantic south to Maine; not into European waters.

Body only moderately elongate; *head* pointed, lower jaw slightly the longer; *teeth* in narrow bands; *lateral line* 1, incomplete; *fins* 48–50 spines, tall fin small and rounded, separate from dorsal and anal fins.

Brownish or bright scarlet with 5 black blotches along the dorsal fin giving coloration resembling the Gunnel.

15 cm.

Over pebble or stony bottoms in cold water at 1–55 m.

Radiated Shanny
Ulvaria subbifurcata (Storer)

Western Atlantic from Newfoundland south to Massachusetts.

Body stout, moderately elongate, *head* moderately pointed, jaws equal; *teeth* small; *eyes* large; *lateral line* complete along middle of flanks but with a branch that extends to above the tip of the pectoral fin; *fins* dorsal fin with 43–44 spiny rays, the dorsal fin is joined to the tail fin by a membrane.

Brownish with pale to yellow belly. There is a dark oval blotch on the front of the dorsal fin. An oblique dark band runs downwards and backwards from the eye.

16 cm.

Amongst seaweeds on rocky shores. Apparently nocturnal, the adults rest inactive on the bottom during the daytime thus avoiding competition with the Gunnel *Pholis gunnellus*, which is diurnal.

Family:

PHOLIDIDAE

Body compressed with very small scales; vertical fins run together, dorsal fin long and low with 75–100 spines. No lateral line. Pelvic with 1 spine and 1 small soft ray.

Butterfish; Gunnel *Pholis gunnellus*

Row of 9–13 black spots circled with white along the base of the dorsal fin. Eastern Atlantic. 25 cm.

Banded Gunnel *Pholis fasciata*

Body elongate, laterally compressed; row of white inverted triangles along back. Western Atlantic. 27 cm.

Butterfish; Gunnel
Pholis gunnellus (L.)

Eastern Atlantic south to the English Channel, north beyond Iceland; English Channel; North Sea; west Baltic.

Body long and flattened from side to side; *head* small; *snout* rounded; *lips* fleshy; *scales* very small; *lateral line* absent; *fins* 1 very long dorsal fin which consists of small spines and commences just behind the gill cover and continues to the base of the tail. 1 anal fin which is about half the length of the dorsal fin. Pelvic fins very reduced and consist of 1 spine and 1 small soft ray. D LXXV-LXXXII; A II, 39–45.

Greyish-brown or brown; there are irregular darker vertical bands on the body but these gradually break up to form a mottled pattern in older fish. There is a row of 9–13 black spots surrounded by a white ring situated along the back at the base of or on the dorsal fin. There is also a dark stripe which runs from the eye to the rear edge of the mouth.

Up to 25 cm.

These fish are common and found from the sea shore down to 40 m and live in a variety of habitats from mud and sand to rocks, frequently amongst kelp (*Laminaria*) holdfasts. They feed on small molluscs and crustacea, worms and other small sedentary animals. Breeding takes place in winter and nests may be made from between tides to 25 m. The egg masses are guarded by either parent though more usually by the female. During this period both the parent fish cease feeding. The skin is very slimy (hence the name Butterfish) and they are extremely difficult to catch.

Banded Gunnel
Pholis fasciata (Bloch & Schneider)

Western north Atlantic, Hudson Bay, Labrador, Newfoundland; north Pacific.

Resembles the Gunnel in body form except that it has 84–91 spines in the dorsal fin.

Ground colour of the body is yellowish-grey, black band runs vertically across the eye and across the top of the head; the black band is bordered by white stripe behind, 10–12 light triangles pointing downwards along the back beside the dorsal fin; the light triangles have dark spots. Some individuals have brillant scarlet sides.

27 cm.

Found in shallow rocky intertidal arctic waters down to about 8 m. Does not appear to frequent tide pools.

Butterfish, *Pholis gunnellus* (Jim Greenfield).

Family:
LUMPENIDAE

Elongate, slightly compressed, no fleshy tentacles on the head, dorsal and anal fins long with a distinct tail fin. Pelvic fins present but with only 1 spiny ray and 3 soft rays.

Snake Blenny
Lumpenus lumpretaeformis (Walbaum)

Cold north Atlantic distribution, in the west from arctic south to Nova Scotia, in the east from the arctic south to the Baltic and North Sea.

Body very long and eel-like, moderately compressed; *head* rather slender and blunt with lower jaw slightly shorter than upper; *teeth* small and conical with a single row in each jaw; *fins* dorsal has 68–85 spines, anal has 47–53 soft rays, tail fin is oval shaped and slightly pointed.

Light pale brown on back with darker brown mottling, under head and front part of body bluish, dorsal and tail fins with oblique pale bars, anal and pectoral fins brown, pelvic fins white.

28 cm.

Found on muddy or hard bottoms at 30–200 m, often in water around 0°C. Believed to form burrows in the mud.

The Slender Eelblenny (*Lumpenus fabricii*) (Valenciennes) is similar. It differs in the tail fin which is more rounded to square-cut and the dorsal fin which has 61–67 spiny rays. Has been observed amongst eel-grass very close to shore, but has been found down to at least 183 m at temperatures between –1.6 and +13°C.

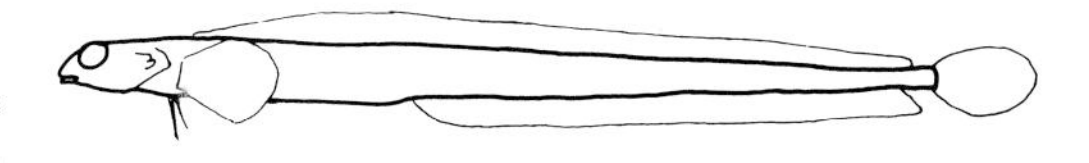

Snake Blenny *Lumpenus lumpretaeformis*
Body slender and eel-like; tail fin distinct, oval-shaped; pectoral fin rounded; lower lip thick.
Cold eastern and western Atlantic.
28 cm.

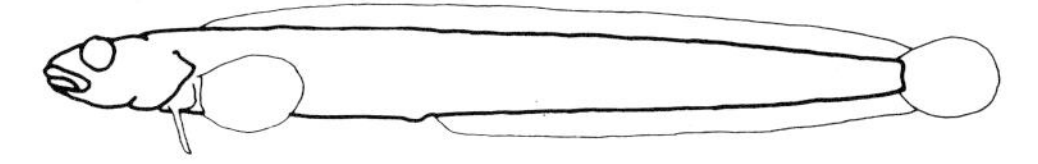

Stout Eelblenny *Anisarchus medius*
Eel-like body; tail fin distinct, rounded; pectoral fin rounded; body light yellowish; dorsal fin rays 58–63.
Cold western Atlantic.
15 cm.

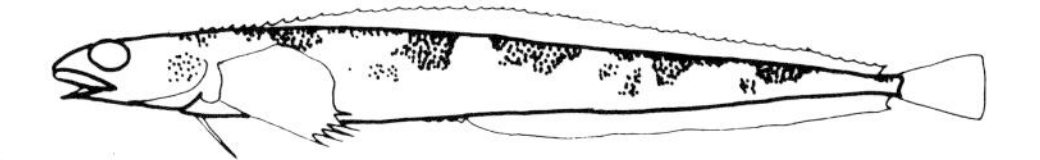

Daubed Shanny *Leptoclinus maculatus*
Eel-like body; tail fin distinct, square-cut; pectoral fin with longer lower rays; body with many dark patches.
Cold eastern and western Atlantic.
15 cm.

Stout Eelblenny
Anisarchus medius (Reinhardt)
= *Lumpenus medius* Reinhardt

Polar distribution, as far south as Gulf of St Lawrence.

Body long, slightly compressed; *mouth* small, terminal; *teeth* small and conical; *fins* dorsal 59–63 spiny rays, tail fin distinct and rounded.

Body yellowish with faint mottlings; in life the dorsal fin has red markings.

15 cm.

On sandy mud at 16–119 m. Usually caught where water temperature is between –1 and +3°C.

Daubed Shanny
Leptoclinus maculatus (Fries)
= *Lumpenus maculatus* (Fries)

Eastern and western Atlantic, south to Cape Cod and Northern Sweden.

Resembles the Stout Eelblenny (*Anisarchus medius*) in body form, but differs in the square-cut tail fin and the pectoral fin which has elongate lower rays, partly free of the fin membrane, which it uses to crawl over the bottom.

Ground colour yellowish with well-marked dark patches on the sides; oblique dark stripes on the dorsal fin, dark bands on the tail fin.

15 cm.

Over sandy mud bottoms often with stones present at 2–250 m, at water temperatures between –1° and +10°C.

Family:

ZOARCIDAE

Elongate blenny-like fish with a long dorsal and a shorter anal fin forming a continuous fringe around the tail. The pelvic fins are set well in front of the pectorals. The head is massive.

Eel Pout; Viviparous Blenny
Zoarces viviparus (L.)

Northern Irish Sea; North Sea; eastern Channel; Baltic; Scandinavian coasts to the White Sea.

Body long, tapering towards the tail; *teeth* conical; *scales* very small embedded in the skin which is slimy; *fins* 1 dorsal fin which is continuous around the tail with the anal fin. There is a distinct dip in the dorsal fin near tail. Pelvic fins small.

Yellowish-green or brown on the back, shading to greyish-brown underneath. There are darker diffuse spots and bands on the back and head and a row of blotches along the sides. There is a row of dark arches on the dorsal fin. The pectoral fins are edged with yellow or orange, but during the breeding season they may be bright red in the male.

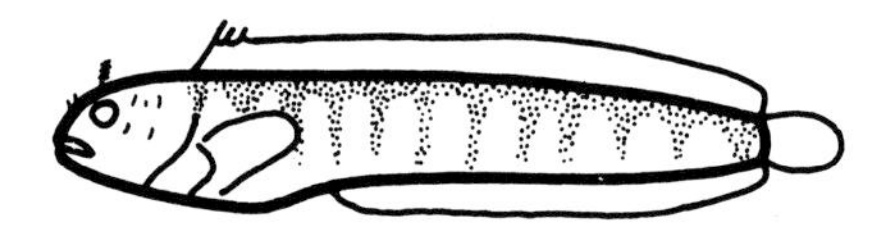

Eel Pout: Viviparous Blenny *Zoarces viviparus*
Dorsal fin continues around tail; distinct dip near tail.
Eastern Atlantic.
46 cm.

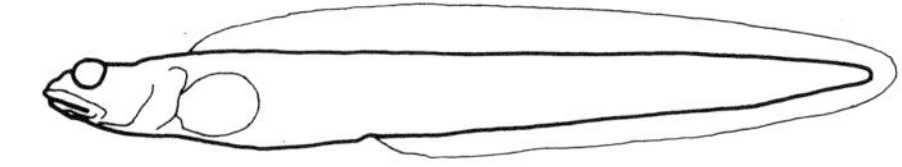

Fish Doctor *Gymnelis viridis*
Long body; soft dorsal and anal fins forming point at tail; no pelvic fins.
Arctic Canada.
25 cm.

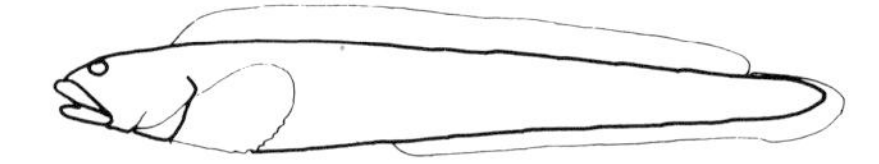

Ocean Pout *Macrozoarces americanus*
Heavy head with thick lips; mottling on back and dorsal fin; dorsal fin dark, edged with yellow.
Western Atlantic.
80 cm

Fish Doctor
Gymnelis viridis (Fabricius)

Circumpolar in arctic waters as far south as Nova Scotia and northern Bering Sea.

Body long and tapering to point at tail, no scales; *head* mouth terminal, horizontal; *teeth* small, conical; *lateral line* faint and straight; *fins* no pelvic fins, dorsal and anal fins with only soft rays, coming to a point at the tail, the tail fin is not separately distinguishable.

Greenish-brown, sometimes banded or mottled, no dark spot on dorsal fin, anal fin dusky in males.

25 cm.

Lives in very cold water, often below O°. Over sandy or muddy bottoms often amongst brown kelp, where it has been observed to make burrows under large stones, from the intertidal down to 100 m.

Eel Pout, *Zoarces viviparus* (David Maitland).

Ocean Pout
Macrozoarces americanus (Bloch & Schneider)

Western north Atlantic from Labrador south to Delaware.

Body long, somewhat compressed, very smooth and covered with mucus; *mouth* large; *teeth* 2 rows of blunt conical teeth in the front of each jaw and one series at the sides; *fins* dorsal fin of more or less uniform height except for the last 16–24 rays which are very short, anal fin long and continuous with the tail fin.

Dull yellow or red brown with a darker mottled pattern on the upper flanks and extending onto the dorsal fin which has a yellow edge, the pectorals are often reddish or orange.

80 cm.

From the intertidal to 200 m, migrating into the deeper waters in autumn and returning in spring. Usually on hard bottoms. Unlike its European relative, it lays eggs and is not viviparous.

Ocean Pout, *Macrozoarces americanus* (Jon D. Witman).

Family:

OPHIDIIDAE

Generally deep-water eel-like fishes with pelvic fins resembling barbels set far forward.

Snake Blenny
Ophidion barbatum L.

Mediterranean and eastern Atlantic from Biscay to Senegal.

Body long, and eel-like, laterally flattened; *head* 1/5 to 1/6 the total body length; *scales* are an elongate oval shape and are embedded in the skin on the upper surface of the fish in a mozaic pattern; *eye* large, having a diameter about 1/4 the head length; *fins* the dorsal, anal and tail fins are united into a single continuous fin. The pelvic fins are set very far forward under the chin; each consists of a double ray and superficially resemble barbels.

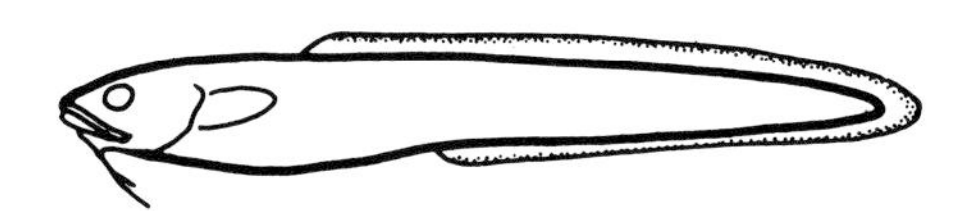

Snake Blenny *Ophidion barbatum*
Pelvic fins resembling barbels under chin; united median fins with dark margin. Mediterranean; eastern Atlantic. 30 cm.

Pinkish above, paler bluish-white below. The united fin has a continuous dark margin.

Up to 30 cm.

Found on sandy or muddy bottoms down to about 150 m. The young may be found in the vicinity of sea-grass (*Posidonia*) meadows.

Family:
CARAPIDAE

The **Pearlfish** have elongated bodies with long dorsal and anal fins merging at the tail. No tail or pelvic fins. The anus opens far forward near the pectoral fin. They live in the body cavity of sedentary echinoderms and molluscs.

Pearlfish
Carapus acus (Brünnich)

Pearlfish *Carapus acus*
Long, slender shape; no tail or pelvic fins; vent opens in front of pectoral fin; lives in sea cucumbers. Mediterranean. 20 cm.

Mediterranean.

Body long, slender and laterally flattened grading gradually in depth from the head to the pointed tail; *jaws* large, extending back past the eye; *vent* situated in front of the pectoral fin; *scales* none; *fins* dorsal and anal fins run most of the length of the body and unite at the tail. In the early larval stages the 1st ray of the dorsal fin is much elongated. No pelvic or tail fins. D approx. 140; A approx. 170.

Silver-white with reddish spots and mottlings.

Up to 20 cm.

The Pearlfish inhabits the body cavity of sea-cucumbers, especially *Stichopus regalis* and *Holothuria tubulosa*. The young enter head first through the anus but the adults enter tail first. Inside they feed mostly upon the gonads of their host, which occasionally eviscerates expelling the fish. At intervals the body of the Pearlfish pulsates in its host presumably to create a current of oxygenated water inside. The Pearlfish is able to live freely outside the body of a sea-cucumber but soon finds a new host, which it recognizes by its mucus and by its long shape.

Family:
STROMATEIDAE

Man-of-War Fish
Nomeus gronovii (Gmelin)

Man-of-War Fish *Nomeus gronovii*
Associated with Man-of-War jellyfish; large irregular bars and blotches; pectoral fin large, black; Western Atlantic, circumtropical; 15 cm.

Western Atlantic, Massachusetts south to Brazil; not in the north-east Atlantic, but otherwise circumtropical.

Body elongate oval, compressed, highest behind gill cover; *scales* thin and easily detached; *mouth* small, terminal; *fins* 2 dorsal fins, pelvic fin large, fan-like and attached almost back to the anal fin, 2 dorsal fins. D IX-XII, 24–28; A I-II, 24–29.

Silvery-white with dark blue mottled pattern consisting of large blotches, spots and short irregular bars extending over entire fish except the pelvic fin, which is mostly black.

15 cm.

Associated with the poisonous siphonophore jelly fish the Portuguese Man-of-War (*Physalia*). Several fish may be associ-

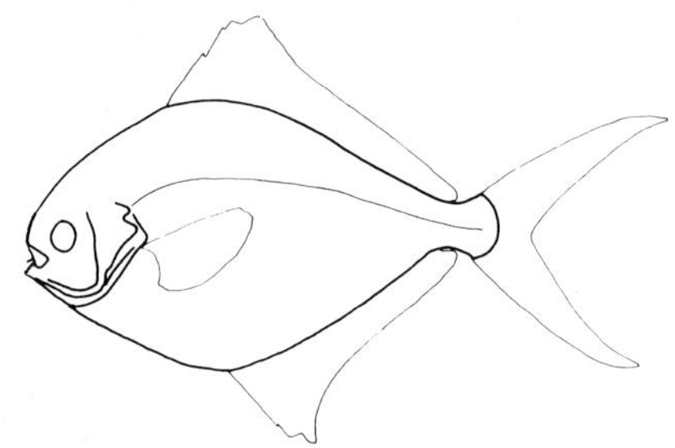

Butterfish *Stromateus fiatola*
Deep compressed body; deeply forked tail; dorsal and anal fins similar. Mediterranean. 50 cm.

Man-of-War Fish, *Nomeus gronovii* (Peter David).

ated with one jellyfish, sheltering under the bell or amongst its long tentacles. The tentacles sting dangerously, but the fish appears to have a limited immunity to the sting. The colouring of the fish mimics the pattern of retracted jellyfish tentacles.

Butterfish
Stromateus fiatola L.

Mediterranean; eastern Atlantic coast south from Spain.

Body deep and compressed; *head* small with blunt front profile; *eyes* small set centrally in the head; *lateral line* bowed upwards following upper profile; *fins* dorsal and anal fins similar in shape, but dorsal fin longer, tail fin deeply forked, pelvic fin present in young but absent in fishes longer than 10 cm. D 42–50; A 33–38.

Bluish shading to whitish below; fish shorter than 10 cm have vertical bars, fish longer than 10 cm have spots on the back.

50 cm.

The striped young are associated with free-floating jelly fish; older fish in schools near the bottom in open water at depths between 12 and 50 m.

Family:
SPHYRAENIDAE

Long cylindrical body with two well-separated dorsal fins. The mouth is large and contains numerous large teeth. The lower jaw extends beyond the upper.

Barracuda
Sphyraena sphyraena (L.)

Mediterranean.

Body long and slender; *jaws* long, reaching back to the front-edge of the eye, lower jaw longest with a small fleshy lobe at the end, gives a streamlined outline when the jaw is closed; *scales* 125–145 along lateral line; *fins* 2 well-separated dorsal fins.

Greenish or brownish above, silvery below; adults have about 24 darker vertical stripes.

50 cm.

Swim in schools sometimes just beneath

Barracuda, *Sphyraena sphyraena* (Christian Petron).

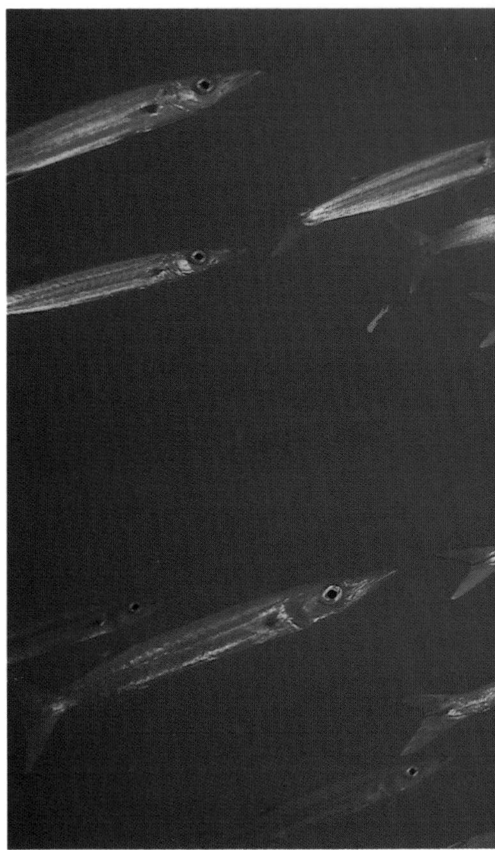

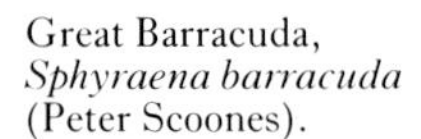

Great Barracuda, *Sphyraena barracuda* (Peter Scoones).

the surface, but may also be much deeper especially over sand.

There are two other species in the eastern Mediterranean. These are *Sphyraena viridensis*, which differs in having a naked edge to the front gill cover; and *Sphyraena chrysotaenia*, which has 150–160 scales along the lateral line, the tips of its 1st dorsal and tail fins are blackfish, and the 2nd dorsal and tail fins are yellowish.

Great Barracuda
Sphyraena barracuda (Walbaum)

Western Atlantic from Massachusetts southwards and in all warm and tropical seas except the eastern Pacific and the Mediterranean.

Body long and slender; *mouth* long, lower jaw longest with no fleshy tip; *teeth* almost upright in the jaw; *lateral line* scales not raised and the same size as neighbouring scales, 75–87; *fins* 2 dorsal fins, pelvic fin originates well in front of 1st dorsal fin, pectoral fin reaches back to the origin of the pelvic fin.

Small fish with longitudinal series of blotches which are often H-shaped, larger fish with 18–22 oblique bands on upper flanks, and irregular dark blotches below. 2nd dorsal fin and tail fin dark with white tips.

170 cm.

Found round wrecks, oil platforms and reefs, often as solitary individuals.

Northern Sennet
Sphyraena borealis DeKay

Western Atlantic south from Cape Cod.

Body long and slender; *head* space behind eyes flat; *mouth* long, lower jaw longest with fleshy tip; *fins* 2 dorsal fins, pelvic fin originates above or slightly behind first dorsal fin, pectoral fin does not reach as far back as the origin of the pelvic fin.

45 cm.

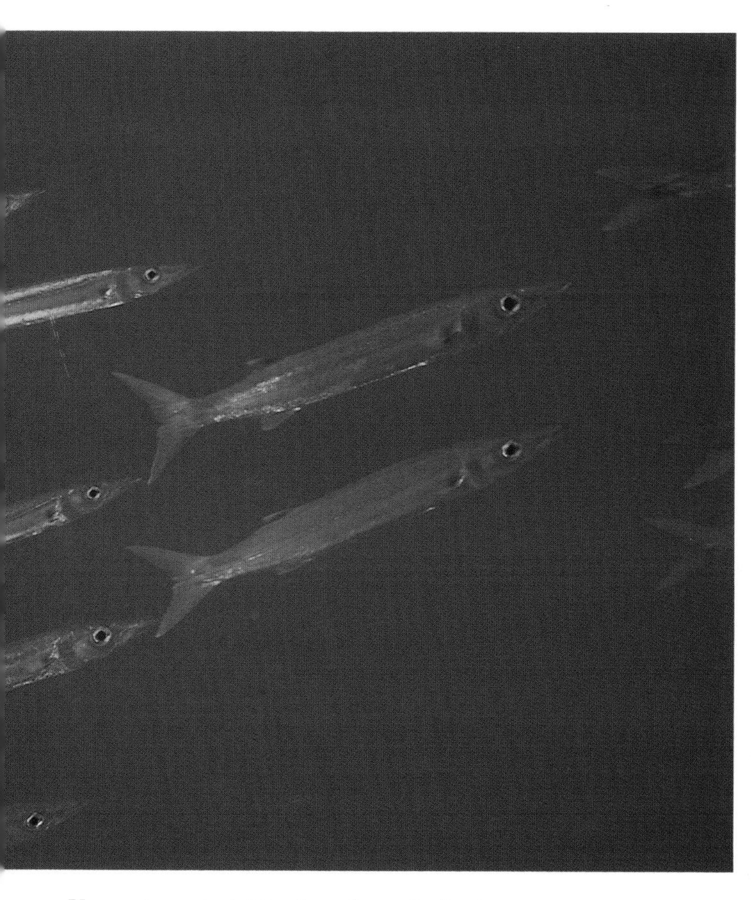

Northern Sennet, *Sphyraena borealis* (James King).

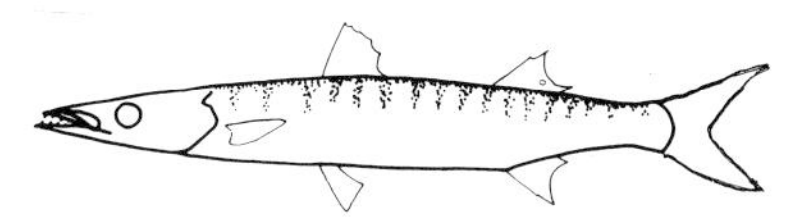

Barracuda *Sphyraena sphyraena*

2 dorsal fins;
short oblique bands in adult.
Mediterranean; eastern Atlantic.

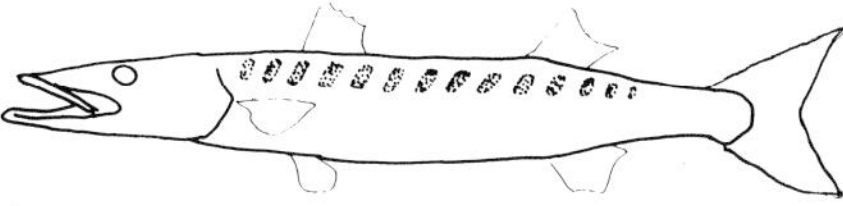

Great Barracuda *Sphyraena barracuda*

2 dorsal fins;
teeth upright in jaws;
short oblique bands in adult.
Western Atlantic.
170 cm.

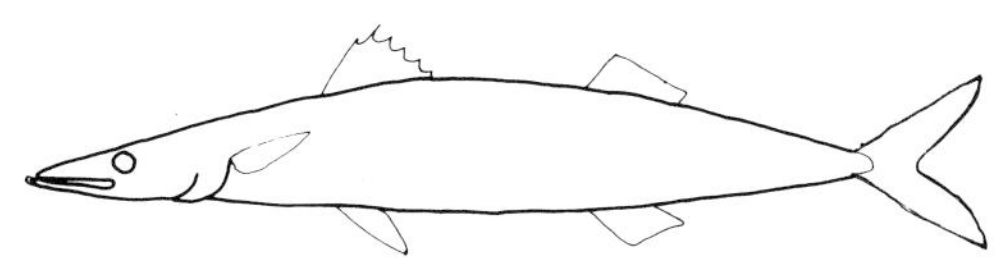

Northern Sennet *Sphyraena borealis*

2 dorsal fins;
fleshy tip to lower jaw;
young with dusky blotches.
Western Atlantic.
45 cm.

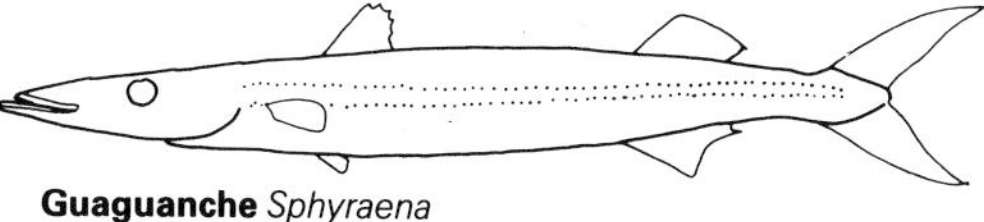

Guaguanche *Sphyraena guachancho*

2 dorsal fins;
3 encircling bands on body;
teeth directed backwards.
Western Atlantic.
60 cm.

Young have dark blotches along the back and on the lateral line.

Guaguanche
Sphyraena guachancho Cuvier

Western Atlantic south from Massachusetts.

Body long and slender; *head* space behind eyes convex; *mouth* long, lower jaw longest; *teeth* directed back; *lateral line* scales along lateral line raised and larger than neighbouring scales 108–114; *fins* 2 dorsal fins, 1st dorsal originates behind pelvic fin, pectoral fin reaches back to the origin of the pelvic fin.

Silvery to brownish-green with two faint longitudinal golden stripes, three dark bands encircle the rear part of the body, margin of pelvic anal fins black, tip of middle tail fin rays black.

60 cm.

An open water species, more common southwards.

Family:
MUGILIDAE

The **Grey Mullets** differ little in general body form. Most species enter brackish water and some penetrate into fresh water. Usually in tight schools and often seen feeding in harbours near sewage outfalls, etc.

Thicklip Grey Mullet
Chelon labrosus (Risso)
9 soft anal fin rays; upper lip very deep with horny papillae; dark stripes along scale rows, no dark spot at base of pectoral fin. Eastern Atlantic, Mediterranean. 60 cm. Usually inshore, entering lagoons.

Thicklip Grey Mullet
Chelon labrosus
9 soft anal finrays; upper lip very deep with horny papillae; dark stripes along scale rows, no dark spot at base of pectoral fin; Eastern Atlantic, Mediterranean. 60 cm.

Golden Grey Mullet
Liza aurata (Risso)
9 soft anal fin rays; adipose eyelid very narrow or absent; golden blotch on gill cover, no dark spot at base of pectoral fin. Eastern Atlantic; Mediterranean. 50 cm. Inshore, enters lagoons and estuaries.

Thinlip Grey Mullet
Liza ramada (Risso)
9 soft anal fin rays; adipose eyelid very narrow or absent; faint dark stripes along scale rows, black spot at base of pectoral fin. Eastern Atlantic; Mediterranean. 50 cm. Inshore, enters lagoons, estuaries and fresh water.

Leaping Mullet
Liza saliens
9 soft anal fin rays; lower half of upper lip smooth; scales on upper half of head and body with 2–8 grooves; no clear markings. Eastern Atlantic; Mediterranean. 35 cm. Inshore, enters lagoons and estuaries.

Boxlip Mullet
Oedalechilus labeo (Cuvier)
11 soft anal fin rays; upper lip wide and bordered by a row of closely packed horny projections; golden stripes along scale rows. Mediterranean. 25 cm. Not entering estuaries.

Striped Mullet; Flathead Grey Mullet
Mugil cephalus (L.)
8 soft anal fin rays; adults with thick transparent adipose eyelid around and over eye; scales on sides streaked with black to form longitudinal stripes; dark blotch at base of pectoral fins. Eastern and western Atlantic. 100 cm. Western Atlantic from Maine south to Brazil; eastern Atlantic south from Bay of Biscay; Mediterranean. Very tolerant of salinity variations; found in very salty sea water to fresh water.

White Mullet
Mugil curema Valenciennes
9 soft anal rays; no stripes; gold spot on gill cover. Western Atlantic. 60 cm. Western Atlantic, Cape Cod south to Brazil. Generally found in fully saline water.

Grey Mullet *sp.* (Christian Petron).

Family:
ATHERINIDAE

The **Sand Smelts** and **Silversides** live in temperate and tropical waters. They are schooling fishes generally near the surface, often in brackish or even fresh water. Divers often mistake these fish for sardines because of their schooling habit and silvery colour. Schools of sand smelts are, however, often almost stationary and show none of the constantly flowing motion so typical of clupeids such as the sardines.

Sand Smelts *Atherina* spp

2 well-spaced dorsal fins; bright silver stripe along flanks;
no lateral line.
Eastern Atlantic; Mediterranean.
Up to 15 cm.

Sand Smelts
Atherina spp

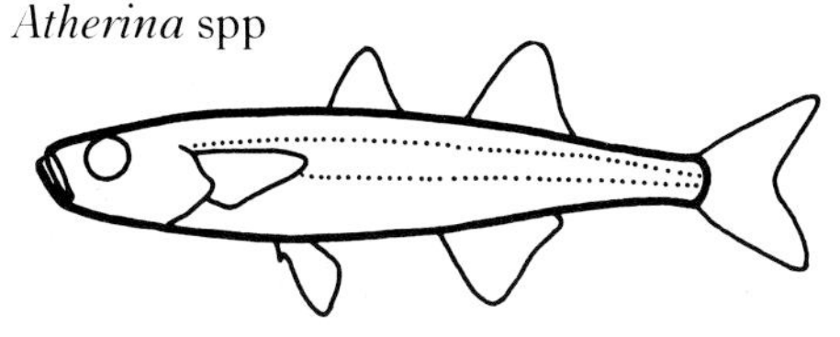

There are three closely similar species in the eastern Atlantic and Mediterranean:

Atherina hepsetus L.
More than 58 scales from gill cover to tail fin. 15 cm. Mediterranean and eastern Atlantic from Spain to Morocco.

Atherina boyeri Risso
44–48 scales from gill cover to tail fin. 9 cm. Mediterranean, isolated patches in southern England, and Atlantic from southern Spain to Morocco.

Atherina hepsetus Cuvier
52–57 scales from gill cover to tail fin. 16 cm. Eastern Atlantic from Scotland south to Morocco, rare in Mediterranean.

Western Atlantic species:

Atlantic Silverside
Menidia menidia (L.)
44–50 scales in the lateral line. 13 cm. Shore waters from the Gulf of St Lawrence to Florida, from Massachusetts to Florida. The Atlantic and Tidewater Silversides occur in the same areas and hybrids exist between them.

Inwater Silverside
Menidia beryllina (Cope)

Scales smooth, 38–40 in the lateral line, scales on back outlined in black. 8 cm. Massachusetts to southern Mexico. Usually limited to the shoreline and tolerant of low salinities.

Rough Silverside
Membras martinica (Valenciennes)

Scales rough to the touch, 43–48 scales in lateral line. 2 rows of spots along back. 9 cm. New York south to Mexico. Usually in more saline waters.

Sandsmelts, *Atherina sp.* (John Lythgoe).

Order:

SCLEROPAREI

Family:

SCORPAENIDAE

The **Scorpion Fishes** are perhaps the most dangerous animal that most Mediterranean swimmers will encounter. At the base of the spiny rays lie venom glands and the poison is directed to the tip of the spine along a canal.

The treatment is to plunge the affected part into very hot water until the pain decreases. Wounds of this nature often go septic and should be thoroughly cleaned and sterilized.

Scorpion Fish

Scorpaena porcus L.

Mediterranean; eastern Atlantic from the Canary Islands to Biscay; very rare as far north as the English Channel.

Head heavily armoured with spines; *eye* large and oval; *appendages* there is a well-developed plume-like appendage above each eye and two smaller ones on the front nostrils. There are no appendages on the chin; *scales* small, at least 55 along the lateral line. D XII, 9–11; A III, 5–6.

Usually brown or reddish-brown with irregular spots and bands which make it extremely difficult to see in the water.

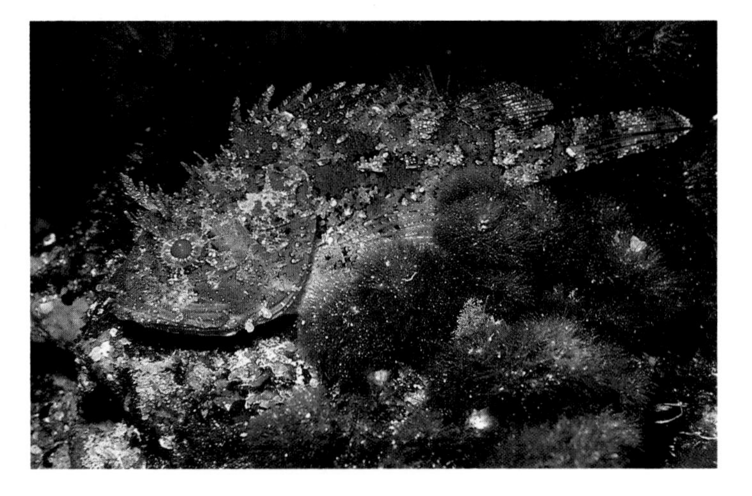

Scorpion Fish, *Scorpaena porcus* (Andy Purcell).

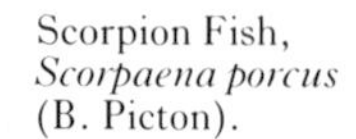

Scorpion Fish, *Scorpaena porcus* (B. Picton).

Up to 25 cm.

During the daytime it usually lies motionless amongst rocks in shallow water. It is extremely difficult to see and a swimmer may inadvertently touch the poisonous spines of the dorsal fin and gill cover (see above for treatment). When disturbed they will swim off at high speed but soon settle again. They are carnivorous and are believed to feed at night. Breed in late spring and summer. The eggs are laid and embedded in a transparent mucous lump.

Scorpion Fish
Scorpaena scrofa L.

Mediterranean; eastern Atlantic from Senegal to Biscay.

Body robust, head armoured with spines; *eye* oval, smaller than in *Scorpaena notata* and *Scorpaena porcus*; *appendages* less well developed than in *S. porcus* except on the chin where they are numerous; *scales* rather large (35–40 along the lateral line). D XII, 9–10; A III, 5–6.

Colour extremely variable, usually reddish-brown with dark and light mottling. In natural light underwater the red colour is not visible and the fish closely resembles the background of rock and weed where it lies. Flash photography shows up the red colour and the fish no longer looks camouflaged. There is often a black spot in the middle of the dorsal fin.

Up to 40 cm, rarely 50 cm.

Like the other Scorpion fishes, they typically rest on the bottom in the daytime. The young fish may be found in shallow water but the adults normally stay between 20 and 200 m. Usually on rocky ground. Their general biology is similar to *S. porcus* but live in rather deeper water. For treatment of stings see above.

Scorpion Fish
Scorpaena notata Rafinesque
= *Scorpaena ustulata* Lowe

Mediterranean and eastern Atlantic from Senegal to the Gulf of Gascony.

Body robust, the head is heavily armoured with spines; *eye* large and nearly round; *appendages* there are no very well developed appendages. There are small ones on the nostrils, a very short one above each eye and none on the chin; *scales* are relatively large (36–40 along the lateral

Scorpion Fish, *Scorpaena scrofa* (Andy Purcell).

Scorpion Fish, *Scorpaena notata* (Andrea Ghisotti).

line). D XII, 9–10; A III, 5.

Brick red with variable mottling. Frequently there are a number of dark spots on the dorsal, anal and tail fins. A constant feature of this fish is a dark blotch between the 8th and 10th dorsal spines. Note that a similar spot is frequent on *Scorpaena scrofa* as well.

15 cm.

Ocean Perch; Golden Redfish

Sebastes marinus (L.)

Cold eastern and western Atlantic; northern North Sea north to Spitzbergen, Gulf of St Lawrence to north Greenland.

Body oval; *head* jaw with rounded protuberance, *eye* large, front gill cover with 5 equal spines, gill cover with 2 spines; *scales* body, head, cheek, snout and lateral line scaly, more than 55 scale rows beneath lateral line. D IV-VI, 13–16; A III, 7–10.

Red or red-brown with dark patch on gill cover.

Young fish in shallow water, the adults in 100–1000 m. Usually in schools.

Small Redfish; Norway Haddock

Sebastes viviparus Kroyer

Cool and cold eastern Atlantic from Ireland; North Sea north to northern Norway; Iceland.

Body oval with rather high shoulder; *head* jaw with rounded protuberance, *eye* large, front gill cover with 5 equal spines, gill cover with 2 spines; *scales* head, cheek, snout and lateral line scaly, less than 55 scale rows beneath lateral line. D IV-VI, 12–15; A III, 6–8.

Golden to red with about 4 dark irregular bands on upper flank.

Usually over rocky bottoms at 10–150 m but are found much deeper. Gregarious but less so than other *Sebastes* species.

Scorpion Fish *Scorpaena porcus*

Plume-like appendages over eye;
no appendages on chin.
Mediterranean.
25 cm.

Scorpion Fish *Scorpaena scrofa*

Numerous appendages on chin.
Eastern Atlantic;
Mediterranean.
40 cm.

Scorpion Fish *Scorpaena notata*

No prominent tentacle over eye;
no appendage on chin;
black spot on dorsal fin.
Eastern Atlantic;
Mediterranean.
15 cm.

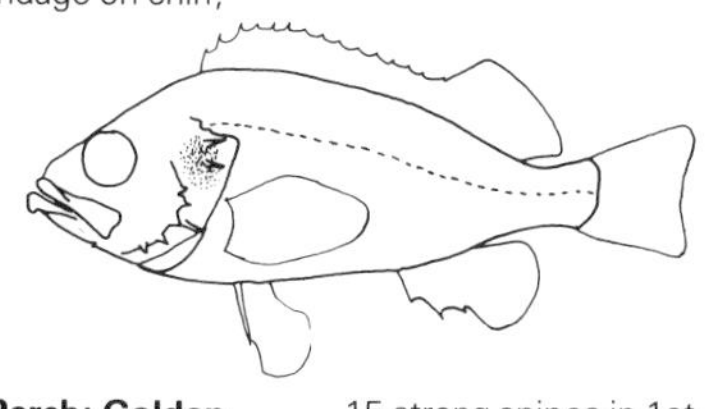

Ocean Perch; Golden Redfish *Sebastes marinus*

5 equal spines on front gill cover;
15 strong spines in 1st dorsal fin;
dusky patch on gill cover.
cold north Atlantic.
50 cm.

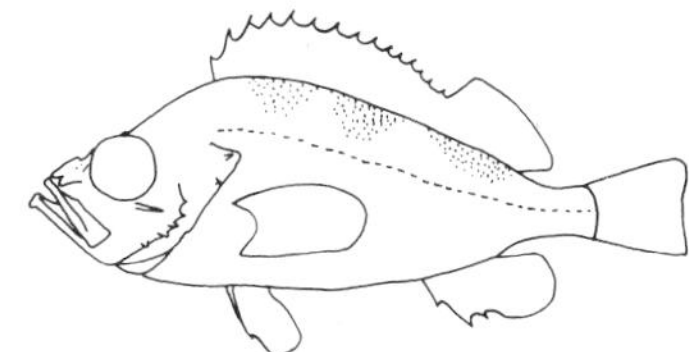

Small Redfish; Norway Haddock *Sebastes viviparus*

5 equal spines on front gill cover;
15 strong spines in 1st dorsal fin;
blotchy dark bands on flanks.
North-eastern Atlantic.
25 cm.

Ocean Perch, *Sebastes marinus* (Jon D. Witman).

Small Redfish, *Sebastes viviparus* (B. Picton).

Family:

TRIGLIDAE

The **Gurnards** and **Sea Robins** have large armoured heads. The most characteristic feature is in the pectoral fins which have a few rays free from the fin membrane and which have taste buds scattered over the surface. The Gurnards delicately probe the bottom with their feelers in their search for food. Very superficially the Gurnards resemble the Red Mullet but differ in having pectoral feelers and not barbels.

Piper
Trigla lyra L.

Mediterranean; eastern Atlantic from northern Scotland to Senegal.

Body tapers steeply from behind the head to the tail; *head* armoured; *profile* steep and concave; *spines* there are 2 bony, toothed lobes protruding in front of the snout. There is a long pointed spine situated immediately above the pectoral fin that is about half the length of the fin. On each side of the dorsal fin is a row of 24–25 short robust spines; *scales* small with the free edge finely toothed; *lateral line* distinct and smooth; *fins* 2 dorsal fins, one anal fin similar to and opposite the 2nd dorsal fin. Pectoral fin long, extending to the 5th ray of 2nd dorsal and anal fin. The 3 lowest rays of the pectoral fin are free and modified into sensory feelers. 1D IX-X; 2D 16–17; A 16–17.

Red in colour, darker on the back, lighter on the flanks and silvery underneath. Fins reddish, the dorsal and pectoral fins often spotted with blue.

Up to 40 cm, occasionally 60 cm.

This species is usually found in deep water at 100–700 m. Occasionally, however, it may penetrate into shallower water and has been caught at around 50 m. Feeds on small crustacea, echinoderms, etc.

Tub Gurnard; Saphirine Gurnard; Yellow Gurnard
Trigla lucerna L.

Mediterranean; Black Sea; eastern Atlantic from Norway to Senegal; English Channel; North Sea.

Body tapers gradually from behind the head to the tail; *head* armoured; *profile* only slightly concave and less steep than in the Piper (*Trigla lyra*); *spines* there are 2 short lobes either side of the snout. The spine above the pectoral fin is short. On each side of the dorsal fin there is a row of

Tub Gurnard, *Trigla lucerna* (Andy Purcell).

Tub Gurnard, *Trigla lucerna* (Peter Scoones).

24–25 small spines; *scales* small on body, 70 larger scales along the lateral line; *lateral line* distinct and smooth; *fins* 2 dorsal fins, 1 anal fin opposite to but rather shorter than the 2nd dorsal fin. Pectoral fin large and extends to the 3rd or 4th ray of the anal fin; the 3 lowest rays are free and modified into sensory feelers. 1D VIII-X; 2D 15–18; A 14–17.

Brilliantly coloured and very variable. The back may be red, reddish-yellow, or brown, or yellowish with brown, or even greenish blotches. The flanks are reddish or yellowish, shading to pink or white on the undersurface. The head is dark red. Pectoral fins have bright blue rims, blue spots and a blue-black blotch on the undersurface with light blue spots on the upper surface.

Up to 65 cm, usually about 30 cm.

This species ranges in depth from 5–300 m, and is found on sand, mud or gravel bottoms. The young fish are frequently encountered near coasts and particularly in or around estuaries and even in fresh water, where they rest on the bottom with their pectoral fins spread out. The adult fish feed mainly on crustacea. Like many of the Triglidae this fish is known to emit noises.

Streaked Gurnard

Trigloporus lastoviza (Brünnich)

Mediterranean; eastern Atlantic north to Scotland, south to Canaries; English Channel.

Body tapers from behind the head to the tail; *head* covered with bony plates; *profile* steep and slightly concave; *snout* blunt, without projecting spines; *spines* there are small spines above the eyes. The spine above the pectoral fin is short but has a very wide base; there are 25 spines along the body either side of the dorsal fins; *skin* arranged in distinct oblique stripes along the body; *fins* 2 dorsal fins, anal opposite and similar to the 2nd dorsal. Pectoral fins reach the 3rd–7th anal fin rays. The lowest 3 rays of the pectoral fin are modified into sensory feelers. 1D IX-XI; 2D 16–17; A 14–16.

Back red, sometimes with darker blotches, sides paler and belly white. Dorsal fins are pinkish and the anal fin is pinkish at the base and yellowish at the tip. Pectoral fins reddish with dark blue spots arranged in rows.

Up to 40 cm.

Most frequently found on muddy or sandy areas near rocks from 45 m. Spawning occurs in the summer. Adult fish feed exclusively on small crustacea.

Grey Gurnard
Eutrigla gurnardus (L.)

Eastern Atlantic from Norway and Iceland to Senegal; Mediterranean; west Baltic; North Sea; English Channel; also Black Sea.

Body tapers from behind the head to the tail; *head* with bony plates; *profile* sloping, almost straight; *snout* with a lobe projecting on either side each with 3–4 small spines; *spines* the spine above the pectoral fin is short. 27 bony spines are present either side of the dorsal fins; *lateral line* 70–77 scales each with a central ridge with a point at the end and toothed free edge; *fins* 2 dorsal fins, anal fin opposite to but shorter than the 2nd dorsal fin. Pectoral fin does not reach the anal fin. The three lowest rays of the pectoral fin are modified into sensory feelers. 1D VII-IX; 2D 18–19; A 17–20.

Variable in colour. The Atlantic form is usually grey, greyish-red or greyish-brown with numerous whitish-yellow spots scattered over the back and sides and the lateral line is also yellowish-white. The 1st dorsal fin has a black patch. The Mediterranean form is usually red or reddish-brown but also has a black patch on the 1st dorsal fin.

Up to 50 cm, the Mediterranean form reaches 30 cm.

Very common in the Atlantic, where they are found in groups on muddy or sandy bottoms or on sandy areas amongst rocks. They inhabit all depths down to 200 m. During the summer there is a migration into shallow water when they may even penetrate into estuaries. They feed predominantly on crustacea but also on small fish. Spawning occurs from January to September, during the later months in the northern ranges and in the early months in the southern ranges. Like the other Gurnards this species is able to emit noises.

Red Gurnard
Aspitrigla cuculus (L.)

Eastern Atlantic north to Norway south to Mauritania; southern North Sea; English Channel; Mediterranean.

Body tapers from behind the head to the tail; *head* covered with bony plates; *profile* fairly steep and somewhat concave; *scales* 65–70 along the lateral line, gradually decreasing in size towards the tail. These scales are very deep and bony but not toothed or spined; *spines* 3–4 small spines project either side of the snout. The spines above the pectoral fin are strong but not large. There are 26–28 spines either side of the dorsal fins; *fins* 2 dorsal fins, anal fin similar to and opposite the 2nd dorsal fin. The pectoral fins extend to about the 3rd ray of the anal fin. 1D VIII-X; 2D 17–18; A 16–18.

Head, back and flanks reddish, undersurface white. Pectoral fins may be greyish, pinkish or yellowish. The tail fin has a pale-coloured base and a dark rear edge.

Grey Gurnard, *Eutrigla gurnardus* (B. Picton).

Red Gurnard, *Aspitrigla cuculus* (Jim Greenfield).

Up to 45 cm, commonly 20–30 cm.

In Atlantic waters this fish is very common and ranges in depths of 5–250 m. In the Mediterranean they are normally caught at 100–250 m, but sometimes in much more shallow water near coasts (30 m) on sandy, muddy, gravel, shell or rocky areas with sandy patches. Spawning occurs during the spring and summer. The adults feed mainly on crustacea but also on fish.

Long-finned Gurnard; Shining Gurnard

Aspitrigla obscura (L.)

Mediterranean; eastern Atlantic north to English Channel.

Body tapers from behind the head to the tail; *head* covered with bony plates; *profile* more or less straight; *spines* there is a bony lobe with 1 distinct spine present on either side of the snout. The spine above the pectoral fin is very short. On each side of the dorsal fin is a row of 27–28 bony plates with toothed margins; *scales* very small on body; large, 68–70, along the lateral line and untoothed; *fins* 2 dorsal fins. The 2nd ray of the 1st dorsal fin is elongate. The 2nd dorsal is similar to the anal fin. The pectoral fin does not extend beyond the 3rd anal fin ray and the 3 lowest rays are modified into sensory feelers. 1D IX-X; 2D 16–18; A 14–18.

The colour is very variable. The back is commonly greyish-pink shading to pearly pink underneath. The lateral line is a silvery-pink. Pectoral fins are bluish, dorsal fins and tail fin reddish. Pelvic and anal fins reddish, sometimes with yellow or white edges.

Up to 35 cm.

Found down to 50 m on rocky and sandy bottoms. Little is known about these fish except that they feed mainly on small crustacea. They are believed to be bioluminescent.

Northern Searobin

Prionotus carolinus (L.)

Western Atlantic.

Head with groove across top; *barbels* none; *scales* chest covered with scales; *fins* with 3 modified rays. Anal has 12 rays; D X + 12–13, A 11–13.

Upper part of body with elongate reddish spots, a dark spot ringed with lighter colour on the 1st dorsal fin between spiny rays 4 and 5, tail and pectoral fins blackish; outer gill (branchiostegal) membranes blackish.

30 cm.

A mainly northern species which frequents sandy bottoms at 15–170 m.

A similar species; *Prionotus martis* (Ginsburg) has two dark spots on the 1st dorsal fin, in between rays 1 and 2, the second between rays 4 and 5.

Leopard Searobin

Prionotus scitulus (Jordan & Gilbert)

Closely related to the Northern Searobin, but its chest is only partly covered with scales and it has a more southerly distribution from North Carolina to Venezuela.

Striped Searobin

Prionotus evolans

Western Atlantic, Nova Scotia south to Florida.

Body short broad and deep; *fins* pectoral fin long, 12–14 adjoining rays and 3 modified rays. D IX-XI, 11–13; A 11–13.

Reddish or greenish-brown above, paler below; there is a narrow black band along the lateral line with a second band running along the middle of the body; pectoral fin with many narrowly separated dark stripes.

45 cm.

Mainly on sandy bottoms from inshore estuaries to about 180 m.

Bighead Searobin
Prionotus tribulus (Cuvier)

Western Atlantic, Chesapeake Bay south to southern Florida.

Body robust; *head* large and broad, the spines on the head are elevated and knife-like, space between eyes deeply concave and at least as wide as the eye diameter; D X, 11–12; A 10–12.

Short oblique bands on upper part of body; 1st dorsal fin with black blotch centred at spines 4–5, tail fin blackish with a light bar near base, pectoral fins with many dark bands and a spot at the top margin, chin is dark.

35 cm.

Common in shallow bays; the young may enter estuaries.

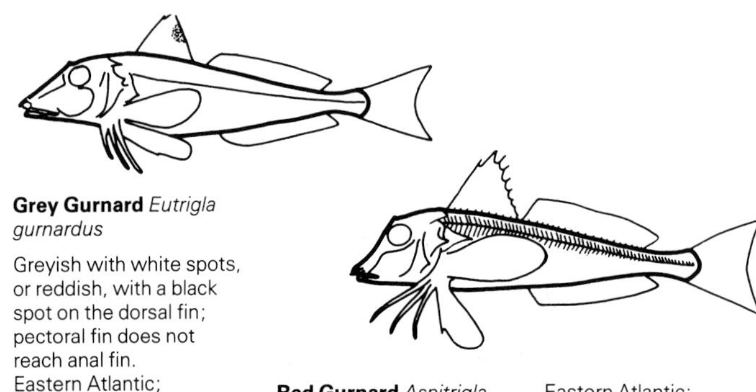

Grey Gurnard *Eutrigla gurnardus*

Greyish with white spots, or reddish, with a black spot on the dorsal fin; pectoral fin does not reach anal fin. Eastern Atlantic; Mediterranean. 40 cm.

Red Gurnard *Aspitrigla cuculus*

Large deep bony scales along the lateral line; red. Eastern Atlantic; Mediterranean. 30 cm.

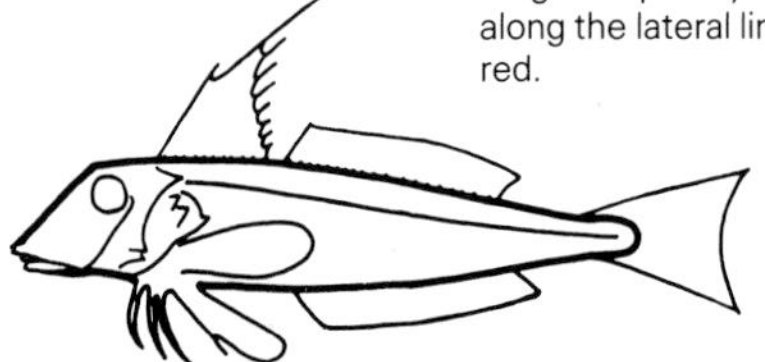

Long-finned Gurnard; Shining Gurnard *Aspitrigla obscura*

2nd ray of 1st dorsal fin elongate; silvery lateral line. Eastern Atlantic; Mediterranean. 35 cm.

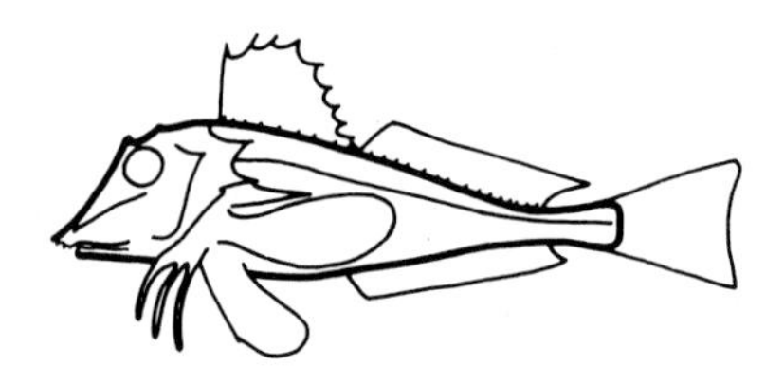

Piper *Trigla lyra*

Large toothed lobes extending either side of the snout; pectoral fins reddish; spines over pectoral fin long. Eastern Atlantic. Mediterranean. 60 cm.

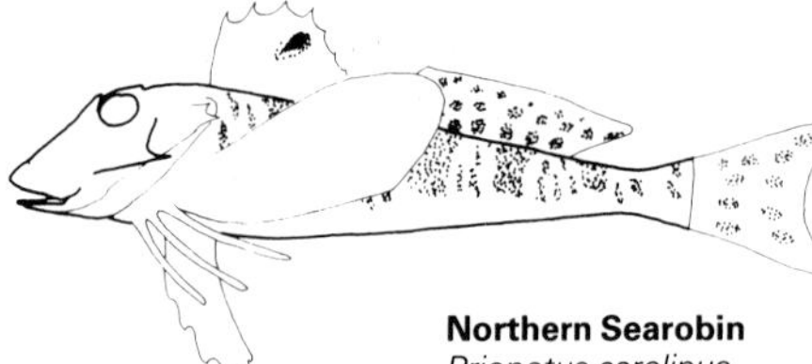

Northern Searobin *Prionotus carolinus*

Dark eyespot near upper edge of dorsal fin; body mottled above. Western Atlantic. 38 cm.

Striped Searobin *Prionotus evolans*

2 stripes along body; dark blotch on middle of 1st dorsal fin. Western Atlantic. 45 cm.

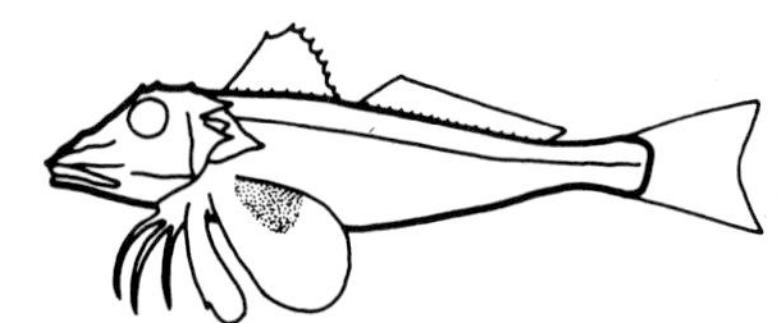

Tub Gurnard; Saphirine Gurnard; *Trigla lucerna*

Spine over pectoral fins short; pectoral fins blue with blue-black blotch on undersurface and light-blue spots on upper surface. Eastern Atlantic; Mediterranean. 65 cm.

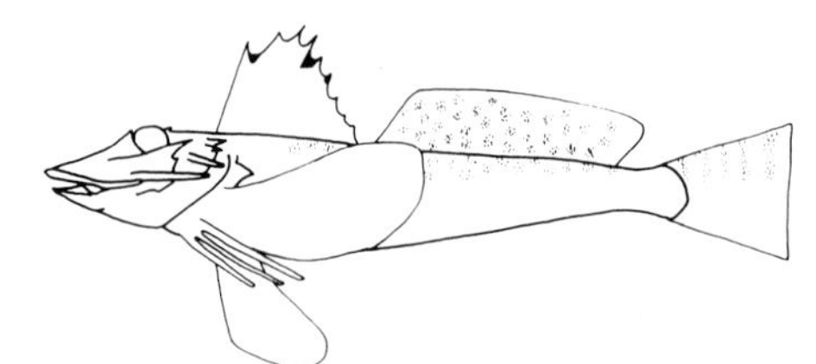

Leopard Searobin *Prionotus scitulus*

2 black spots on 1st dorsal fin; body with brown spots. Western Atlantic. 25 cm.

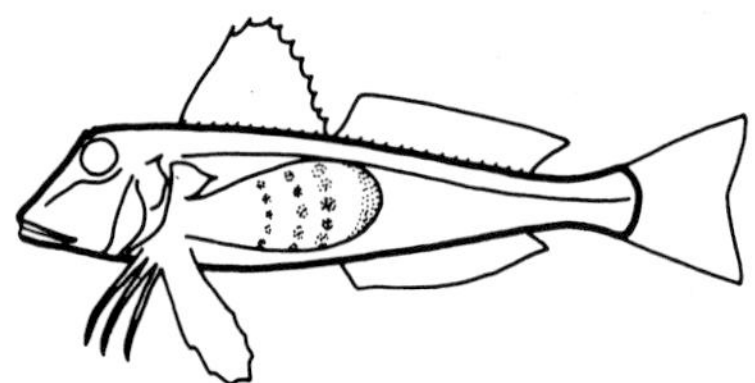

Streaked Gurnard *Trigloporus lastoviza*

Skin arranged in distinct oblique stripes along the body; head and body red, pectoral fin with rows of blue spots. Eastern Atlantic; Mediterranean. 40 cm.

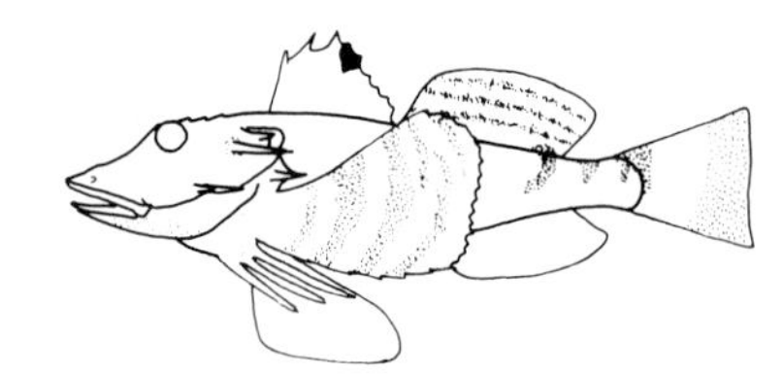

Bighead Searobin *Prionotus tribulus*

Pectoral fin banded, dark; head big and broad. Western Atlantic. 35 cm.

Family:
COTTIDAE

The head is wide, the gill cover has at least 1 spine; skin scaleless but often with hard plates of varying size embedded in it.

Father Lasher; Shorthorn Sculpin; Bull Rout
Myoxocephalus scorpius (L.)

Eastern Atlantic south to Biscay, western Atlantic south to New Jersey.

Head wide, flattened from above and with bony crests; *jaws* no fleshy barbels at the corners of the jaws; *spines* 2 spines present on the front gill cover, the upper longer than the lower but less than the eye diameter; *skin* smooth, without scales but with bony plates embedded in it and spines either side of the lateral line; *lateral line* smooth; *gill cover* the gill covers are joined by a skin membrane which forms a flap across the body of the fish; *fins* 2 dorsal fins, 1 anal fin. Pectoral fin large, pelvic fin with 3 rays. 1D VII-XI; 2D 14–17; A 10–14.

This fish is very variable in colour. Frequently greyish or brownish with darker blotches on the head and back with lighter spots on the flanks. Belly yellowish in the male and orange in the female. The fins are light with darker spots and bars. During the breeding season the males are coppery with lighter spots.

Up to 30 cm.

In the northern ranges of its distribution this fish is very common and found near the shore line. In the more southern water they are found rather deeper at 2–60 m. They inhabit a variety of bottoms including mud, sand, shell, gravel and seaweed and also penetrate into estuaries. Adult fish feed mainly on crustacea and fish. Spawning takes place during winter. The eggs are laid in small masses near rocks and seaweed.

Fatherlasher, *Myoxocephalus scorpius* (Jon D. Witman).

Long-spined Sea Scorpion
Taurulus bubalis (Euphrasen)

Eastern Atlantic south to Portugal; North Sea; English Channel; rare in Mediterranean.

Head large with bony crests; *jaws* with 1 or 2 fleshy barbels at their corners; *spines* there are 3 or 4 spines present on the front gill cover, the uppermost is longest, more than the eye diameter, and pointed; *skin* smooth without scales; *lateral line* consists of a row of 32–35 spiny plates; *gill cover* the skin membrane extending from the gill covers is joined to the body of the fish and does not form a flap; *fins* 2 dorsal fins, 1 anal fin, pectoral fin large, pelvic fin with 3 rays. 1D VII-IX; 2D 10–14; A 8–10.

Variable in colour but frequently brown with greenish and light blotches. Undersurface yellowish or silvery, particularly in the breeding male. They may sometimes be reddish in colour when at first glance they may be mistaken for a Scorpion Fish.

Males 8–12 cm, females 10–25 cm.

Very common in shallow water and rarely found below 30 m. They live amongst rocks and particularly in areas where there is plenty of seaweed or seagrass. Their diet is very varied and includes

Sea Scorpion, *Taurulus* ***bubalis*** (Pete Atkinson).

Sea Scorpion, *Taurulus* ***bubalis*** (Jim Greenfield).

Sea Scorpion, *Taurulus* ***bubalis*** (Jim Greenfield).

small crustacea, worms and small fish, which they catch by making short forays from their lair. Breeding occurs during spring and the eggs are laid in a mass amongst the rocks and frequently in small crevices.

Norway Bullhead
Micrenophrys lilljeborgi Collett

West coast of Scandinavia to west Iceland; Irish Sea.

This species is very similar to *Myoxocephalus bubalis* but may be distinguished by a row of bony spines above and parallel to the lateral line; the pelvic fins only have 2 rays and a maximum size of 60 cm. There are 4 spines on the front gill cover but the uppermost is only equal to the eye diameter.

Light brown or yellowish with 4 darker, greyish, transverse bands on body and 1 on head. During breeding male fish have a red band behind the head and red patches along the flanks.

Up to 60 cm.

Found from 20 m on coarse shell and gravel bottoms. Breeding occurs during spring in its southern range and summer in the more northern latitudes. The eggs are laid amongst stones.

Arctic Hookear Sculpin
Artediellus uncinatus (Reinhardt)

Western Atlantic from Labrador south to Cape Cod; eastern Pacific; arctic Alaska.

Head somewhat flattened with fleshy tabs, an upwardly hooked spine on front gill cover; *teeth* few or absent on the palate, 0–3 vomerine teeth; *scales* absent from body; *lateral line* with 20–30 pores; *fins* dorsal fins close but separate, ray tips elongate in males.

Body light-coloured with darker blotches; there are round white spots on body, head and dorsal fin, the white spots are particularly evident on the dorsal fin of the males.

6 cm.

Found on a variety of hard or soft bottoms at depths of 13–183 m.

Atlantic Hookear Sculpin
Artediellus atlanticus Jordan & Evermann

Western Atlantic from 77°N to Cape Cod; eastern Atlantic from Barents Sea south to Ireland; not North Sea or English Channel.

Resembles the Arctic Hookear Sculpin (*Artediellus uncinatus*), but has teeth on the palate and 16–39 vomerine teeth.

The dorsal fin has oblique dark stripes and there are no white spots.

10 cm.

Usually in deeper water than the Arctic Hookear Sculpin, and prefers mud bottoms.

Arctic Staghorn Sculpin
Gymnocanthus tricuspis (Reinhardt)

Western north Atlantic, arctic south to Maine; eastern Atlantic arctic; generally circumpolar.

Head and chin without fleshy tabs; *fins* 2 separate dorsal fins; 3 blunt *spines* on front gill cover, the top spine broad with branchlets at tip; *teeth* none on vomer; *scales* none except below pectoral fins.

Back dark, flanks with 2 lines of irregular blotches, dorsal fin with dark and light stripes; tail, anal and pectoral fins light, the pectoral fin has a yellow tip with 4–5 bands of dark spots.

26 cm.

On rock or sandy bottoms at 2–174 m. Prefers water between −1.8 and 5°C.

Sea Raven
Hemitripterus americanus (Gmelin)

Western Atlantic from Newfoundland south to Chesapeake Bay.

Body heavy in front part tapering back to the tail; *head* large, ornamented with many bony humps and ridges above, and many fleshy tabs on the head and lower jaw; *front gill cover* with 2 short spines; *fins* dorsal fins separated by a short space, the front dorsal fin has irregular spines with fleshy tabs at the tips, skin with prickles which are larger on the back and near the lateral line.

Reddish-brown, marbled and mottled with darker markings.

60 cm.

Usually found at 2–100 m on rocky bottoms. Inhabits sea water that is almost cold enough to freeze, and also water as warm as 16°C. Feeds on almost any bottom-living invertebrate it can find.

Spatulate Sculpin
Icelus spatula Gilbert & Burke

Western Atlantic from Labrador south to Cape Cod; Russian Arctic, Bering Sea in north Pacific.

Head upper spine on front gill cover forked at tip, no fleshy tabs on the head or chin; *lateral line* usually complete with a row of 40 bony plates along the lateral line, these plates have no small spine below the pores.

Brown or brownish-green with darker mottling, spots and bars on dorsal fins.

11 cm.

A coldwater species living in water generally below zero. Usually caught over soft bottoms.

The Twohorn Sculpin (*Icelus bicornis*) (Reinhardt) differs from the Spatulate Sculpin in having 38 plates along the lateral line, and each plate has a small spine below as well as above the pore. Found at depths greater than 40 m.

Grubby
Myoxocephalus aenaeus (Mitchill)

Western Atlantic, Newfoundland south to New Jersey.

Head broad, covered with smooth skin, there is a ridge on top of the head bearing 1–2 short spines over each eye and 2 short spines between the nostrils; *front gill cover* with 3 short spines, the bottom pointing downward; *lateral line* well-marked with no bony plates, set high on the body.

Varies according to the bottom, with darker bars blotches and shadings.

15 cm.

A coastal species which is tolerant of a wide range of temperatures and salinity, although it does not appear to enter brackish water. In protected situations it may live as shallow as 1 m on sand, gravel and mud; in slightly deeper water it is known to occur on rocky bottoms as well as eel-grass beds. Can be abundant on eel-grass beds. Feeds on a wide variety of bottom-living invertebrates as well as on small fish.

Sea Raven, *Hemitripterus americanus* (Gilbert van Rijckevorsel).

Longhorn Sculpin
Myoxocephalus octodecemspinosus (Mitchill)

Western Atlantic from Newfoundland south to Virginia.

Body naked except for a series of smooth plates along the lateral line; *head* with blunt profile, heavy, front gill cover with 3 spines, the upper one at least 3 times longer than the 2 below, with a sharp naked tip; *fins* no space between dorsal fins.

Brown or greenish-brown with 3–4 irregular blotch-like crossbars. 1st dorsal fin dark with darker mottling, 2nd dorsal with 3–4 oblique bars.

40 cm.

Found in coastal waters, moving further offshore in the summer. When molested, it erects the spines on its gill cover.

Longhorn Sculpin, *Myoxocephalus octodecemspinosus* (Jon D. Witman).

Arctic Sculpin
Myoxocephalus scorpioides (Fabricius)

Western Atlantic from Labrador to the St Lawrence; northern Pacific.

Body high at shoulder tapering back to narrow tail stalk; *head* upper head spines modified into fleshy tabs, warty bumps on head; 3 short *spines* on front gill cover, at least one pointed downward; *lateral line* almost straight with few, if any, armoured scales; *fins* no more than 16 pectoral rays.

Generally blackish in males, lighter in females with indistinct bands and blotches on body and fins, white blotches or stripes on dorsal and anal fins.

30 cm.

Usually less common than the Shorthorn Sculpin (*Myoxocephalus scorpius*), it lives near the bottom in shallow water, usually in the intertidal.

Four-horned Sculpin
Triglopsis quadricornis (L.)

All coasts of the northern hemisphere northwards of Denmark. Not around the British Isles. Baltic and freshwater lakes.

Body rather tadpole shaped; *head* large, flat and wide, there are 4 conspicious bony warts on the top of the head; *scales* absent, there is a row of granular bumps between the lateral line and the back; *gill cover* the front gill cover has 3–4 spines of which the uppermost is the longest, equalling the diameter of the orbit; *fins* 2 dorsal fins, the 1st is low, ⅔ height of the 2nd dorsal.

Greyish-brown above tinged with red on the gill covers, yellowish on the sides, whitish below.

20 cm, rarely up to 60 cm.

Able to tolerate salt, brackish and fresh water. Its distribution throughout the cold waters of the northern hemisphere and its presence in land-locked lakes suggest that it was once associated with glacial conditions and remained in place when the ice retreated.

Moustache Sculpin
Triglops murrayi Gunther

Cold eastern and western Atlantic distribution: western Atlantic from arctic Canada south to Cape Cod; eastern Atlantic from arctic south to Denmark and Scotland.

Body with regular diagonal skin folds below lateral line; *head* large and blunt with jaws about equal, front gill cover with 4 small simple spines; *scales* 45–49 broad plate-like scales along the lateral line; *fins* 2 dorsal fins with distinct space between them, 20–27 rays in anal fin.

Brownish with irregular blotches and cross-bars, usually 3 dark bars on the tail fin; there is a characteristic dark band above the mouth and below the eye that resembles a moustache.

12 cm.

Usually found on soft bottoms, 18–110 m, but known down to at least 300 m.

The Ribbed Sculpin (*Triglops pingeli*) (Reinhardt), is similar to the Moustache Sculpin except that the snout is flattened and square-cut at the end, there are no cross-bars on the tail and no black spot on the front tip of the first dorsal fin.

Usually found at 10–100 m but often goes deeper. May live at sea temperatures below 0°C.

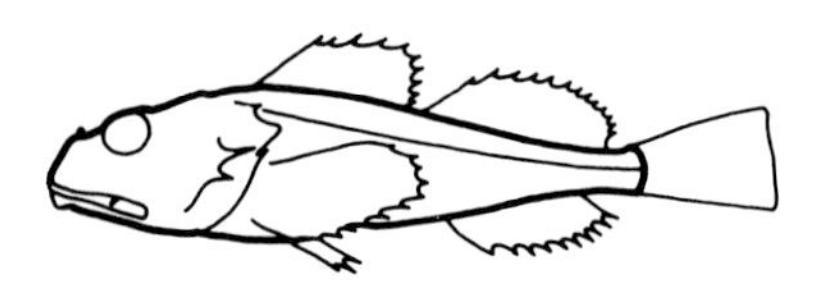

Father Lasher; Bull Rout
Myoxocephalus scorpius

2 spines on front gill cover;
no fleshy barbel at corner of jaws;
smooth lateral line.
30 cm.

Atlantic Hookear Sculpin
Artediellus atlanticus

Front gill cover with upwardly hooked spine skin naked;
dorsal fin with oblique dark stripes.
Cold eastern and western Atlantic.
10 cm.

Grubby *Myoxocephalus aenaeus*

No gap between dorsal fins;
body naked above lateral line;
no bony plates on lateral line;
3 spines on front gill cover, bottom one pointed downward.
Cool western Atlantic.
15 cm.

Long-spined Sea Scorpion *Taurulus bubalis*

No prickles along bases of dorsal fins;
a few sharp prickles on each lateral line scale;
upper front gill cover spine long.
Eastern Atlantic; Mediterranean.
15 cm.

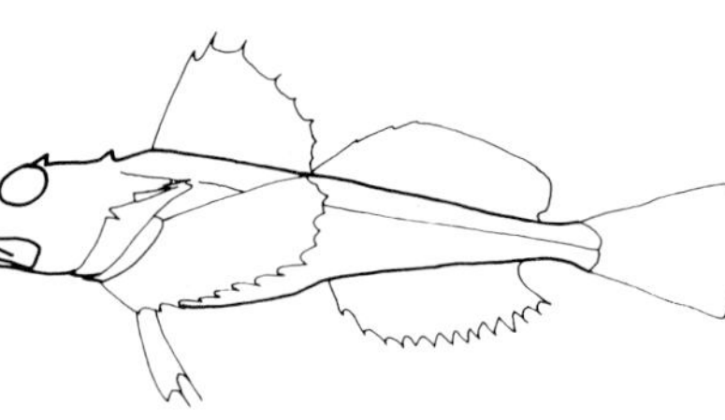

Longhorn Sculpin
Myoxocephalus octodecemspinosus

No gap between dorsal fins;
body naked above lateral line;
no bony plates on lateral line;
3 spines on front gill cover, top spine long with naked tip.
Western Atlantic.
40 cm.

Arctic Staghorn Sculpin
Gymnocanthus tricuspis

Upper spine on front gill cover with spinelets at tip;
Pectoral fin with yellow tip and 4–5 bands of dark spots;
Cold eastern and western Atlantic.
26 cm.

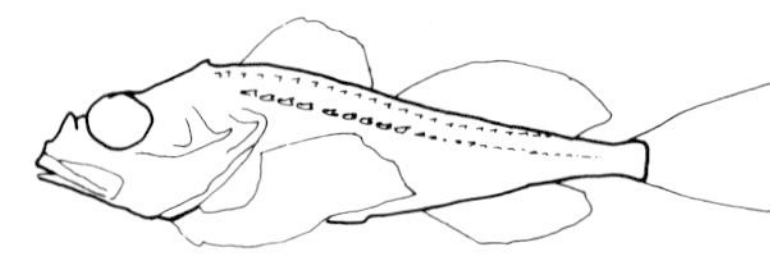

Norway Bullhead
Micrenophrys lilljeborgi

Prickles along bases of dorsal fins;
lateral line scales each with 1 sharp spine;
front gill cover with 4 spines,
upper long and strong.
Cold and cool eastern Atlantic.
7 cm.

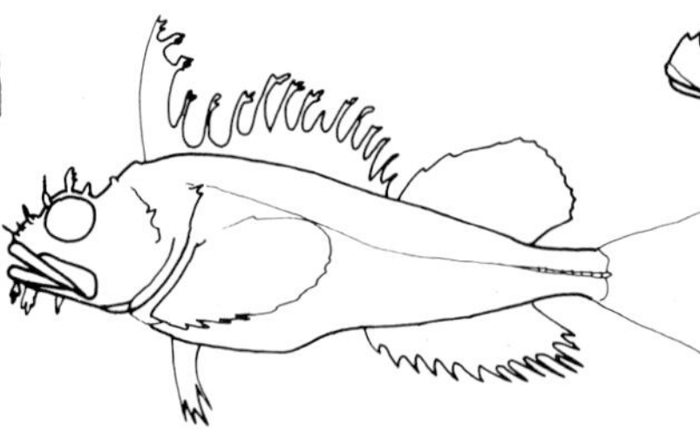

Sea Raven *Hemitripterus americanus*

Head with many bony humps and ridges; fleshy tabs on head;
dorsal spines with ragged tips, irregular.
Cool western Atlantic.
60 cm.

Arctic Sculpin *Myoxocephalus scorpioides*

No gap between dorsal fins;
body naked above lateral line;
no bony plates on lateral line;
head with fleshy tabs over eyes.
Cold western Atlantic.
20 cm.

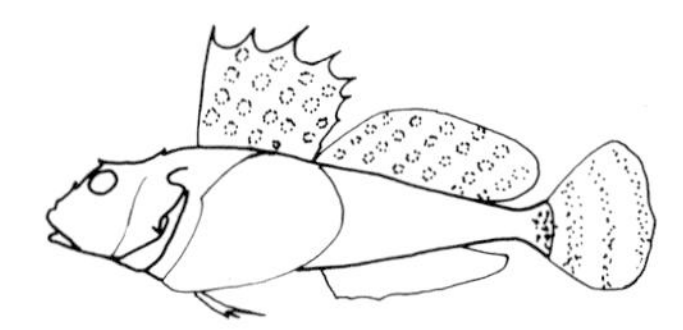

Arctic Hookear Sculpin
Artediellus uncinatus

Front gill cover with upwardly hooked spine skin naked;
dorsal fin with small white blotches.
Cold western Atlantic.
6 cm.

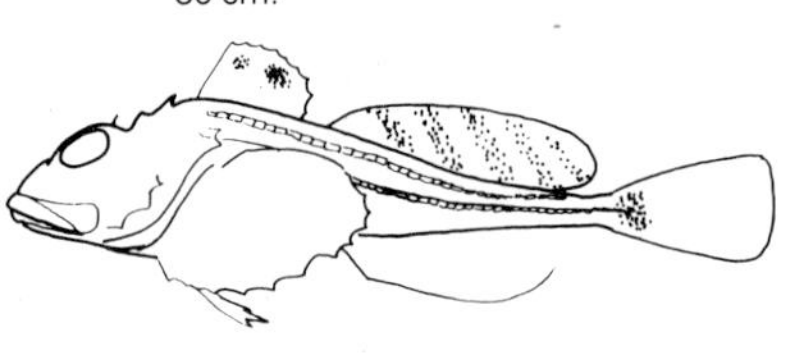

Spatulate Sculpin *Icelus spatula*

Row of bony plates on back;
lateral line with 40 bony plates;
upper gill cover spine with forked tip;
lateral line reaching tail fin.
Cold eastern and western Atlantic.
11 cm.

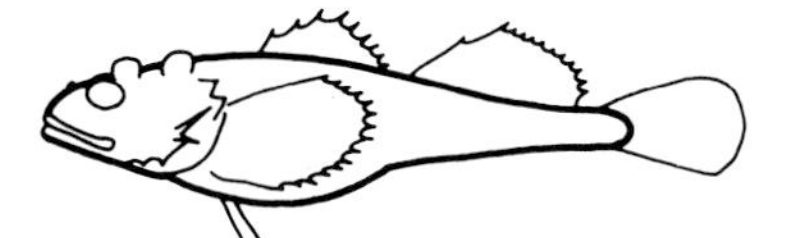

Four-horned Sculpin
Triglopsis quadricornis

Gap between dorsal fins,
2nd dorsal fin highest;
body above lateral line
with bony plates;
4 bony lumps on head.
Eastern and western cold
and cool Atlantic.
25 cm.

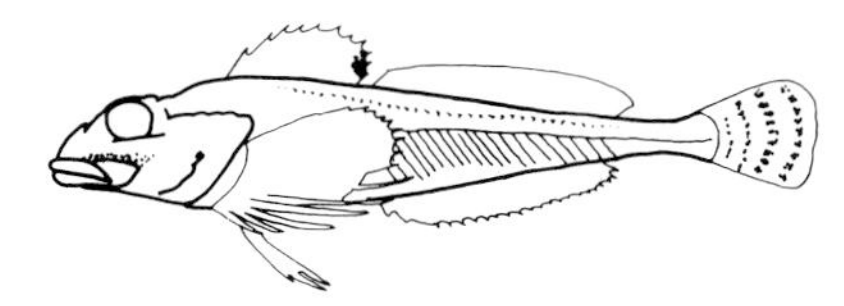

Moustache Sculpin *Triglops murrayi*

Body with diagonal skin folds
below lateral line;
armoured plates on lateral line;
'Moustache' – like mark above mouth
below eye.
Cold eastern and western Atlantic.
12 cm.

Family:
AGONIDAE

A group of temperate and coldwater bottom-living fishes. The body is completely covered with bony plates. There are 2 dorsal fins, the head is spiny and the mouth is on the underside.

Pogge; Hook-Nose; Armed Bullhead
Agonus cataphractus (L).

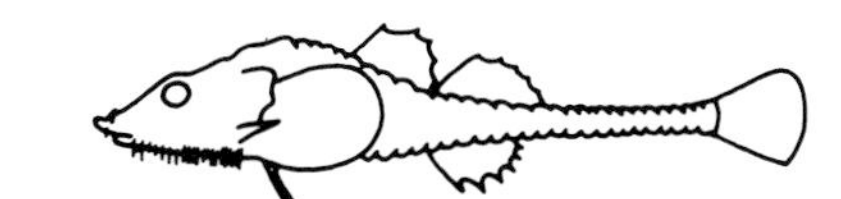

Pogge; Armed Bullhead
Agonus cataphractus

Armoured body;
small barbels under head.
15 cm.

Eastern Atlantic north beyond Iceland and Norway, south to English Channel; North Sea; western Baltic.

Head very wide and triangular when seen from above; *body* very deep behind the head, curves steeply to a long slender rear portion. Completely covered with hard bony plates; *profile* slightly concave; *snout* with one pair of spines on either side; *barbels* there are a large number of short barbels present on the lower surface of the head; *fins* 2 dorsal fins set close together. 1D V-VI; 2D 6–8; A 6–7.

Brown with 4–5 darker saddles along the back. Undersurface whitish. The pectoral fin may be orange.

Up to 15 cm.

A very common fish found from shallow water to below 500 m on mud, sand or stones but rarely seen by divers, perhaps because it lies buried in the sand. In the southern range of its distribution it is frequently encountered in estuaries during the winter. Spawning occurs during autumn and winter; the yellow eggs are laid in clumps, frequently amongst kelp (*Laminaria*) holdfasts. Its diet is very varied and includes small crustacea, echinoderms and other shellfish.

Pogge, *Agonus cataphractus* (B. Picton).

Family:
CYCLOPTERIDAE

Lump Sucker
Cyclopterus Lumpus L.

Both sides of north Atlantic from the Arctic sea South to Portugal, and from Hudson Bay south to Maryland.

Body massive, tall and rounded; *head* bony, armoured with bony plates of which there are 4 enlarged rows, 1st along the back, the 2nd running backwards from the eye to the tail fin, the 3rd from the angle of the mouth to the tail fin and the 4th from the pectoral fin base to the anal fin base; *fins* 2 dorsal fins in the young. In the adult the 1st dorsal becomes hidden by thick skin. The pelvic fins are united into a sucker. 1D 6–8; 2D I, 10; A I, 10.

Bluish or greyish; the males in the breeding season are orange or brick-red below. The young are green.

Male up to 50 cm, females up to 60 cm.

A bottom-living fish found from very shallow water down to 300 m and more. The well-developed sucker is used to attach the fish to rocks and other solid objects. In late winter and spring they come into shallow water sometimes even into the intertidal zone to breed. The eggs are pink and laid in a large mass. They take between one and two months to hatch, during which time they are guarded with great tenacity by the male who fans water through the egg mass to keep them oxygenated. The female meanwhile has returned to deep water. The larvae live in the plankton.

The eggs ae dyed, preserved and sold as 'caviar' substitute.

Another Lump Sucker, *Eumicrotremus spinosus* (Fabricius), is sometimes caught in the arctic western Atlantic at depths less than 25 m. Its body is covered with very large conical tubercles.

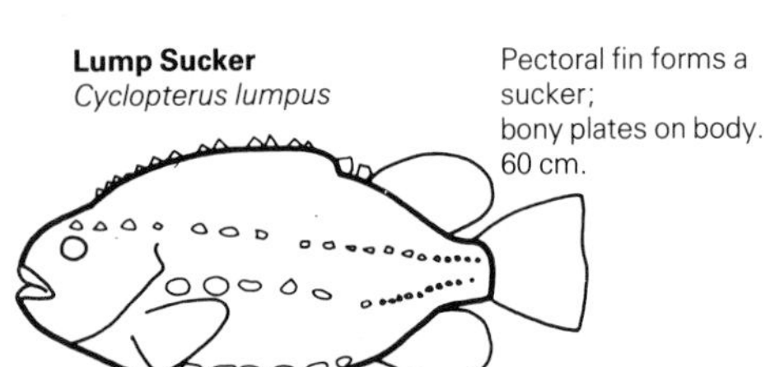

Lump Sucker
Cyclopterus lumpus

Pectoral fin forms a sucker;
bony plates on body.
60 cm.

Lump Sucker, *Cyclopterus lumpus* (Andy Purcell).

Lump Sucker, *Cyclopterus lumpus* (Jim Greenfield).

Lump Sucker, *Cyclopterus lumpus* (Phil Lobel).

Family:
LIPARIDIDAE

Tadpole-shaped fish of gelatinous consistency. The pelvic fins are modified to form a sucker. The skin is naked.

Sea Snail, Snailfish
Tadpole-shaped body with sucker; flaccid skin; anal fin joined to tail fin.

Sea Snail; Unctuous Sucker
Liparis liparis (L.)

Northern Atlantic from the Arctic Ocean to English Channel; sometimes in Baltic.

Body rounded in front, laterally compressed behind; *head* broad and blunt-snouted; *sucker*, there is a well-marked round sucker set forward on the belly; *skin* loose and slimy extending over the fins, there are no scales; *fins* the pelvic fin is modified into a sucker; pectoral fins rounded, nearly meeting under the head. The last ray of the anal fin is joined to the tail fin by a membrane. D 33–36; A 27–29.

Variable in colour, usually of some drab shade that matches the bottom. Sometimes with indistinct streaks and lines.

Up to 12 cm, sometimes 18 cm.

Usually in shallow water, but its range extends from just beneath the surface to 100 m amongst stones and in estuaries. It breeds in winter in shallow water, usually near the mouths of rivers. The eggs are laid in clumps amongst hydroids or amongst fine algae. It feeds on small animals, chiefly crustacea.

Montagu's Sea Snail
Liparis montagui (Donovan)

Resembles the Sea Snail (*Liparis liparis*) except that the the anal fin is not joined by a membrane to the tail fin. D 28–30; A 22–25.

Up to 9 cm.

Lives in shallower water than *Liparis liparis*, it extends downwards from mid-tide level to 30 m or more. Breeds in spring and early summer.

Atlantic Snailfish
Liparis atlanticus (Jordan & Evermann)

Western north Atlantic from Labrador south to Connecticut.

Body elongate, compressed behind, skin loose with no scales but scattered prickles; *fins* dorsal with 31–35 rays with a distinct notch or valley deepest between about the 5th and 6th rays; the pelvic fins are fused to form an adhesive disc, main part of pectoral fins broad and fan-like, but forming a secondary frilled lobe under the head.

Usually brown or reddish-brown, there may be lighter bars.

14 cm.

A shallow water and tide pool fish rarely deeper than about 2 m. It uses its adhesive disc to attach itself to the underside of small boulders.

Other Snailfish have been confused with the Atlantic Snailfish. The Gulf Snailfish (*Liparis coheni*) (Able) has a dorsal fin with no notch, and the pectoral fins have a

Sea Snail, *Liparis liparis* (Linda Pitkin).

shallow notch at the 8–9th rays above the ventral insertion. Caught at 2–210 m in the Bay of Fundy and Gulf of Maine.

The Inquiline Snailfish (*Liparis inquilinus*) (Able) has a dorsal fin notched at the 5–7th rays, head broader than high, pectoral fin notched at the 8–11th rays from bottom. This species attaches itself to the inside of the upper valve of the sea scallop (*Plactopecten magellanicus*) during the daytime, but comes out at night to feed. Several Snailfish may shelter within one scallop. Depth range at least 5–197 m. North-west Atlantic from Nova Scotia south to Cape Hatteras.

The Kelp Snailfish (*Liparis tunicatus*) (Reinhardt) has a dorsal fin with no notch, the pectoral fin has a deep notch at the 7–9th rays from bottom of fin; like other Snailfish it is brown, but this species may have small black spots. Arctic Atlantic and Pacific coasts of the New World. Usually in less than 50 m, within 5 km of the shore. It is most often found associated with kelp (*Laminaria*) forests where it uses its ventral sucker to attach itself to kelp fronds. Seems to prefer temperatures from –1.5 to +2.1°C.

Family:

DACTYLOPTERIDAE

The **Flying Gurnards** are separated from the Gurnards (Triglidae) by the primitive characteristics of their head bones.

Flying Gurnard
Dactylopterus volitans

Mediterranean; eastern Atlantic north to Brittany, south to Angola, western Atlantic from Massachusetts south to Argentina.

Head large with a corselet type structure extending onto the body; *snout* rounded; *eyes* large; *fins* the pectoral fins are very large and make this fish unmistakable. They are divided into two sections, the front section is short and consists of 6 rays and is attached by the base only to the 2nd much larger, section. The 1st and 2nd rays of the 1st dorsal fin are separate. 1D, II + IV; 2D I + 8; A 6.

Head, back and flanks greyish or brownish often with darker spots and blotches, and lighter spots. Undersurface pinkish-white. The pectoral fins are brown with bright blue spots and stripes arranged in a regular pattern.

Up to 50 cm.

Live on sandy or muddy bottoms, usually around 10–30 m. They feed mainly on small bottom-living animals predominantly crustacea. Spawning occurs during summer. The very large pectoral fins are usually folded back when the fish is resting on the bottom but when disturbed the fins open and the fish 'fly' through the water.

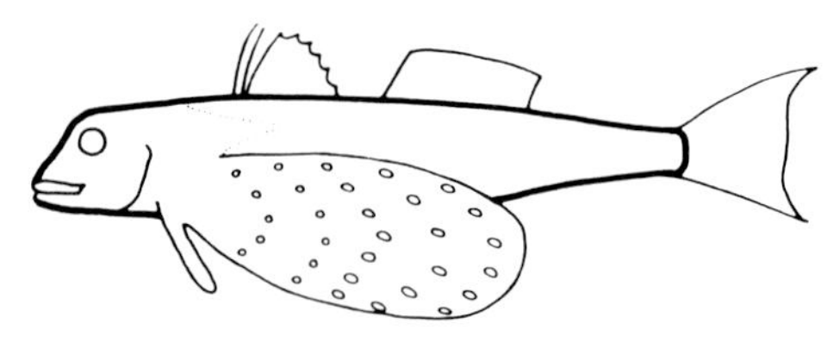

Flying Gurnard
Dactylopterus volitans
Very large pectoral fins with bright blue spots and stripes.
Eastern and western Atlantic.
50 cm.

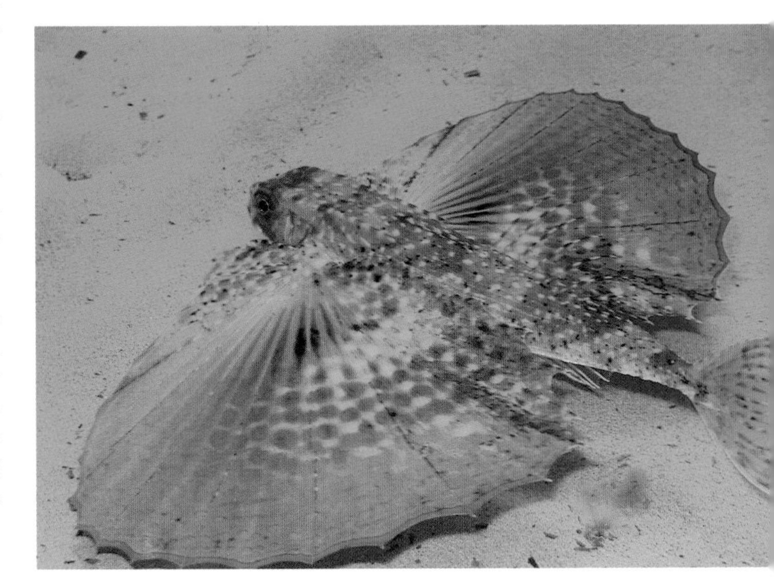

Flying Gurnard, *Dactylopterus volitans* (Andy Purcell).

Order:
HETEROSOMATA

This order contains the well-known **Flatfish**. The adult fishes lie on their side with both eyes on the uppermost side. The dorsal and anal fins are long and many-rayed. The upper 'eyed-side' is coloured usually in such a way as to camouflage the fish on the bottom; the under 'blind-side' is usually white.

The eggs are pelagic but hatch in a few days. The larvae are symmetrical like conventional fishes and swim in the upper layers of the sea. However, when they have reached a length of about 2 cm, one eye moves over the top of the head taking up station next to the other. At the same time the dorsal fin extends forward onto the head and the body becomes flattened. At about the time that the eye migrates, the fish comes to rest on the bottom to begin a fundamentally bottom-living existence.

Family:
SCOPHTHALMIDAE

Flatfishes with eyes on the left side, the pelvic fins on both the eyed side and blind side elongate.

Turbot
Psetta maxima L.

Eastern Atlantic; west Baltic; Mediterranean. There is a related species *Scophthalmus maeoticus* (Pallas) in the Black Sea.

Body very deep and rounded; *eyes* situated on the left side of the head and the lower eye is slightly in front of the upper; *mouth* large and situated on the left of the eyes; *profile* there is a slightly indentation at the origin of the dorsal fin; *skin* there are no scales on the skin but the eyed side has a number of bony tubercles scattered over the surface; *fins* the dorsal fin commences just in front of the lower eye. The first few rays are slightly branched and the tips of the rays are free from the fin membrane. Neither the dorsal nor the anal fin extends under the tail. D 57–72; A 43–56.

The colour of the eyed-side is variable and depends greatly on the colour of the sea-bed on which the fish is living. It ranges from light grey-brown to dark chocolate-brown with yellowish, light or dark brown, blackish or greenish spots. The fins are speckled, including the tail fin which is densely covered with spots. Blind side whitish.

Up to 80 cm, occasionally up to 100 cm.

The Turbot is found on sandy, muddy, shell and gravel bottoms from very shallow water to below 80 m, and also inhabits brackish water. The younger, smaller fish are commonly in the shallow water and the older, larger, specimens in the deeper water.

Breeding occurs during spring and summer. In northern European water the fish spawn over gravel bottoms at depths of 10–80 m. At about 25 mm, when they are about 4–6 months old, the young fish take up a bottom-living existence. They are carnivorous and feed mainly on other fish. The Turbot is of great commercial value, the majority of the fish being caught in the central North Sea.

Turbot, *Psetta maxima* (Leo Collier).

Brill
Scophthalmus rhombus (L.)

Eastern Atlantic to central Norway; North Sea; English Channel; western Baltic; Mediterranean.

Body oval; *eyes* situated on the left side of the head. The lower eye is slightly in front of the upper; *mouth* large and situated on the left of the eyes; *profile* smooth; *skin* covered with scales, which are rather large on the eyed-side. There are no tubercles on the skin; *fins* the first rays of the dorsal fin are partly free from the fin membrane and branched. Neither the dorsal nor the anal fin continues under the tail. D 73–83; A 56–62.

The colour of the eyed side is very variable and ranges from sandy to grey brown to dark brown with many small irregular darker spots and other lighter spots scattered over the surface. The fins are lighter, and the tail fin is slightly spotted. Blind side white, sometimes with occasional darker blotches.

Up to 70 cm.

The Brill may be found from very shallow water down to about 70 m on sandy, muddy or gravel bottoms. They also penetrate into areas of brackish water. The larger, older fish are usually found in the deeper water and the younger, smaller specimens in the shallow water. Breeding occurs during spring and summer in the Atlantic, North Sea and English Channel and during February to March in the Mediterranean. The adults feed mainly on fish. They are edible but not of great commercial importance. Hybrids between the Brill and the Turbot are sometimes found.

Topknot
Zeugopterus punctatus (Bloch)

Eastern Atlantic south to Biscay; English Channel; North Sea.

Body very deep, oval (height more than ½ length); *profile* smooth; *eyes* situated on the left side of the head one above the other; *mouth* large, situated on the left of the eyes; *scales* toothed on the eyed side and smooth on the blind side; *fins* dorsal fin commences very far forward on the snout. Both the dorsal and anal fin clearly continue under the tail stalk where they form distinct lobes. The pelvic fins are joined to the anal fin. D 88–102; A 67–76.

Eyed side brown with darker blotches. There is a dark spot just behind the curve of the lateral line and dark bars extending from each eye.

Up to 25 cm.

Usually found from shallow water down to 37 m. Although fairly common they are extremely difficult to find as they have the habit of clinging to the bottoms and sides of large rocks where they are almost invisible. Spawning occurs in late winter.

Bloch's Topknot; Eckström's Topknot
Phrynorhombus regius (Bonnaterre)

Eastern Atlantic north to Shetland Isles; Mediterranean.

Body deep and oval; *profile* with a distinct notch in front of the upper eye; *eyes* on the left side of the head. They are large and situated one above the other or with the lower eye slightly in front of the upper. There is a bony ridge separating the 2 eyes; *mouth* large and situated on the left of the eyes; *scales* small with toothed free margins on the eyed side and less so on the blind side. 72–80 scales along the lateral line; *fins* the dorsal fin commences in front of the eyes and just behind the notch present in the profile. The 1st ray of the dorsal fin is longer than the others. Both the dorsal and the anal fins continue onto

Topknot, *Zeugopterus punctatus* (B. Picton).

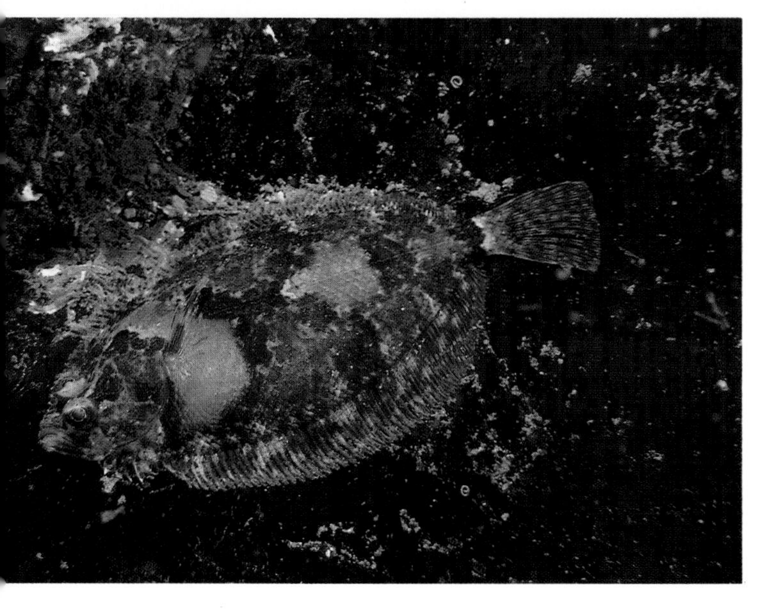

Bloch's Topknot (top), *Phrynorhombus regius* (B. Picton).

Norwegian Topknot, (bottom), *Phrynorhombus norvegicus* (Jim Greenfield).

the blind side of the tail stalk. D 73–80; A 60–68.

Eyed side reddish-brown with darker spots and blotches. There is one large dark spot, often with a lighter centre situated on the lateral line towards the tail. The dorsal and anal fins have dark spots which are more or less regularly arranged.

Up to 20 cm.

This species is not common in the Atlantic waters. It lives from 9–55 m on a variety of bottoms both sandy and rocky. In the Mediterranean they have been caught from between 100–300 m. It is possible that these fish cling to the sides of rocks in a similar fashion to the Topknot (*Zeugopterus punctatus*). Spawning occurs in spring and summer.

Norwegian Topknot
Phrynorhombus norvegicus (Günther)

Eastern Atlantic south to English Channel; north to Iceland.

Body oval (depth much less than ½ body length); *profile* smooth; *eyes* situated on the left side of the head. They are very close together, one above the other; *jaws* situated on the left of the eyes. They extend just beyond the front edge of the eyes; *scales* large, finely toothed on the eyed side, less so on the blind side; *fins* dorsal fin starts in front of the eyes and both the dorsal and anal fins clearly continue under the tail stalk where they form distinct lobes. Pelvic fin separate from anal fins. D 76–84; A 58–68.

Eyed side sandy-brown with darker blotches. The dorsal and anal fins have irregular dark markings.

Up to 12 cm.

This species is the smallest and most northern of the Topknots. Very little is known about its habits except that they live in rocky areas and are usually found down to 50 m, sometimes down to 170 m. Spawning occurs in spring and early summer and the young fish take up a bottom-living existence at about 13 mm.

Megrim; Sail-Fluke
Lepidorhombus whiffiagonis (Walbaum)

Eastern Atlantic north to Iceland south to Morocco; North Sea; western English Channel; western Mediterranean.

Body long, slender, oval; *snout* rather long; *eyes* situated on the left side of the head. They are large, oval, close together and separated by a bony ridge. The lower eye is clearly situated in front of the upper; *jaws* on the left of the eyes. The lower jaw is longer than the upper and has a small bony spine at the tip; *scales* small, 95–109 along the lateral line; *fins* the dorsal fin starts immediately in front of the front eye. The dorsal and anal fins only just continue under the tail stalk. The longest rays of the dorsal and anal fins are found towards the rear of the fins. D 80–94; A 61–74.

The eyed side is sandy-brown with darker blotches. The eye iris is greenish-yellow and the pupil is circled with bright yellow. Blind side white.

Up to 61 cm.

Found on sandy bottoms from 10–600 m, but most commonly from 50 m, in the Atlantic and rather deeper in the Mediterranean. Only very rarely are they found in shallow water or near coasts. They feed on large quantities of many varieties of fish but also on crustacea. Breeding occurs in water between 50–200 m deep from March to June. When the young fish reach about 19 mm they take up a bottom-living existence.

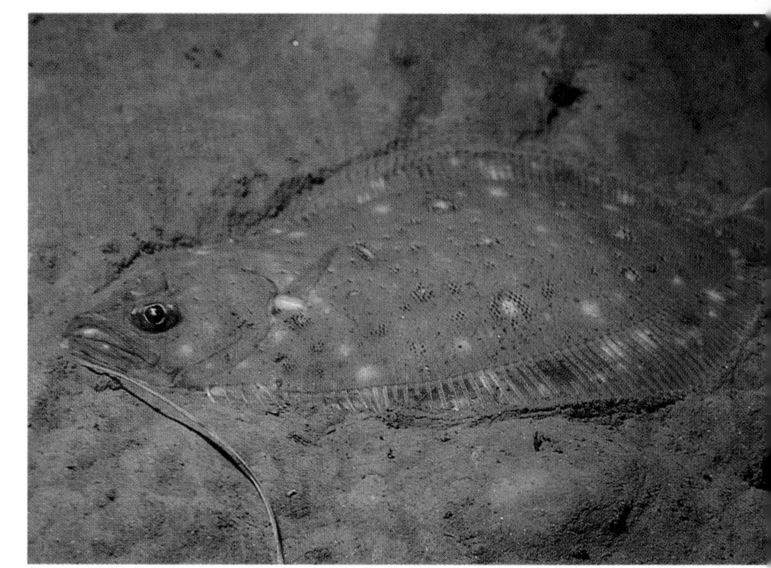

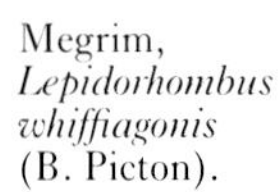

Megrim, *Lepidorhombus whiffiagonis* (B. Picton).

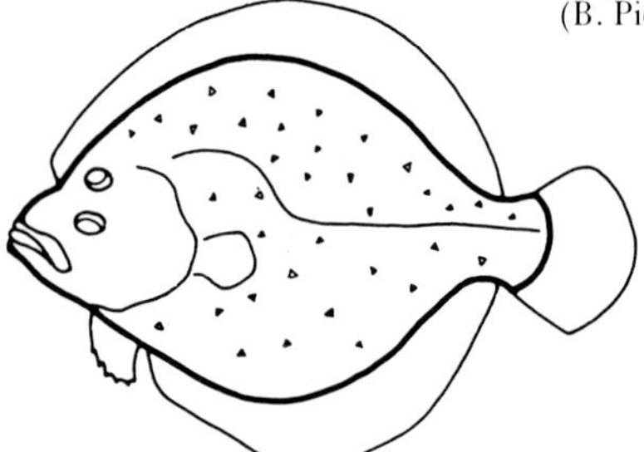

Turbot *Psetta maxima*

Skin without scales but with scattered tubercles; tail fin heavily speckled. Eastern Atlantic; Mediterranean. 80 cm.

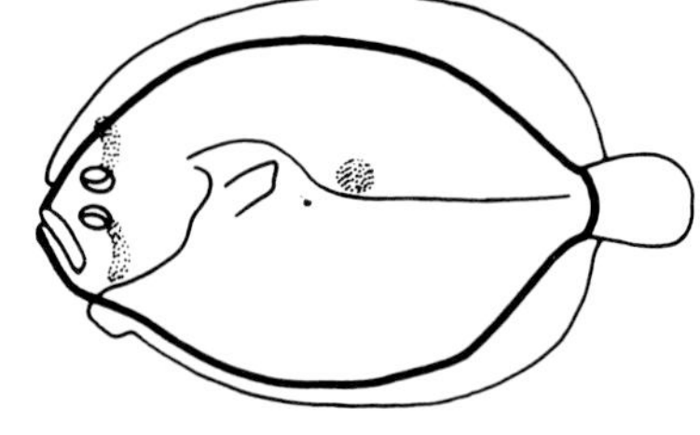

Topknot *Zeugopterus punctatus*

Pelvic fin attached to anal fin; dorsal and anal fins form lobes under tail stalk; dark spot behind curve of the lateral line; dark bars extending from eyes. Eastern Atlantic. 25 cm.

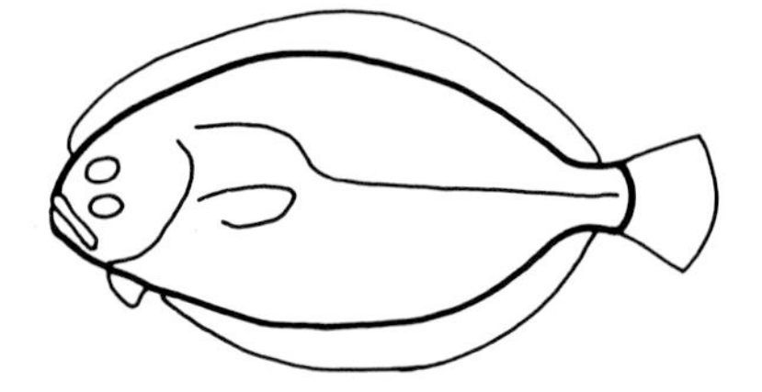

Norwegian Topknot *Phrynorhombus norvegicus*

Dorsal and anal fins form lobes under tail stalk; profile smooth; body slender. Eastern Atlantic. 12 cm.

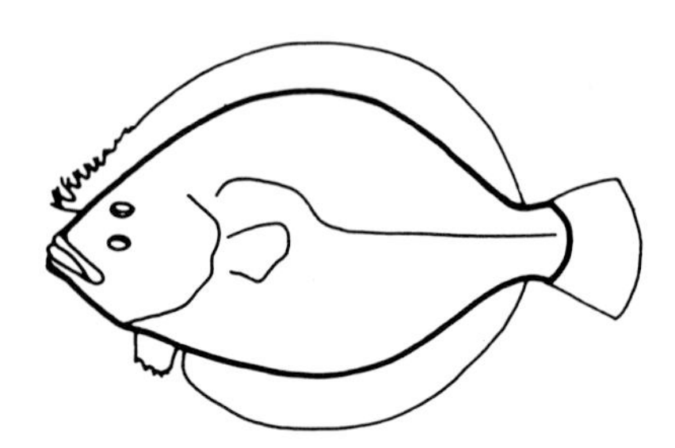

Brill *Scophthalmus rhombus*

Skin with scales and no tubercles; tail fin only slightly spotted; first rays of the dorsal fin branched. Eastern Atlantic; Mediterranean. 70 cm.

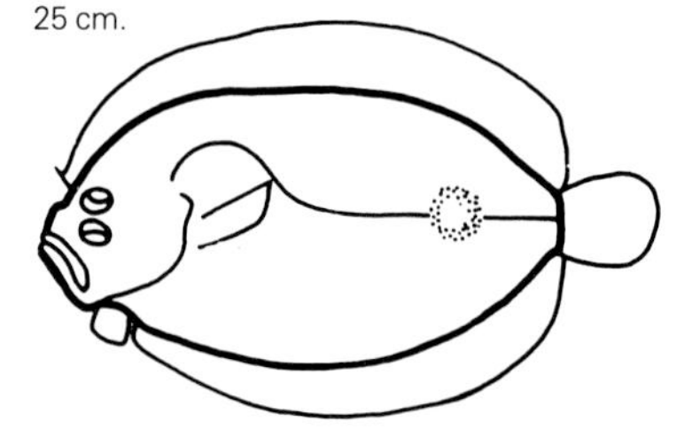

Bloch's Topknot; Eckström's Topknot *Phrynorhombus regius*

Dorsal and anal fins form lobes under tail stalk; 1st ray of dorsal fin elongate; distinct notch in profile. Eastern Atlantic; Mediterranean. 20 cm.

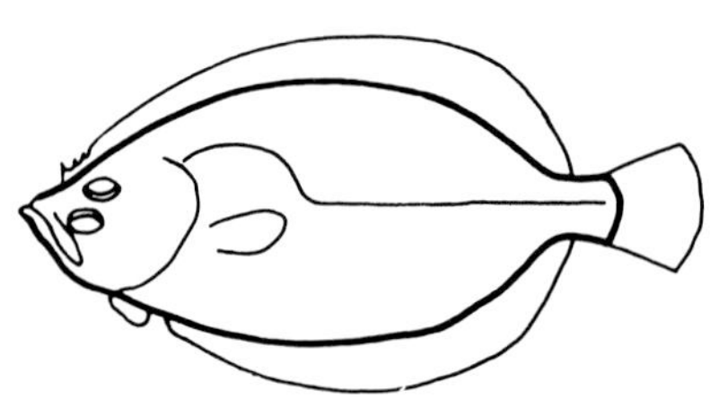

Megrim; Sail-Fluke *Lepidorhombus whiffiagonis*

Slender body; lower eye clearly in front of upper; lower jaw prominent. Eastern Atlantic; Mediterranean. 61 cm.

Family:
BOTHIDAE

Flatfishes with eyes on the left side. Pelvic fin base on the eyed side is longer than that on the blind side.

Wide-eyed Flounder
Bothas podas (Delaroche)

Mediterranean; eastern Atlantic north to Azores, south to Angola.

Body deep oval; *eyes* situated on the left side of the head. The lower eye is in front of the upper and they are separated by a very wide space. (In the female the lower eye is not completely in front of the upper and the space separating the eyes is not as great as in the male.) The male fish has a spine directly in front of the lower eye and another on the snout; *mouth* small and situated on the left of the eyes; *scales* small, 82–91 along the lateral line; *fins* the dorsal fin commences in front of the eyes. D 85–94; A 65–70.

The colour of the eyed side is very variable and depends greatly on the colour of the bottom where the fish lives. They may be greyish, or brownish with lighter spots, or uniform in colour. There are 1 or 2 distinct spots on the lateral line towards the tail. Blind side bluish-white.

Up to 20 cm.

Usually found in shallow sandy areas where they feed on small animals. Breeding occurs during the summer.

Scaldfish
Arnoglossus laterna (Walbaum)

Eastern Atlantic north to Norway, south to Morocco; Mediterranean; North Sea; English Channel.

Body oval, slender; *eyes* situated on the left side of the head. The eyes are separated by a bony ridge and the lower eye is slightly in front of the upper; *mouth* situated on the left of the eyes; *scales* very easily detached from the skin. 51–56 along the lateral line; *lateral line* has a branch over the pectoral fin; *fins* dorsal fin starts in front of the eyes. The first few rays are partly free from the fin membrane. The pelvic fin on the blind side is very much smaller than that of the eyed side. Tail fin rounded. D 84–98; A 63–75.

Sandy brown or greyish with darker patches. The fins may have small dark spots.

Up to 19 cm, usually around 10 cm.

The Scaldfish are most common in the Atlantic, Channel and North Sea, where they are found on muddy or sandy bottoms at 10–200 m. In the Mediterranean they have been caught from 40–1000 m, but are most frequently found at 100–300 m. They feed on small bottom-living animals.

Spawning occurs in spring and summer.

Scaldfish
Arnoglossus thori Kyle

Mediterranean; eastern Atlantic from northern Ireland to Senegal; western English Channel.

Body oval but fairly deep; *eyes* situated on the left of the head, close together, the lower eye slightly in front of the upper; *mouth* situated on the left of the eyes. Small, it extends only to the first ⅓ of the lower eye; *scales* large 49–56 along the lateral line; *fins* the dorsal fin starts in front of the eyes, the first few rays are longer and not attached to the fin membrane. The 2nd ray is elongate and has a membrane attached to it which gives the appearance of a fringe. The pelvic fin on the blind side is much smaller than that on the eyed side. D 81–91; A 62–67.

Eyed side brownish or greyish with darker blotches and spots. Two of these blotches, one behind the curve of the lateral line and another towards the tail on the lateral line are more conspicuous than the others. There is also a narrow black stripe at the base of the tail fin and black spots scattered over the fins. There may be a dark spot on the blind-side of the pelvic fin.

Up to 18 cm.

Found on sandy, muddy and rough ground from 15 m to below 100 m. Spawning occurs from March to July and possibly also during the winter in the

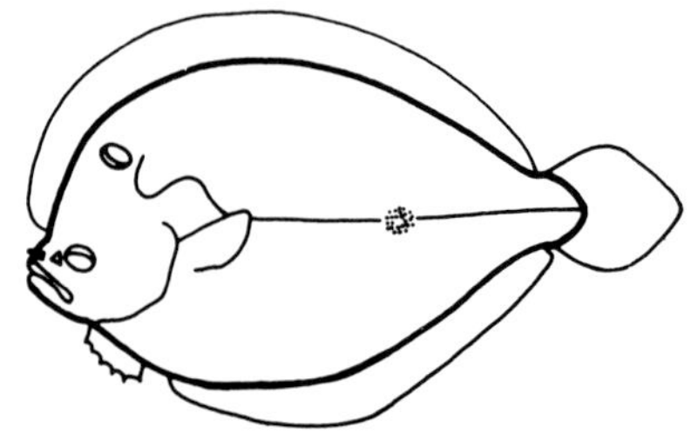

Wide-eyed Flounder *Bothas podas*

Eyes widely separated; 1 or 2 dark spots on lateral line towards tail. Male fish with spine in front of lower eye and another on snout. Mediterranean. 20 cm.

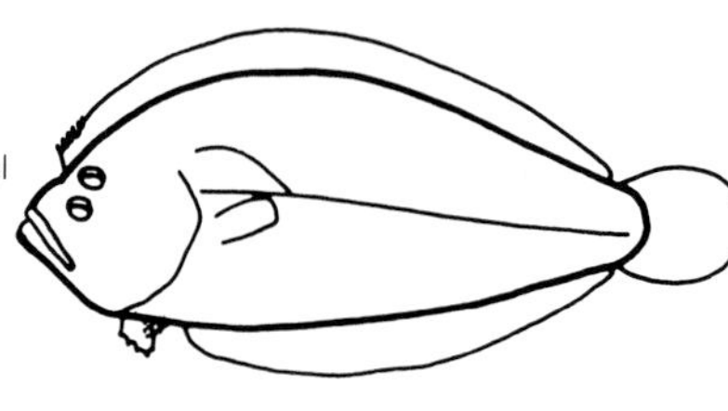

Scaldfish *Arnoglossus laterna*

Scales very easily removed; 1st rays of the dorsal fin not elongate but partly free from fin membrane. D 84–98; A 63–75. Eastern Atlantic; Mediterranean. 19 cm.

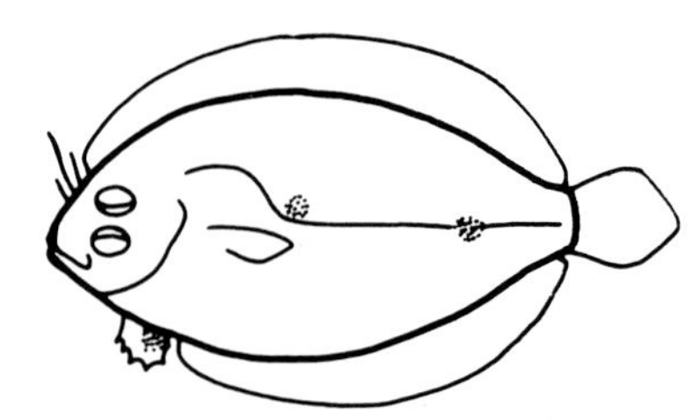

Scaldfish *Arnoglossus thori*

1st rays of the dorsal fin not attached to fin membrane; 2nd ray of dorsal fin elongate. Eastern Atlantic; Mediterranean. 18 cm.

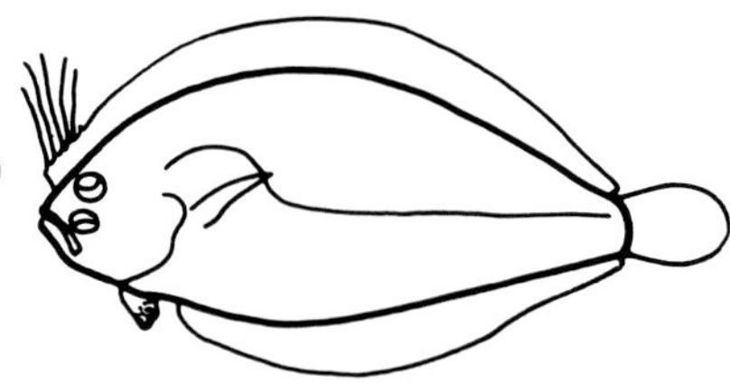

Scaldfish *Arnoglossus imperialis*

2nd–5th or 6th rays of the dorsal fin elongate, particularly in adult male. Eastern Atlantic; Mediterranean. 25 cm.

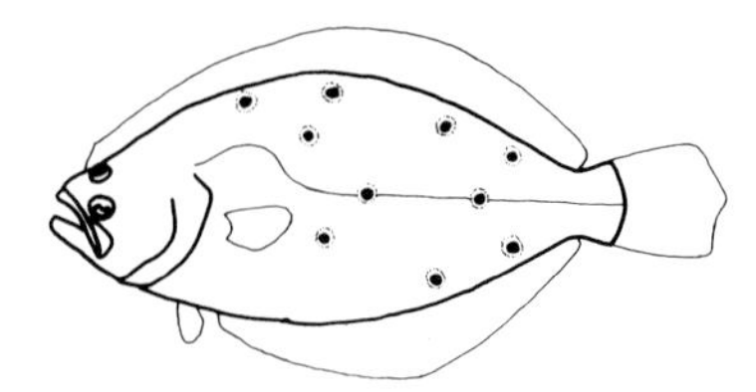

Summer Flounder *Paralichthys dentatus*

Eyes on left side; 10–14 dark spots ringed with white; 85–94 dorsal fin rays; Western Atlantic. 100 cm.

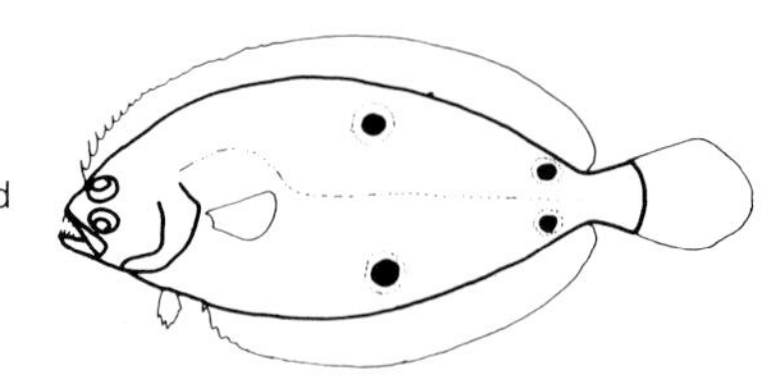

Fourspot Flounder *Paralichthys oblongus*

Eyes on left side; 4 black spots ringed with white; 72–81 dorsal fin rays. Western Atlantic. 40 cm.

Mediterranean. The adults feed on small bottom-living animals, mainly invertebrates.

Scaldfish

Arnoglossus imperialis (Rafinesque)

Eastern Atlantic north to Scotland, south to West Africa; western Mediterranean. *Body* oval; *eyes* situated on the left side of the head; the lower eye slightly in front of the upper; *mouth* situated on the left of the eyes; *jaws* lower jaw with a slight protuberance at the tip; *scales* large, detached very easily and often missing, 58–63 along the lateral line; *fins* the dorsal fin commences in front of the eyes. In the male the 1st ray is fairly short but the next 5 or 6 rays are thickened and elongated and for the majority of their length free from the fin membrane. In females the first rays of the dorsal fin are slightly longer than the following rays and are free from the fin membrane at their tips only. D 95–106; A 74–82.

Eyed side sandy or greyish with darker blotches. Male fish have a black spot on the pelvic fin.

Up to 25 cm.

This species is found on mud and sand in fairly deep water at 60–350 m.

Summer Flounder

Paralichthys dentatus (L.)

Western Atlantic from Maine (rarely Nova Scotia) south to South Carolina.

Body compressed; *jaws* lower slightly longer than upper; *eyes* on left side, closely spaced; *lateral line* arched over pectoral fin; *fins* dorsal has 85–94 rays, anal 60–73 rays, tail fin rounded.

Eyed side varies according to the background; 10–14 dark spots with paler borders scattered over the surface, mostly near base of dorsal and anal fins; blind side is whitish.

100 cm.

Usually found at 20–200 m on sandy or muddy bottoms.

Fourspot Flounder
Paralichthys oblongus (Mitchill)

Western Atlantic from Bay of Fundy south to South Carolina.

Body compressed; *mouth* terminal and large, lower jaw longer than upper; *teeth* 4–5 canines near front of jaw; *eyes* on left side, close together but separated by sharp bony ridge; *lateral line* arched over pectoral fin; *fins* dorsal fin has 72–81 rays, anal fin has 60–67 rays, tail fin rounded.

Eyed side mottled grey with 4 (occasionally 6) black eyespots with paler ring, 2 on the tail stalk, 2 just behind mid-point of the body; blind side pale grey.

40 cm.

Usually caught some way offshore between 15 and 350 m.

Windowpane
Scophthalmus aquosus (Mitchill)

Western Atlantic from the Gulf of St Lawrence south to Florida.

Body compressed and almost round in outline; *mouth* large, its angle under front edge of pupil; *eyes* on left side moderately spaced; *lateral line* arched over pectoral fin; *fins* dorsal has 63–73 rays first, 12–15 rays free and branched at their tips, anal has 46–54; tail fin rounded; pectoral fins large, inserted on sides behind gill opening.

Eyed side made up of many dark spots, each of which is itself composed of smaller dark spots; blind side white, sometimes with dark blotches.

40 cm.

From shallow water down to about 70 m, and able to tolerate a wide range of temperatures.

Fringed Flounder
Etropus crosssotus Jordan & Gilbert

Western Atlantic from Chesapeake Bay south to French Guiana.

Head upper jaw very short, only 25 per cent of head length and angle stretching back only to the front margin of the eye;

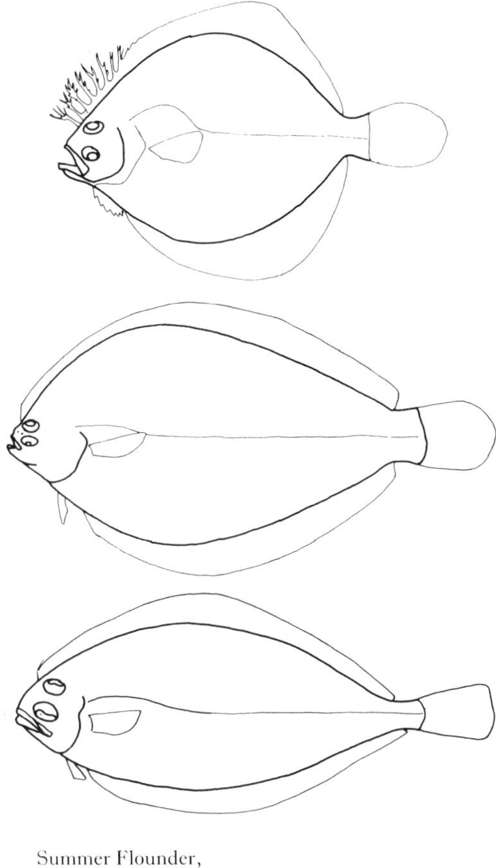

Windowpane
Scophthalmus aquosus

Eyes on left side;
body very flat and almost round;
front dorsal rays branched and fringe-like;
pelvic fin very broad.
Western Atlantic.
40 cm.

Fringed Flounder
Etropus crossotus

Eyes on left side;
upper jaw only 25 per cent head length;
no large spots or bars on body or fins;
lateral line without high arch over pectoral fin.
Western Atlantic.
18 cm.

Bay Whiff *Citharichthys spilopterus*

Eyes on left side;
2 dark spots on tail stalk;
light spot under pectoral fin;
jaw moderate or long extending to middle of lower eye;
no spines on head.
Western Atlantic.
15 cm.

Summer Flounder, *Paralichthys dentatus* (Roy Manstan).

eyes on left side; *scales* easily dislodged, does not have secondary scales developed on top of the main scale row; *lateral line* no high arch over pectoral fin; *fins* pectoral fin present on blind side, pelvic fin short, dorsal fin has 75–85 rays, anal fin has 58–68 rays.

No spots on body, fins with dusky blotches, tail fin often edged with black.

18 cm.

A shallow water species usually in 10–80 m.

Bay Whiff

Citharichthys spilopterus Günther

Western Atlantic from New Jersey south to Brazil.

Head with no spines, lower jaw moderate or long, the angle extending back to the pupil of the lower eye, longer than 35 per cent of head length; *eyes* on left side, diameter 25 per cent of head length; *gill rakers* long and slender, 9–13 on lower limb; *fins* dorsal 75–84 rays, anal 56–63 rays, pectoral fin with short base and present on blind side.

Eyed side variable according to colour of bottom, two dark spots on the tail stalk and a light spot at the base of the pectoral fin, otherwise the body is not covered with regular patterns of spots or blotches.

15 cm.

Found close inshore down to about 150 m, comes into shallower water in the warmer months.

Family:

PLEURONECTIDAE

The eyes are on the right side but very occasionally the eyes may be on the left side. This 'reversal' is relatively common in the **Flounder** (*Platichthys flesus*). The front gill cover has a free edge, the dorsal fin begins above the upper eye and the mouth is terminal. The egg yolk contains no oil droplet.

Dab

Limanda limanda (L.)

Atlantic from the White Sea to Biscay; North Sea and western Baltic.

Body oval; *mouth* terminal, directed to the right of the eyes reaching back to the first 1/3 of the eye; *head* 1/4 body length; *lateral line* strongly arched over the pectoral fin; *scales* finely serrated along the margins giving a rough feel to the eyed-side, the blind-side has slightly rounded rear margins. D 65–81; A 50–64.

Yellowish-brown on the eyed side, often with indistinct blotches and small dark spots. Frequently found with scattered orange spots. Pectoral fin orange. The blind side is white.

Up to 20 cm, rarely 40 cm.

One of the most common flatfishes of the north-east Atlantic. It prefers sandy shoals and banks from just beneath the surface to 150 m or more. There appear to be no well-defined spawning grounds but the young fish generally live on the bottoms only a metre or so beneath the surface. Spawning occurs from late winter to early summer, the earlier spawning being in the southern parts of its range. Whilst the Dab chiefly eats crustacea it will take almost any small animal it can find including small fish, echinoderms and molluscs. It feeds in a similar way to the Lemon Sole but strikes

Dab, *Limanda limanda* (B. Picton).

Plaice, *Pleuronectes platessa* (B. Picton).

obliquely at its prey. This feeding method seems to be less suitable for catching polychaetes but is better for catching most other organisms.

Flounder
Platichthys flesus (L.)

Northern Atlantic from the White Sea to Gibraltar; western Mediterranean. A related form *Platichthys flesus italicus* in the Adriatic.

Body oval; *jaws* terminal, directed to the right of the head, does not reach to beneath the eye; *lateral line* gently curved over the pectoral fin; *scales* small, without serrated margins. There is a row of hard warts running along the base of the dorsal and anal fins, similar warts either side of the lateral line; *fins* tail fin square-cut. D 52–67; A 35–46.

Greenish or brownish, sometimes mottled with light or dark spots. Some fishes have large dull orange spots. Blind side white.

Up to 30 cm, sometimes 50 cm.

The Flounder is particularly tolerant of differences in salinity and may be found high up river estuaries or even in completely fresh water. Alternatively they may be found in the sea down to 50 m or more. It is most common in areas of rather low salinity such as the Baltic.

Plaice
Pleuronectes platessa L.

Northern Atlantic from the White Sea to the Bay of Cadiz; western Mediterranean.

Body oval; *jaws* terminal, directed to the right and reaching back to beneath the first ⅓ of the eye; *lateral line* gently curved over the pectoral fin; *scales* have smooth margins; no bony warts at the base of the dorsal and anal fins. 4–7 warts between the beginning of the lateral line and the eyes. D 65–79; A 48–59.

Eyed side brown with large conspicuously orange or orange-yellow spots. Blind side white, sometimes with dark blotches.

Up to 60 cm, rarely up to 90 cm.

The Plaice is a very important commercial species, especially in the North Sea. It is also frequently seen by divers especially on sand and gravel bottoms. The orange spots on the back are very conspicuous underwater but when the nature of the bottom allows it, the Plaice hides itself by flapping its fins in the bottom and thus covering itself with a deposit of bottom sediment. Where there is a tidal current the Plaice orientates itself, pointing upstream and presses its dorsal and anal fins firmly onto the bottom.

In northern European waters the Plaice is found from the shore line down to about 100 m or somewhat more; in the western

Mediterranean it lives in much deeper water down to 400 m. Like other members of the family it is tolerant of brackish water and may move some way into estuaries to feed, but it is not nearly so tolerant of fresh water as the Flounder (*Platichthys flesus*).

Lemon Sole
Microstomus kitt (Walbaum)

Northern Atlantic from the White Sea to Biscay; North Sea; English Channel.

Body broad, oval; *head* small, only 1/5 body length; *jaws* small, directed to the right and not reaching the front of the eye; the lower jaw does not extend beyond the upper; *lateral line* gently curved over the pectoral fin; *scales* no bony warts, the scales have smooth margins, the skin is smooth to the touch; *fins* tail rounded. D 85–97; A 69–76.

Eyed side warm brown or yellow brown, mottled with darker brown, dull yellow or dull orange. Blind side white.

Up to 40 cm, rarely 70 cm.

The Lemon Sole may be found on coarse sand or gravel bottoms from a few metres down to over 200 m. There are no well-defined spawning areas. Spawning occurs from early spring to late summer; the earlier times being in the more southern waters. The food consists largely of bottom living polychaete worms. Do not generally eat hard-shelled creatures but will bite off the protruding soft parts of shellfish.

Yellowtail Flounder
Limanda ferruginea (Storer)

Western Atlantic from Labrador south to Chesapeake Bay.

Body oval compressed; *head* less than ¼ total length with concave upper profile; *eyes* close-set, separated by a high narrow ridge; *jaws* small, angle reaching midway between snout and eye, lower jaw projecting; *lateral line* arched above pectoral fin. D 73–91; A 55–68.

Eyed side including fins brown or greenish-brown with many rust-coloured spots, lemon yellow on margins of dorsal, anal and tail fins and on tail stalk; blind side white with yellow on tail stalk.

60 cm.

Found on offshore shallow-water sand and mud banks at depths between 27 and 364 m, but usually at 30–100 m. A valued food fish, and supports a dedicated fishery on the Canadian Grand Banks.

Smooth Flounder
Liopsetta putnami (Gill)

Western Atlantic from northern Labrador south to Rhode Island.

Body oblong and rather thick; *head* small less than ¼ body length; *jaws* small, the angle of the jaw in front of the eyes; *scales* smooth in females, rough in males, no scales in the area between the eyes, which are separated by a ridge; *lateral line*

Lemon Sole, *Microstomus kitt* (Jim Greenfield).

straight; tail fin rounded; D 53–59, A 35–41.

Brownish or brownish-grey or almost black, mottled brown or with darker spots; blind side white. Fins may have darker spots or blotches.

30 cm.

An inshore species of shallow water where it inhabits soft mud or silt bottoms. Is more tolerant of low salinities and warm temperatures than the Winter Flounder (*Pseudopleuronectes americanus*). Of little commercial importance although they may be caught in large numbers in smelt nets in winter. An indifferent food fish.

Winter Flounder

Pseudopleuronectes americanus (Walbaum)

Western Atlantic from southern Labrador to Georgia.

Body oval with wide tail stalk; *head* very small, less than 1/5 body length, no mucous pits on the underside of the head; *mouth* small angle in front of the eyes; *eyes* on right side of head, right eye slightly in front of the left, separated by a scaled area which is about as wide as the eye diameter; *scales* rough on eyed side, smooth on blind side; *lateral line* almost straight. D 60–76; A 44–58.

Varied according to individual and bottom colour, sometimes with spots or mottling; blind side white often with spots, and rarely dark coloured.

45 cm.

Lives on muddy or moderately hard bottoms in shallow water from about 2 m to 40 m, but has been caught as deep as 143 m. Present in inshore water as well as offshore banks. Tolerant of a wide range of temperatures including the freezing inshore waters of Newfoundland. In the more southern part of their range Winter Flounder move offshore in summer, in the northern part of their range where they encounter very cold temperatures, they move offshore in winter.

Very good to eat, they are often fished for by sports anglers, but are caught only incidentally by commercial fisheries.

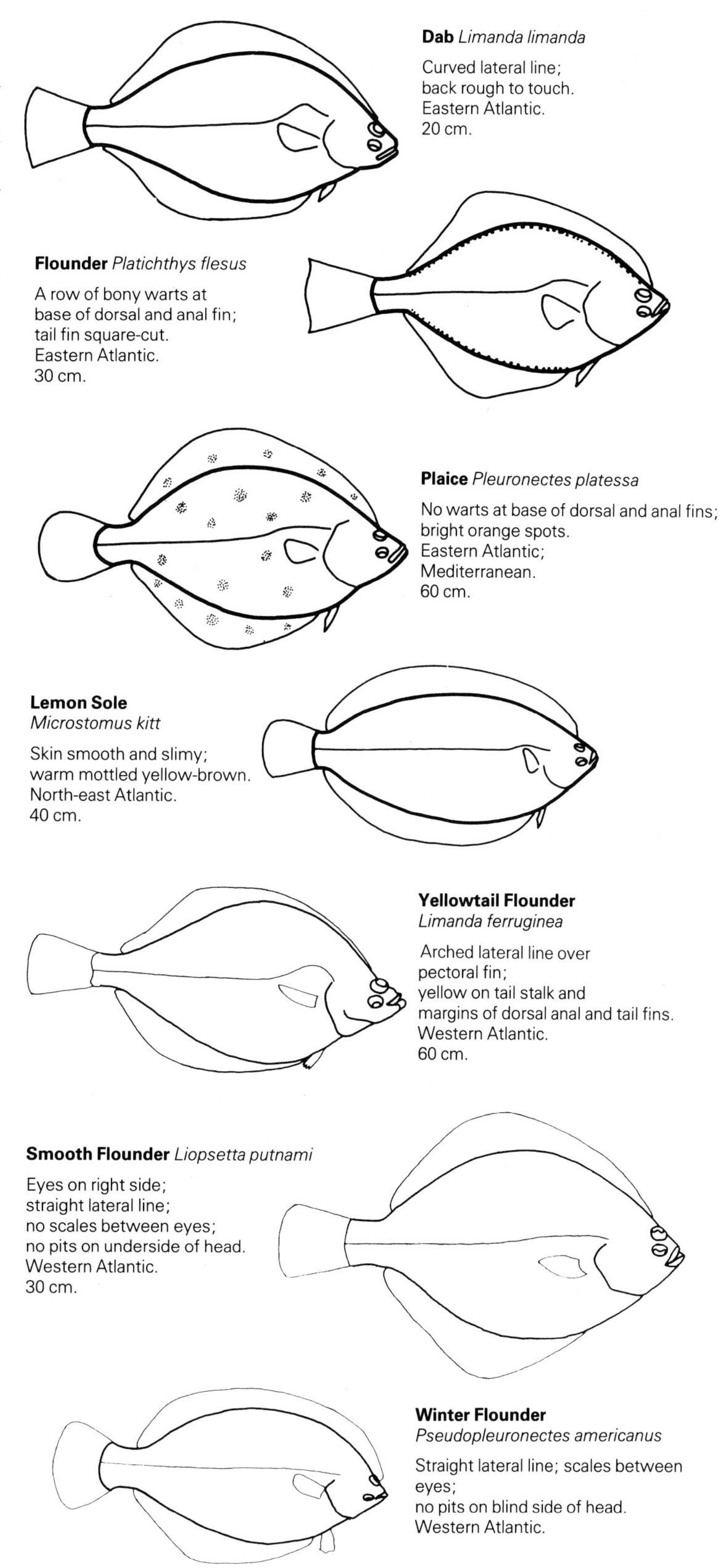

Dab *Limanda limanda*
Curved lateral line;
back rough to touch.
Eastern Atlantic.
20 cm.

Flounder *Platichthys flesus*
A row of bony warts at base of dorsal and anal fin;
tail fin square-cut.
Eastern Atlantic.
30 cm.

Plaice *Pleuronectes platessa*
No warts at base of dorsal and anal fins;
bright orange spots.
Eastern Atlantic;
Mediterranean.
60 cm.

Lemon Sole *Microstomus kitt*
Skin smooth and slimy;
warm mottled yellow-brown.
North-east Atlantic.
40 cm.

Yellowtail Flounder *Limanda ferruginea*
Arched lateral line over pectoral fin;
yellow on tail stalk and margins of dorsal anal and tail fins.
Western Atlantic.
60 cm.

Smooth Flounder *Liopsetta putnami*
Eyes on right side;
straight lateral line;
no scales between eyes;
no pits on underside of head.
Western Atlantic.
30 cm.

Winter Flounder *Pseudopleuronectes americanus*
Straight lateral line; scales between eyes;
no pits on blind side of head.
Western Atlantic.

Family:

SOLEIDAE

Flatfishes with oval, much-compressed bodies, with small eyes on the right side. The snout is rounded and the mouth is small.

All but one of the species included here live in the eastern Atlantic or Mediterranean. All live in shallow water, but some are found in much deeper water. In general they prefer soft muddy bottoms and are thus only rarely seen by divers. Soles are carnivorous, feeding on bottom-living invertebrates. The Dover Sole is a much valued food fish.

Solenette
Buglossidium luteum (Risso)
= *Solea lutea* (Risso)

Eastern Atlantic north to Scotland; English Channel; North Sea; Mediterranean.

Body oval and fairly elongate; *snout* rounded with the upper jaw slightly elongated and rounded to form a 'beak' *eyes* on the right side of the head. They are small and the diameter of the upper eye is less than the space between it and the front of the head; *scales* 55–70 along the lateral line; *fins* 1 dorsal fin which commences opposite the lower edge of the upper eye. There is a small membrance at the base of the dorsal and anal fins. The pectoral fins are small and much reduced on the eyed side. D 65–78; A 50–63.

The colour of the eyed side is variable but frequently yellowish or light brown either with or without darker blotches and spots. The dorsal and anal fins are sandy but every 5th or 6th (occasionally every 4th or 7th) ray is dark for the majority of its length.

Up to 13 cm, it is the smallest of the Soles.

This fish is common and may be found from 9–250 m, either on, or half buried in, sand. Spawning occurs in early summer in the English Channel and during spring in the Mediterranean. The young fish take up a bottom-living existence at about 12 mm in length. Adult fish feed on small bottom-living animals. They have little commercial importance.

Six-eyed Sole
Dicologoglossa hexophthalma

Mediterranean; eastern Atlantic.

Body oval, nostril on blind side not enlarged, front nostril on eyed side with backward-pointing tube; *fins* pectoral fin present on blind side.

Dark reddish-brown bands run across the body with 6 eye spots at the margins of the body.

20 cm.

Shallow water.

Bastard Sole
Microchirus azevia (Capello)

Body oval; *eyes* set very close and near to the top profile of the head; front nostril on blind side not enlarged, front nostril on eyed side a backward-pointing tube; *fins* tail fin completely separate from dorsal and anal fins, pectoral fin on blind side reduced to 2–3 rays.

No conspicuous cross-bands or eye spots.

30 cm.

Found on sand and mud from the shoreline down to at least 250 m.

Four-eyed Sole
Microchirus ocellatus (L.)

Mediterranean.

Body oval and elongate; *eyes* on the right side of the head; *snout* rounded; *scales* 70–75 along the lateral line; *fins* 1 dorsal fin which commences at a point in front of the upper eye. The dorsal and anal fins are not attached to the tail fin. The pectoral fin on the blind side is smaller and has fewer rays than that on the eyed side. D 63–73; A 50–57.

Eyed side variable in colour. Greyish-yellow, light brown or chocolate brown with large dark spots on the front half of the body. The rear half has 4–5 dark spots each with a yellow centre and surrounded by a

Four-eyed Sole, *Solea ocellata* (Guido Picchetti).

yellow ring. The base of the tail has a dark stripe and the tip of the tail fin has a fine white stripe. Blind side white.

Up to 20 cm.

Usually found from 100–300 m either on or half buried in mud or sand or amongst sea-grass (*Posidonia*). Solitary fish are occasionally found in much shallower water.

Thickback Sole

Microchirus variegatus (Donovan)

Mediterranean; eastern Atlantic north to the English Channel and occasionally to Scotland.

Body oval and rather thick; *eyes* on the right side of the head. The upper eye is slightly in front of the lower eye and has a diameter greater than the space between the eye and the edge of the head; *snout* rounded; *scales* 70–92 along the lateral line; *nostrils* small and tubular on the blind-side; *fins* 1 dorsal fin which commences at a point level with the top edge of the upper eye. The anal and dorsal fins are well separated from the tail fin. The pectoral fin is small and much reduced on the blind side. D 63–77; A 51–67.

The colour of the eyed side is variable and ranges from greyish-yellow or reddish-brown with about 5 darker cross-bars which extend onto the dorsal and anal fins. There may also be other less clear, narrower stripes. The dorsal and anal fins may have dark rays scattered throughout their length. The rear edge of the tail fin has a dark stripe and the pectoral fin is dark. The blind side is whitish.

Up to 22 cm, usually up to 14 cm in the Mediterranean.

Found on fine coarse sand. In the Atlantic they are most common between 35–90 m, but extend 10–300 m. In the Mediterranean they range 80–400 m, rarely shallower. Spawning occurs in the English Channel in spring and early summer at a depth between 55–75 m, and in the Mediterranean from February. The adults feed on small bottom-living animals and are themselves an excellent food fish.

Sand Sole; French Sole

Solea lascaris Günther

Eastern Atlantic north to Britain, occasionally along the west coast of Britain; Mediterranean.

Body oval and elongate; *snout* rounded with the upper jaw slightly longer than the lower; *eyes* on the right side of the head; *nostrils* the front nostril on the blind-side is large (nearly the same diameter as the eye) and resembles a rosette; *scales* 96–140 along the lateral line; *lateral line* straight along the body but steeply curved above the eyes; *fins* both the dorsal and anal fins have a membrane at their rear ends. The pectoral fin on the blind side is only slightly smaller than that on the eyed side. D 71–90; A 55–75.

The colour of the eyed side is variable, greyish, yellowish or reddish-brown with small, irregular, darker spots scattered over the surface. The pectoral fin has a black spot circled with white at the extreme tip.

Up to 40 cm.

Found on or half buried in sand down to 48 m. It may be very common locally.

Sole; Dover Sole

Solea vulgaris (Quensel)

Eastern Atlantic north to Scotland; southern North Sea; English Channel; west Baltic; Mediterranean.

Body oval and elongate; *eyes* on the right side of the head; *snout* rounded; *mouth* small, semicircular in shape and situated on the lower edge of the body; *nostrils* small and tubular on the blind side; *scales* 116–165 along the lateral line; *fins* 1 long dorsal fin which commences at a point midway between the upper eye and the tip of the snout; 1 anal fin similar to, but shorter than, the dorsal fin. Both the dorsal and anal fins are attached to the tail fin by a membrane. The pectoral fin on the blind side is slightly smaller than that on the eyed side. D 75–93; A 59–79.

Eyed side brownish or greyish-brown with large, dark irregular blotches. The pectoral fin on the eyed side has a black spot at its tip which seems to intensify after death. The dorsal and anal fins have a white edge. Blind side whitish. The colours of these fish vary greatly with the bottom and in some areas the fish markets are able to tell from which ground the fish were caught.

Up to 50 cm, usually only up to 30 cm in the Mediterranean.

Found from shallow water down to 183 m either on, or half buried in, sand or mud. Frequently found in estuaries. During the winter there is a migration to deep offshore water, usually between 70–130 m. Feeding occurs mainly at night with peaks at dusk and dawn. The food consists of small bottom-living animals. During the spring in the Atlantic the fish migrate to specific spawning areas which lie at a depth of 40–60 m. A number of these spawning grounds in the North Sea, Irish Sea and English Channel are known. Spawning occurs in winter in the Mediterranean. The young fish live in mid-water until they reach a length of between 15–18 mm, when they start their bottom-living life. The Sole is of great commercial value and an excellent food fish.

Klein's Sole
Solea kleinii (Bonaparte)

Mediterranean.

Body oval and elongate, gradually tapering towards the tail; *snout* rounded; *eyes* on the right side of the head, the upper eye slightly in front of the lower; *fins* 1 dorsal fin which commences on the snout level with the upper edge of the upper eye. D 72–91; A 57–72.

Eyed side very variable in colour, usually brownish with darker patches and small lighter spots. The edge of the dorsal and anal fins are dark. The pectoral fin has a black base, an orange centre and a white rear edge.

Up to 32 cm.

Found on sandy or muddy bottoms from 20–100 m.

Whiskered Sole
Monochirus hispidus Rafinesque.

Mediterranean; eastern Atlantic from Portugal to Morocco.

Body oval but tapers rather less than the other soles; *snout* rounded; *eyes* on the right side of the head; *scales* large and rather 'hairy'; *fins* the dorsal fin commences above the upper eye. There is no pectoral

Dover Sole, *Solea vulgaris* (B. Picton).

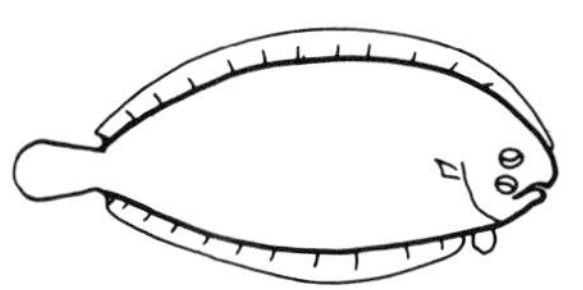

Solenette *Buglossidium luteum*

Lateral line with no branch over eye;
tail fin joined to anal and dorsal fins by thin membrane;
pectoral fin on blind side reduced;
every 4th to 9th ray of anal and dorsal fins dark.
Eastern Atlantic; Mediterranean.
15 cm.

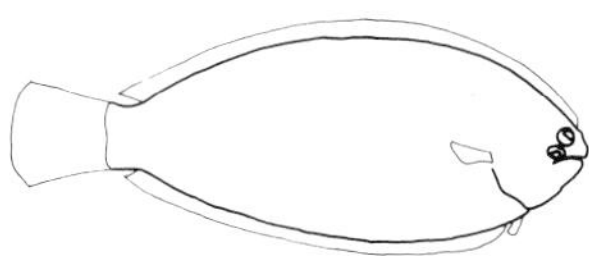

Bastard Sole *Microchirus azevia*

No eye spots or dark cross-bands;
pectoral fin on blind side reduced;
tail fin separate from dorsal and anal fins.
Eastern Mediterranean; eastern Atlantic.
30 cm.

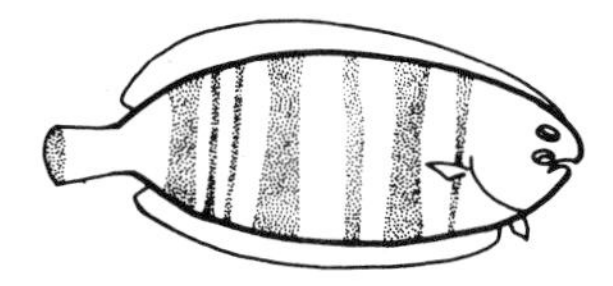

Thickback Sole *Microchirus variegatus*

Dark cross-bands on body extending onto fins;
tail fin separate from dorsal and anal fins;
pectoral fin on blind side reduced to 2–3 rays.
Eastern Atlantic; Mediterranean.
20 cm.

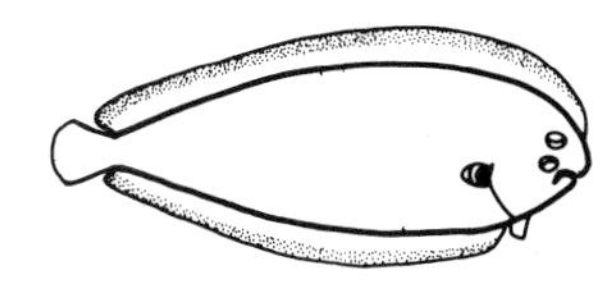

Klein's Sole *Solea kleinii*

Black, orange and white pectoral fin;
anal and dorsal fins with dark edges.
Mediterranean; Eastern Atlantic.
30 cm.

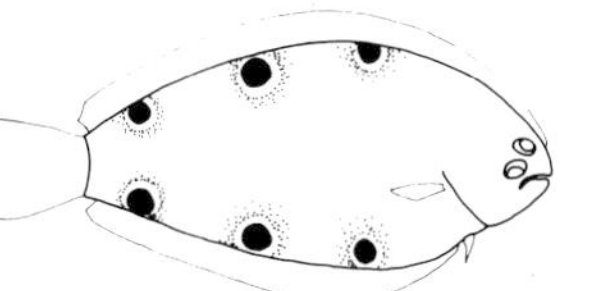

Six-eyed Sole *Dicologoglossa hexophthalma*

6 eyed spots along margins of body.
Mediterranean; eastern Atlantic.
20 cm.

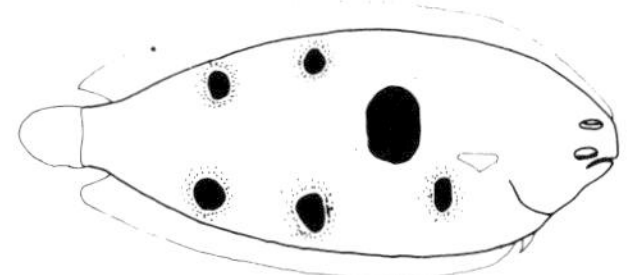

Four-eyed Sole *Microchirus ocellatus*

Large blotch in mid-body plus 4 dark spots with light rings.
Mediterranean; eastern Atlantic.
20 cm.

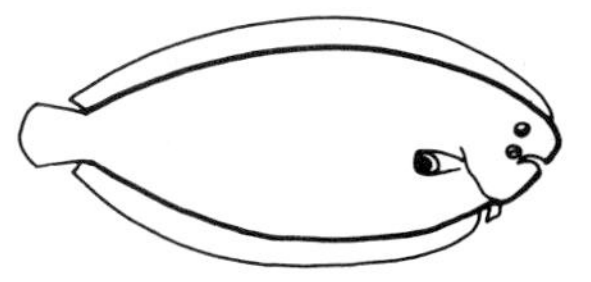

Sand Sole *Solea lascaris*

Body yellow-brown with dark spot;
pectoral with dark spot with yellow border;
front nostril on blind side rosette-like.
Eastern Atlantic; Mediterranean.
40 cm.

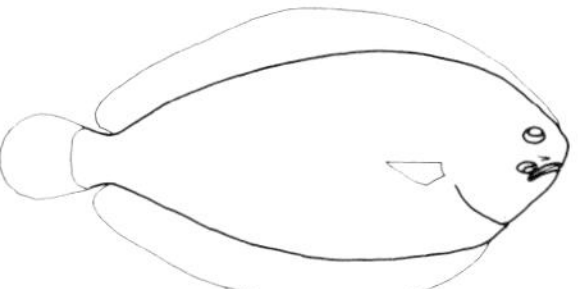
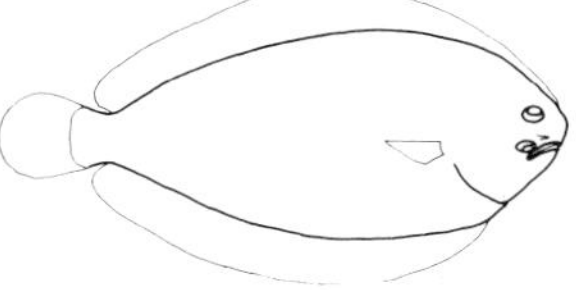

Whiskered Sole *Monochirus hispidus*

No pectoral fin on blind side;
front nostril on eyed side a very long tube;
dark marbled pattern with chevron on tail stalk.
Mediterranean; eastern Atlantic.
20 cm.

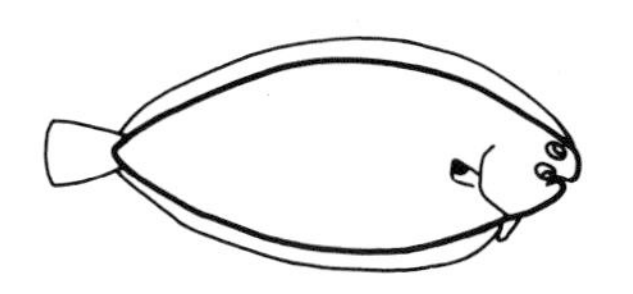

Sole; Dover Sole *Solea vulgaris*

Membrane joins dorsal, tail and anal fins;
pectoral fin with black spot;
front nostril on blind side not enlarged;
pectoral fin on blind side well developed.
Eastern Atlantic; Mediterranean.
70 cm.

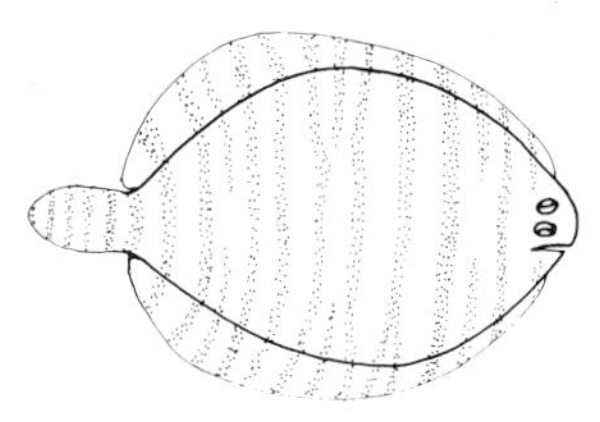

Hogchoker *Trinectes maculatus*

No pectoral fins;
about 8 dark vertical stripes;
tail fin spotted.
Western Atlantic.
15 cm.

fin on the blind-side. D 50–58; A 40–45.

Eyed side greyish or yellowish-brown with a dark marbled patterning which continues onto the dorsal and anal fins as a row of regular spots. There is a V-shaped stripe at the base of the tail fin.

Up to 14 cm.

This species may be found from 10–250 m on sand, mud or sea-grass (*Posidonia*) bottoms. Spawning occurs in late summer.

Hogchoker

Trinectes maculatus (Bloch & Schneider)

Western Atlantic from New York to the south Atlantic.

Greenish brown with about 8 vertical bands that may vary in width; tail fin with elongate spots; blind side may have numerous round dark spots.

15 cm.

Ascends up estuaries and even penetrates fresh water above tide water.

Order:

DISCOCEPHALI=ECHENEIFORMES

The dorsal fin is transformed into a sucking disc on the top of the head.

Family:

ECHENEIDIDAE

The only family in the order.

Common Remora; Shark Sucker
Remora remora (L.)

Mediterranean; all warm oceans.

Body only slightly elongate and fairly stout; *head* large; *sucker* there is a large sucker situated along the top of the head and the front part of the body. The sucker is about twice as long as wide and has between 17–19 pairs of ridges; *mouth* large and ends at the front edge of the eye; *jaws* the lower jaw is longer than the upper and rounded; *fins* the dorsal and the anal fins are opposite and similar to each other. Tail fin slightly curved. D 24–27; A 23–25.

Uniform dark brown or greyish-brown. Pectoral fins lighter.

Up to 64 cm.

Frequently found attached by the sucker on the top of the head to sharks and other large fish and mammals. They feed on the small crustacean parasites which infest the skin of the host, but they supplement their diet with other free-living small crustacea and fishes.

Spawning occurs in the early summer in mid-Atlantic and in the late summer and autumn in the Mediterranean. Newly hatched fish live a free-swimming life until they reach about 3 cm in length, when they attach themselves to a host fish, often another Remora.

The Spearfish Remora (*Remora brachyptera*, Lowe) differs in having 27–34 dorsal fin rays and less than 24 gill rakers and preferring billfishes as hosts. Western Mediterranean and neighbouring Atlantic, elsewhere in all warm seas.

The Marlinsucker (*Remora osteochir*, Cuvier), differs from the Spearfish Remora in having 20–26 finrays. Probably only on billfishes. Mediterranean and all warm seas.

Common Remora, *Remora remora* (Doug Perrine).

Shark Sucker, *Echeneis naucrates* (Herwath Voigtmann).

Shark Sucker; Remora
Echeneis naucrates (L.)

Mediterranean and all warm oceans.

Body longer and more slender than *Remora remora*; *snout* pointed; *sucker* there is a large sucker situated along the top of the head and the front of the body. The sucker is three times as long as wide with 20–24 pairs of ridges; *jaws* the lower jaw is much longer than the upper and has a fleshy, flexible tip; *fins* the 2nd dorsal and anal fins are opposite and similar to each other, tail fin square-cut or slightly rounded. Pectoral fin pointed. D 32–40; A 31–39;

The back and underside are grey and the sides have a central, longitudinal black stripe flanked on each side by a white stripe.

Up to 65 cm, occasionally up to 1 m.

Usually attaches to sharks, but will attach to a variety of large mammals including scuba divers.

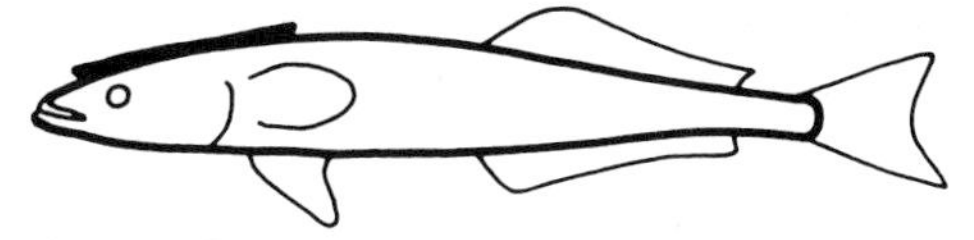

Common Remora; Shark Sucker *Remora remora*

Uniform brown or greyish-brown; 17–19 pairs of ridges on the sucker; tail fin slightly forked. 64 cm.

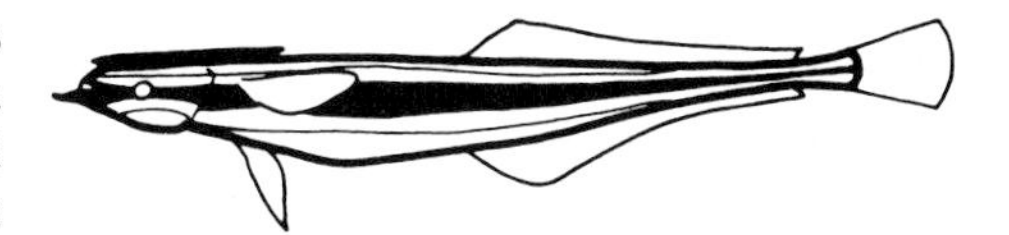

Shark Sucker; Remora *Echeneis naucrates*

Grey with a black longitudinal stripe flanked with white. 20–24 pairs of ridges along the sucker; tail fin square-cut. 65 cm.

Order:

PLECTOGNATHI=TETRAODONTIFORMES

Skeleton not completely bony with few vertebrae. The gill opening is a narrow slit situated in front of the pectoral fins. The bones of the upper jaw mostly fused sometimes forming a beak modified for grazing and scraping. The teeth may be distinct or absent.

Family:

BALISTIDAE

The teeth are implanted in sockets in the bone. The dorsal fin has 3 spines, the 1st is very strong and can be locked into position by a bony knob protruding forward from the base of the 2nd spine.

Trigger Fish
Balistes capriscus Gmelin

Eastern Atlantic from North Sea south to Angola; Mediterranean; western Atlantic from Nova Scotia to Argentina.

Body diamond-shaped, deep but narrows just before the tail fin; flattened laterally, *snout* angular; *mouth* very small; *teeth* large and pointed; *lips* relatively large and fleshy; *gill slit* simple and situated immediately in front and above the pectoral fin; *scales* large and arranged in a mozaic pattern. There are about 2 large spines behind the gill slit; *fins* 2 dorsal fins, the 1st fin consists of 3 spines. The 1st spine is large with a rough edge, the 2nd is smaller and close to the 1st. The 3rd spine is small and well separated from the 1st and 2nd spines. The 2nd dorsal fin and anal fin are similar in size and shape. Tail fin large with the outermost rays elongate. The pelvic fin consists of a large rough moveable spine. 1D III; 2D 27–28; A 25–27.

The colour is very variable and may be greenish, grey or brownish with green, blue or violet reflections on the back. The dorsal and anal fins sometimes have whitish or violet-black or bluish markings.

Up to 41 cm, usually around 25 cm.

Not strong swimmers, they swim by characteristic undulating movements of the dorsal and anal fins. Usually found in rocky or seaweed-covered areas near coasts. The adults probably feed on small crustacea and molluscs which they find on rocks and on the sea bed. Spawning in the Mediterranean occurs during the summer when the sea reaches a temperature around 21°C. A hollow nest is excavated from the sand of the sea bed and in this the egg mass is laid. The adult fish guard the egg mass and ensure that it is aerated. The eggs hatch after about 2 days.

Trigger Fish, *Balistes capriscus* (Andy Purcell).

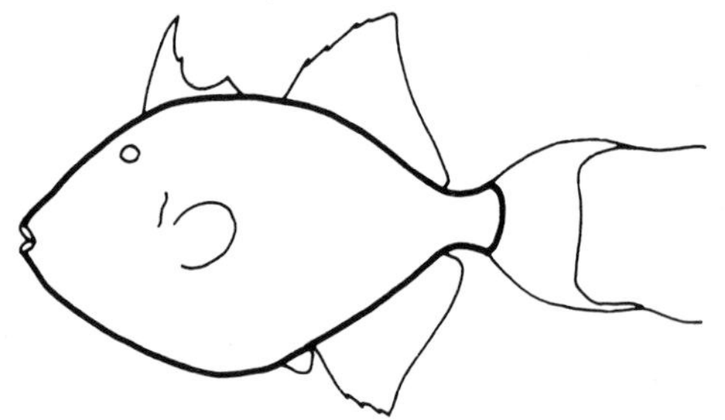

Trigger Fish *Balistes capriscus*

Body diamond-shaped and laterally flattened; outer rays of tail fin elongate.
25 cm.

Family:
DIODONTIDAE

Striped Burrfish
Chilomycterus schoepfi (Walbaum)

Cape Cod south to Brazil.

Body a little broader than deep, capable of inflation, but not as much as porcupine fish; the body is covered with low triangular *spines*, each with 3 roots and are not depressable.

Young are dark green with close-set wavy lines and spots; larger fish are lighter in colour and have wide wavy parallel lines on back and sides; the body has spots with lighter surrounds.

25 cm.

Puffer Fish inflate their body when alarmed. The closely related Puffer Fishes which occur too far south to be included in our area are able to inflate their body into an almost perfect sphere. The Striped Burrfish can inflate its body to a considerable extent, but does not become spherical.

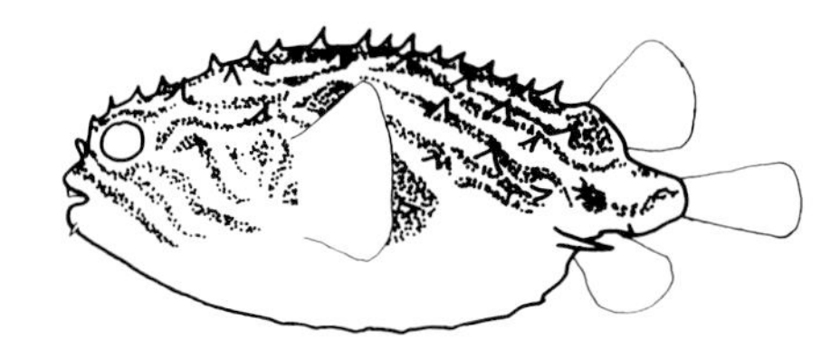

Striped Burrfish
Chilomycterus schoepfi

Inflatable spiny body; spines do not fold down; stripes on body. Eastern Atlantic. 25 cm.

Family:
MOLIDAE

Body disc-shaped with a leathery skin and no scales. In the adults the tail fin resembles a wavy frill on the blunt rear part of the body.

Sunfish
Mola mola (L.)

Throughout all the tropical and temperate oceans of the world.

Body almost rectangular and laterally flattened. There is no tail stalk; *snout* blunt; *mouth* small; *teeth* 1 beak-like tooth in each jaw; *gill slit* situated immediately in front of the pectoral fin; *skin* very thick and rough but without scales; *fins* the dorsal and anal fins are short but tall. The tail fin is very narrow and attached to both the dorsal and anal fins, the rear edge undulates. D 16–20; A 14–18.

The back, fins and base of the tail fin are greyish, brownish or greenish; the flanks and undersurface are lighter.

Up to 300 cm.

They feed on a large number of animals ranging from jellyfish, crustacea, echinoderms, fish and algae. Very little is known about their breeding habits but it is possible that the female may produce more than 300 million eggs.

Sunfish, *Mola mola* (Norbert Wu).

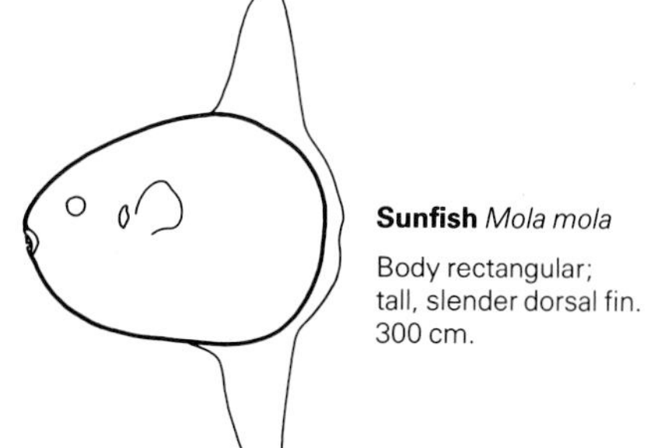

Sunfish *Mola mola*

Body rectangular; tall, slender dorsal fin. 300 cm.

Order:

XENOPTERYGII=GOBIESOCIFORMES

Family:

GOBIESOCIDAE

Scaleless fishes with flattened bodies. There is a sucker modified from the pelvic fin. This is surrounded by a fold of skin supported in the front of the rays of the pectoral fin. One family only.

Blunt-nosed Clingfish; Blunt-nosed Sucker

Gouania wildenowi (Risso)

Mediterranean.

Body flattened sideways; *profile* curved; *head* rather flattened from above and wider than the body; *snout* blunt; *eyes* small, separated by a space about 3 times the diameter of the eye; *gill slits* very prominent; *sucker* small and oval and situated under the front part of the body; *tentacle* there is a slender branched tentacle in front of each eye; *fins* the dorsal and anal fins are continuous around the tail. The dorsal fin does not reach half way along the body. D 14–19; A 10–12.

Unlike all the other species of clingfish this one is uniform in colour, greyish-yellow, light brown or greenish-brown on the back and lighter on the belly.

From 4–7 cm.

Very little is known about this fish except that it is found attached to stones and rocks from very shallow water.

Small-headed Clingfish; Small-headed Sucker

Apletodon dentatus dentatus (Facciola) = *Lepadogaster microcephalus* Brook

Mediterranean; western English Channel and northern Biscay; south-west coast of Scotland.

Body flattened sideways; *head* flattened from above; *sucker* situated under the front part of the body; *snout* rounded; *cheeks* the male has enlarged rounded cheeks; *eyes* fairly large, separated by a space about twice the diameter of the eye; *nostrils* the front nostril has a small appendage on the rear edge; *mouth* large; *teeth* there is a row of 4–8 small incisor-type teeth in the front of the jaw and 1–3 larger canine-type teeth at each side. The teeth are characteristic of this species; *fins* the simple dorsal and anal fins are short, opposite each other and nearly equal in size. They are separated from the tail fin which is rounded. D 5–6; A 5–7.

Greenish, sometimes brownish, with lighter and darker irregularly arranged spots. The male usually has less of these spots and has been described as having a violet patch in the centre of the dorsal and anal fins and underneath the throat. The female has a light-coloured throat.

Up to 5 cm.

Found down to 25 m often amongst kelp (*Laminaria*) and sea-grass (*Zostera*). Breeding occurs from May to September and the eggs are laid amongst the holdfasts of seaweeds.

Two-spotted Clingfish; Two-spotted Sucker

Diplecogaster bimaculata bimaculata (Bonnaterre)

Eastern Atlantic north to Norway; western Channel; Mediterranean, Black Sea.

Body round but flattened sideways near the tail; *head* flattened from above; *profile* slopes from the tip of the snout to the back of the head; *snout* rather triangular; *sucker* situated under the front part of the body; *nostrils* the front nostril has a small appendage on the rear edge; *mouth* rather large; *jaws* upper jaw slightly longer than the lower; *teeth* small and pointed and arranged in patches; *fins* the dorsal and anal fins are both separated from the tail fin. The anal fin is shorter than the dorsal. D 5–7; A 4–6.

This fish is extremely variable in colour, usually reddish on the back and sides and yellowish underneath with yellow, blue and brown spots scatted over the surface. The yellow spots may be arranged in patterns. The male has a purple spot circled with yellow behind the pectoral fin.

Up to 5 cm.

This clingfish is found in deeper water than the other species; down to 55 m in British waters and down to 100 m in the Mediterranean. Usually they are found amongst sea-grass or on coarse bottoms, often where there are plenty of shells. Breeding occurs during spring and summer. The eggs are laid in masses on the undersurface of shells or under stones and are guarded by the adults, particularly the male.

Shore Clingfish; Cornish Sucker

Lepadogaster lepadogaster lepadogaster (Bonnaterre)
= *Lepadogaster gouanii* Risso

Eastern Atlantic north to Scotland; western Channel; Mediterranean.

Body flattened sideways and back curved; *head* flattened from above and triangular in shape; *profile* slopes steeply from the snout to behind the back of the head; *sucker* situated underneath the front half of the body; *jaws* prominent and shaped rather like a duck bill. The upper jaw completely overlaps the lower; *lips* fleshy; *nostrils* with a large fleshy appendage on the rear edge of the front nostril; *fins* the dorsal fin larger than the anal; both these fins are attached to the tail fin but there are distinct dips in the membrane. D 16–19; A 9–10.

Very variable in colour, yellowish, reddish or brownish with irregular brownish spots. There are 2 bright blue spots outlined in dark-brown, red or black on the back at the base of the head.

Up to 8 cm.

Found in very shallow water, from between the tides to about 1 m. They are frequently found in rock pools amongst stones and seaweeds. Breeding occurs during spring and summer. The eggs are golden in colour and are laid in masses which are fastened to the bottom of stones and guarded by the parent fish.

Blunt-nosed Clingfish; Blunt-nosed Sucker
Gouania wildenowi

Uniform colour; 1 fin which continues around the tail; branched tentacle in front of eye. 7 cm.

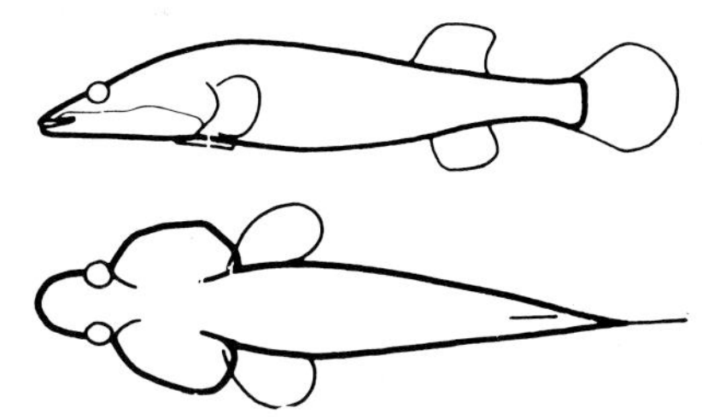

Small-headed Clingfish; Small-headed Sucker
Apletodon dentatus

Dorsal and anal fin separate from tail fin; greenish or brownish with white and dark spots; large canine type teeth at sides of jaws; 5 cm. Female: light coloured throat. Male: large cheeks.

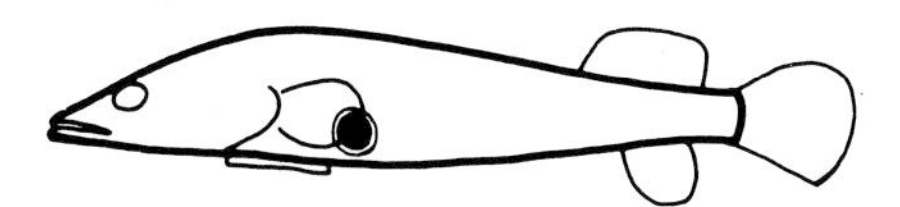

Two-spotted Clingfish; Two-spotted Sucker
Diplecogaster bimaculata

Dorsal and anal fins separate from tail fin; anal fin smaller than dorsal; reddish with yellow markings; no canine-type teeth; 5 cm. Male fish have a purple spot circled with yellow behind the pectoral fin.

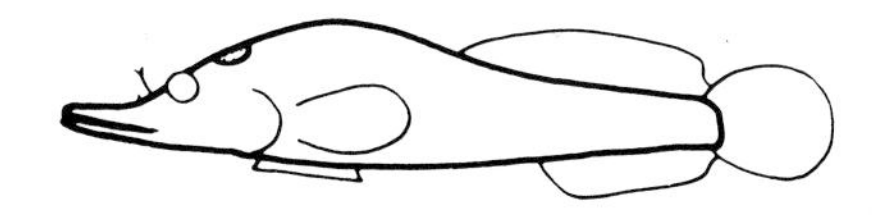

Shore Clingfish; Cornish Sucker *Lepadogaster lepadogaster*

Dorsal and anal fins attached to tail fin but distinct; large blue eye spots at the base of the head on the back. 8 cm.

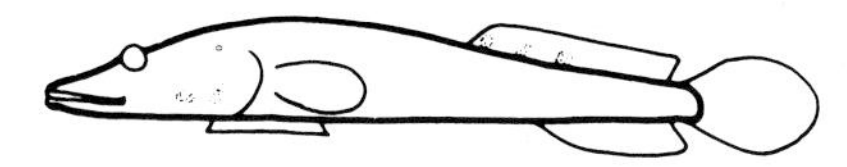

Connemara Clingfish; Connemara Sucker
Lepadogaster candollei

Dorsal and anal fins not attached to the tail fin; dorsal fin long; male fish with red spots on cheeks and at base of dorsal fin. 7.5 cm.

Connemara Clingfish; Connemara Sucker

Lepadogaster candollei Risso
= *Mirbelia decandollei* Canestrini

Eastern Atlantic north to Scotland; western English Channel; Mediterranean; Black Sea.

Body flattened sideways in the rear half of the body; *profile* slopes gently from the tip of the snout to the back of the head; *sucker* situated under the front part of the body; *snout* rather long and pointed; *jaws* shaped rather like a duck-bill; *nostrils* only a very small appendage present on the front nostril edge; *fins* dorsal fin long and not attached to the tail fin. D 13–16; A 9–11.

The colour of this fish is extremely variable. The ground colour may be reddish, brownish or greenish with red, brown or white dots, spots or stripes. Large specimens, probably the males, have 3 red spots at the base of the dorsal fin and other red spots on the head.

Up to 7.5 cm.

Connemara Clingfish, *Lepadogaster candollei* (B. Picton).

These fish are usually found on shores and around the low-water mark amongst kelp, sea-grass and stony bottoms. Eggs are laid under shells or stones, where they are fixed by filaments on the eggs. They are laid down to about 30 m in depth and guarded by the parent fish.

Order:

BATRACHOIDEFORMES

Family:

BATRACHOIDIDAE

Oyster Toadfish

Opsanus tau (L.)

Western Atlantic from Cape Cod south to Cuba.

Body robust, naked; *head* flattened especially in older individuals, mouth large with very strong jaws, broad flap above eye, lower jaw with tentacles. D III, 26; A 24.

Dusky or olive-brown with irregular dark cross bands; pale cross bars across pectoral and tail fins.

Close to shore amongst rocks and weed. A hardy fish that can be abundant especially in the northern part of its range.

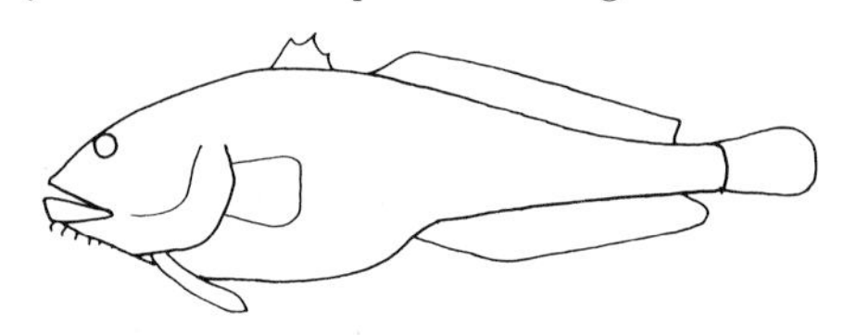

Oyster Toadfish
Opsanus tau

Body robust, naked;
Head flattened;
3 spiny rays in first dorsal fin.
35 cm.

Oyster Toadfish, *Opsanus tau* (Roy Manstan).

Order:
PEDICULATI=LOPHIIFORMES

The dorsal fin composed of a few flexible rays of which the 1st is placed on the head and generally ends in some sort of tassel or bulb; in some deep-sea species this bulb is luminous. In some deep-water forms the male is dwarfed and lives a parasitic existence on the female.

Sub-order:
LOPHIOIDEA

Pelvic fins present (toothed), males not dwarfed, frontal bones in contact for most of their length.

Family:
LOPHIIDAE

The only family in the sub-order.

Angler Fish; Monkfish
Lophias piscatorius (L.)

Throughout Atlantic; Mediterranean; English Channel; North Sea.

Body front part of body flattened from above but gradually tapers to a normal fish-shaped rear half; *head* large, wide and flattened from above; *mouth* very large, semicircular in shape; *jaws* lower jaw longer than upper; *teeth* curved, irregularly sized and spaced; *eyes* small and situated on top of the head; *gill slits* 2 large openings immediately behind the pectoral fins; *skin* there are no scales but a number of fringed lobes which run around the sides of the head and the body. The skin is very loose; *fins* there are a number of isolated rays which run down the mid-line. The 1st ray, which is the longest, is situated just behind the lip of the upper jaw and ends in a branched fleshy lobe. The 2nd ray is shorter but fringed and situated in front of the eye. The 3rd ray is found some way behind the other 2. There are 2 dorsal fins, the 1st with 3 rays joined by a very short membrane. The 2nd is longer, opposite and similar to the anal fin. The pectoral fins are large; the pelvic fins are small and situated on the undersurface of the head. D I + I + I + III; 2 D 10–13; A 9–11.

Brownish, reddish or greenish-brown with darker blotches. Undersurface white, except for the rear edge of the pectoral fin which is black.

Up to 1.98 m.

Angler Fish, *Lophius piscatorius* (Bob Soames).

Common, range from 18 to 550 m. They are normally encountered either on or half buried in sandy and muddy bottoms, but may also be found amongst seaweeds and sandy areas between rocks. They feed on smaller fish (flatfish, haddock, dogfish, etc.) which they attract by twitching the fleshy lobe at the end of the 1st dorsal ray. They do, however, feed on practically any other bottom-living animals, and may even make short trips off the bottom and have been recorded as attacking sea birds on the surface.

Spawning occurs during late winter, spring and summer in very deep water, possibly as deep as 180 m. The eggs are laid in large ribbon like sheets up to 9 m long and 90 cm wide. After a free-living existence the young take up a bottom-living life at about 5–6 cm. The Angler Fish is of no commercial value.

The **Goosefish** *Lophius americanus* that is found in the western Atlantic from the Gulf of St Lawrence south to Florida is apparently a distinct species from the eastern Atlantic's *Lophius piscatorius*.

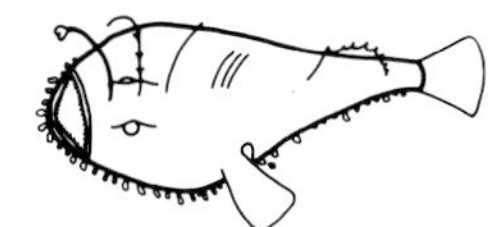

Angler Fish; Monkfish; Goosefish
Lophia sp.

Large flattened body;
very large semicircular mouth;
large spine with a fleshy lobe at the tip just behind the upper lip.
198 cm.
Mediterranean, eastern Atlantic.
Goosefish
Western Atlantic.

Index of Scientific Names

Index of Common Names